Trees, Knots, and Outriggers

Studies in Environmental Anthropology and Ethnobiology

General Editor: **Roy Ellen**, FBA

Professor of Anthropology, University of Kent at Canterbury

Interest in environmental anthropology has grown steadily in recent years, reflecting national and international concern about the environment and developing research priorities. This major new international series, which continues a series first published by Harwood and Routledge, is a vehicle for publishing up-to-date monographs and edited works on particular issues, themes, places or peoples which focus on the interrelationship between society, culture and environment. Relevant areas include human ecology, the perception and representation of the environment, ethno-ecological knowledge, the human dimension of biodiversity conservation and the ethnography of environmental problems. While the underlying ethos of the series will be anthropological, the approach is interdisciplinary.

Volume 1
The Logic of Environmentalism: Anthropology, Ecology and Postcoloniality
Vassos Argyrou

Volume 2
Conversations on the Beach: Fishermen's Knowledge, Metaphor and Environmental Change in South India
Götz Hoeppe

Volume 3
Green Encounters: Shaping and Contesting Environmentalism in Rural Costa Rica
Luis A. Vivanco

Volume 4
Local Science vs. Global Science: Approaches to Indigenous Knowledge in International Development
Edited by Paul Sillitoe

Volume 5
Sustainability and Communities of Place
Edited by Carl A. Maida

Volume 6
Modern Crises and Traditional Strategies: Local Ecological Knowledge in Island Southeast Asia
Edited by Roy Ellen

Volume 7
Traveling Cultures and Plants: The Ethnobiology and Ethnopharmacy of Human Migrations
Edited by Andrea Pieroni and Ina Vandebroek

Volume 8
Fishers And Scientists In Modern Turkey: The Management of Natural Resources, Knowledge and Identity on the Eastern Black Sea Coast
Ståle Knudsen

Volume 9
Landscape Ethnoecology: Concepts of Biotic and Physical Space
Edited by Leslie Main Johnson and Eugene Hunn

Volume 10
Landscape, Process and Power: Re-Evaluating Traditional Environmental Knowledge
Edited by Serena Heckler

Volume 11
Mobility and Migration In Indigenous Amazonia: Contemporary Ethnoecological Perspectives
Edited by Miguel N. Alexiades

Volume 12
Unveiling the Whale: Discourses on Whales and Whaling
Arne Kalland

Volume 13
Virtualism, Governance and Practice: Vision and Execution in Environmental Conservation
Edited by James G. Carrier and Paige West

Volume 14
Ethnobotany in the New Europe: People, Health and Wild Plant Resources
Edited by Manuel Pardo-de-Santayana, Andrea Pieroni and Rajindra K. Puri

Volume 15
Urban Pollution: Cultural Meanings, Social Practices
Edited by Eveline Dürr and Rivke Jaffe

Volume 16
Weathering the World: Recovery in the Wake of the Tsunami in a Tamil Fishing Village
Frida Hastrup

Volume 17
Environmental Anthropology Engaging Ecotopia: Bioregionalism, Permaculture, and Ecovillages
Edited by Joshua Lockyer and James R. Veteto

Volume 18
Things Fall Apart? The Political Ecology of Forest Governance in Southern Nigeria
Pauline von Hellermann

Volume 19
Sustainable Development: An Appraisal Focusing on the Gulf Region
Edited by Paul Sillitoe

Volume 20
Beyond the Lens of Conservation: Malagasy and Swiss Imaginations of One Another
Eva Keller

Volume 21
Trees, Knots, and Outriggers: Environmental Knowledge in the Northeast Kula Ring
Frederick H. Damon

Trees, Knots, and Outriggers

Environmental Knowledge in the Northeast Kula Ring

Frederick H. Damon

berghahn
NEW YORK · OXFORD
www.berghahnbooks.com

First edition published in 2017 by

Berghahn Books

www.berghahnbooks.com

©2017 Frederick H. Damon

Library of Congress Cataloging-in-Publication Data

Names: Damon, Frederick H.. author.
Title: Trees, knots, and outriggers (Kaynen Muyuw) : environmental knowledge
in the northeast Kula Ring / Frederick H. Damon.
Other Titles: Kaynen Muyuw
Description: First edition. | New York : Berghahn Books, [2016] | Series:
Studies in environmental anthropology and ethnobiology ; volume 21 |
Includes bibliographical references and index.
Identifiers: LCCN 2016022585| ISBN 9781785332326 (hardback : alk. paper) |
ISBN 9781785332333 (ebook)
Subjects: LCSH: Traditional ecological knowledge—Papua New Guinea—
Woodlark Island. | Human ecology—Papua New Guinea—Woodlark Island.
| Muyuw (Papua New Guinean people) | Ethnology—Papua New Guinea—
Woodlark Island.
Classification: LCC GN671.N5 D36 2016 | DDC 305.8009953—dc23
LC record available at https://lccn.loc.gov/2016022585

British Library Cataloguing in Publication Data

A catalogue record for this book is available from the British Library

ISBN 978-1-78533-232-6 (hardback)
ISBN 978-1-78533-320-0 (paperback)
ISBN 978-1-78533-233-3 (ebook)

To Grace, Nancy, and Kate

Contents

List of Figures, Graphs, Maps, and Tables   viii

Acknowledgments   x

PART I. Among the Scientists: New Perspectives on the Massim

Introduction. Changes and Last Chapters   3

Chapter 1. Return to the Garden: *Gwed,* Locating Intentions, and Interpretive Puzzles   36

PART II. Toward an Ethnography of Trees

Chapter 2. The Trees: Classificatory Forms, Landscape Beacons, and Basic Categories   81

Chapter 3. The Forest and Fire, Tasim, *Inverted Landscapes and Tree Meanings*   121

Chapter 4. A Story of *Calophyllum*: From Ecological to Social Facts   180

PART III. Synthesizing Models

Chapter 5. *Vatul*: A Life Form and a Form for Life   247

Chapter 6. Geometries of Motion: Trees and the Boats of the Eastern Kula Ring   296

References   353

Index   364

Figures, Graphs, Maps, and Tables

Figures

2.1.	Tree Parts and Terms	114
3.1.	Multivariate ENSO Index	148
4.1.	Keel Design and Its Consequences	233
4.2.	*Wag/Bab* Shape	234
5.1.	Muyuw Fishing Net Structure	280
6.1.	*Kaynikw* Structure	309

Graphs

1.1.	pH vs. Fallow and Forest Type	70
1.2.	P vs. Fallow and Forest Type	71
1.3.	K vs. Fallow and Forest Type	72

Maps

0.1.	The Voyage of 2002	9
1.1.	Okaibom *takulamwala*	48
1.2.	Extensive versus Intensive Resources Bases	50
2.1.	Fractal Representation Of Base/Tip Contrasts	116

Tables

0.1.	GPS Readings from Ole to Panamut; Panamut to Nasikwabw; and Nasikwabw to Waviay	10
1.1.	1995–96: *Gwed* [*Rhus taitensis*] Samples Sorted by Ascending ΔN	65
1.2.	1998 *Gwed* [*Rhus*] Harvest Experiment	65
1.3.	*Tabnayiyuw* ("Tab.," *D. Papuanum*) Comparisons	67
1.4.	Surface/Subsurface C&N Readings	68

1.5.	Code Key for Soil Types	69
1.6.	pH and Fallow Forest Types	69
2.1.	Muyuw Life Forms	89
2.2.	Grouped Trees and Their Western Determinations	95
2.3.	Grouped Trees and Their Western Determinations	96
2.4.	Grouped Trees and Their Western Determinations	98
2.5.	Grouped Trees and Their Western Determinations	99
2.6.	Grouped Trees and Their Western Determinations	99
2.7.	Grouped Trees and Their Western Determinations	101
2.8.	Analysis of Variation of Tree Features	102
2.9.	Grouped Trees and Their Western Determinations	105
2.10.	Grouped Trees and Their Western Determinations	106
3.1.	Wabunun Yam House	159
3.2.	Kaulay, Central Muyuw, Yam House	160
3.3.	Mwadau, Western Muyuw, Yam House	162
3.4.	Iwa Yam House, *Bweyma*	163
3.5.	Ritual Firewood Trees by Place	169
4.1.	Muyuw/Western Identifications	190
4.2.	*Ayniyan/Aynikoy* Comparison	211
4.3.	*Lavanay Anageg/Kemurua* (2002) Measurements	230
5.1.	Vines	251
5.2.	Vine Names	257
5.3.	Finger Names for String Figure Teaching	286
5.4.	String Figure Moves	287
6.1.	*Lavanay Eyalyal* Measurements	310
6.2.	*Talapal* Measurements	310
6.3.	*Kaynik* Measurements	311
6.4.	*Lavanay Nedin* Measurements	313
6.5.	Sail (*Alit'*)/Rudder (*Kavavis*) Dynamics	318
6.6.	*Lavanay Kavavis* Dimensions	319
6.7.	Star Courses (*Kut*)	348

Acknowledgments

As an analytical ethnography, this book combines anthropological traditions with several of the earth sciences, botanical systematics, geochemistry, ecology, and recent work in climatology concerning El Niño Southern Oscillation, ENSO. The work testifies to the willingness of the University of Virginia to sponsor serious interdisciplinary research when its outcome is far in doubt. Support through an initial Dean's Grant, a sequence of Summer Grants, and a final grant from Arts, Humanities, and Social Sciences Research is greatly appreciated, as is a critical 2009 American Philosophical Society Franklin Research Grant that enabled a late exploration of the outrigger canoe form that, in the last analysis, carries this work.

In the first analysis, however, are a set of scientists who gave me the time and attention to learn from their endeavors. Many of these people are named in the text as my arguments unfurl. I note two to represent many others. One is H. Hank Shugart, a forest ecologist and ecological modeler, who, long before I sought him out, realized there needed to be more interaction among natural scientists and social and historical scientists/scholars if we are to make sense of and be engaged with our complex world. Shugart often talked with me about ecological models as I picked them up from his work and the historical ecologists who also play an enormous role in this book. We attended each other's seminars and taught one course together. Everything I have published since 1998 shows his influence; this has also been a two-way street (see Shugart, 2014). The second scientist is Stephen A. Macko, the geochemist who tutored me in stable isotopes and helped me think through some of my data; our joint labors continue. My engagement with these two people, and what they represent, has been as taxing and as stimulating as any engagement with another culture can be. And as hoped, we have managed to send graduate and undergraduate students between our departments.

During the research that resulted in this book, Roy Wagner was a continuing presence; his bountiful enthusiasm for outrigger canoe technicalities testifies to his fertile mind. My teaching also never strayed far from issues addressed herein. That teaching also animated the research for and writing of this book. Two masters guided my way. One is Steve Lansing with his splendid contributions to Austronesian (Balinese) and ecological studies, *Priests and Programmers* (1991) and *Perfect Order* (2006). The less than perfect order in this book derives in part from the question I see Lansing as having defined: how do we describe the complex models inherent in many non-Western social systems? Lansing's passage from Balinese temples to computer simulation helped me move from Kula Ring forests and fields to the ways by which outrigger canoes model a regional social system's complexity. The second master is Anthony Wallace. His brilliant historical ethnographies of the nineteenth-century United States, *Rockdale* and *St. Clair*, profoundly and elegantly discuss interrelationships among a society's technical and social forms. While teaching Wallace's books and conducting the research reported here, "Materiality" and "Science and Technology Studies" became significant subjects in anthropology and related disciplines. A contribution my effort makes to these endeavors partly comes by means of Wallace's patient brilliance.

An aspect of Wallace's work leads to one attempt in this study. The summation of a section of *Rockdale* and decade in the nineteenth century (1825–35), "Thinking about Machinery," describes the thought of what Wallace calls the "mechanician." This refers to a category of people enmeshed in the early Industrial Revolution in the United States (and world); I have increasingly come to think he characterizes the thought found in Kula Ring, if not far beyond it. "The thinking of the mechanician," Wallace writes, "in designing, building and repairing tools and machinery had to be primarily visual and tactile … and this set it apart from those intellectual traditions that depended upon language, whether spoken or written." If Wallace has correctly typified two orders of thought, then a problem in anthropology is that the anthropologist is a language-dependent person whereas many of the people he or she describes have been visual and tactile people. "The product of the mechanician's thinking," Wallace writes, "was a physical object, which virtually had to be seen to be understood; descriptions of machines, even in technical language, are notoriously ambiguous and extremely difficult to write, even with the aid of drawings and models. ... If one visualizes a piece of machinery, however, and wishes to communicate that vision to others, there is an immediate problem. Speech (and writing) will provide only a garbled and incomplete translation of the visual image. One must make the thing—or a model, or at the least a drawing—in order to ensure that one's companion has approximately the same visual experience as oneself" (237–38).

As I pursued trees and the things Kula Ring people did with them I became increasingly convinced that much of their understanding was visual and tactile. Late in my 1996 research I began to consider more seriously the pictures I was taking, then mostly of flora; as the years moved along and my boat investigations intensified, increasingly these pictures were about what people turned their flora into, their outriggers. This passion deepened as my canoe advisers—Adrian Horridge, Geoffrey Irwin, Pierre-Yves Manguin, and Erik Pearthree—encouraged, advised, and suggested I was dealing with a unique and complex sailing tradition. Realizing how important these pictures were to the audiences I have been presenting this material to since 2000, in 2013 I decided that a photographic component needed to accompany this text. Through the University of Virginia's renowned Sciences, Humanities, & Arts Network of Technological Initiatives (SHANTI), Rafael Alvarado connected me to Renee Reighart, a digital scholarship and services librarian. Since the spring of 2013 we have assembled an online photographic essay to accompany this text. Organized by chapters and their sections, as this work goes to press there are approximately three hundred captioned images meant to augment and extend this written account. These may be viewed at https://pages.shanti .virginia.edu/Trees_Knots__Outriggers/. More will be added to this study and additional photographic, online essays are planned.

Beyond the University of Virginia's context a number of individuals and institutions must be thanked. I begin with Maurice Godelier, who has served as my root to French Anthropology since he first brought me to Paris in 1982. Leaving aside his influential writings, several times he has welcomed me to his village and sponsored sympathetic audiences for my intensely ethnographic presentations. He introduced me to André Iteanu. Defined in relation to Dumont's work, Iteanu's voice—one knowledgeable about wind and boats—backgrounds this study. Initial versions of chapters 5 and 6 were delivered in front of groups composed of people Iteanu assembled from the Dumont *equip*, ERASME. I hope the honor I received from being associated with that group is modestly repaid in what attends here. Long ago, also through Godelier, I met Pierre Lemonnier. But it was not until 2000 that we connected concerning my boat material, some incorporated in Lemonnier (2012), much more in this volume. Then I only knew that it was important, but not really how or how to go about pursuing it. Lemonnier's gentle encouragement and critical friendship—along with his associates like Ludovic Coupaye—have been pivotal for finishing this book.

On my ways to and from Papua New Guinea, Canberra and the Australian National University have been stopping-off points since the very beginning of this project in 1991. Part of this association was through my long-term personal and professional relationships to Chris Gregory and Mark Mosko. But increasingly ANU became for me the institutional center of the Pacific and Austronesian studies with which this book is aligned. The people I have been fortunate to meet and discuss my developing material with—Michael Bourke, Shirley Campbell, James J. Fox, Bill Gammage, Robin Hide, Adrian Horridge, Andrew Pawley, Kathryn Robinson, Malcom Ross—are masters who kindly steered and encouraged this interloper. In time Fox's Austronesian Project will be recognized as one of the great intellectual contributions of the late twentieth century, and this work is a modest contribution to it.

I initiated this project in 1991 when I first went to Taiwan and then Kunming in Yunnan Province, the intended place in China for the other half of my planned research. Ho Ts'ui-p'ing, then my student, now my friend, colleague, and adviser, was my introduction to things Chinese. And what an introduction it has been, if first to Yunnan with its many likenesses to the Austronesian world, now to the scholarship of China in general and Fujian Province in particular. If in the Malinowskian tradition I look at China as if it was the Kula Ring, thanks to Ho Ts'ui-p'ing I reciprocate the perspective. And by the aid of the artists and scholars I met through her, I number among my associates Ken Dean, Stephan Feuchtwang, David Gibeault, Zhang Bin, and, far from the least, Wang Mingming. Wang Mingming and his legion of students have kindly brought another foreigner into their midst, increasing our understanding not only of East and West, but the passage between East Asia and the Austronesian world. That influence leads into this work's closing chapters. Ho Ts'ui-p'ing also introduced me to Susanne Küchler's work on knots (2001) after my 2002 return to Muyuw. That enabled what now appears as chapter 5 in this book. Our conversations also brought me to Taiwan for part of a year in 2004, allowed me to complete chapter 2, and gave me a more sustained understanding of forests, patches, attentions to detail, and modeling procedures—arguably great themes in the cultural systems that swirl across the Indo-Pacific. As my ties to China continued the concluding chapter in this book was first drafted in Quanzhou in association with the Quanzhou Maritime Museum and its delightful staff.

Muyuw, what some call Woodlark Island, is in Milne Bay Province in Papua New Guinea. Many provincial government officials, especially in Alotau, have said yes and given enthusiasm to this study. Standing for many of them is Mr. Titus Philemon, Governor of Milne Bay Province as this book goes to press and a politician whose name has been on the lips of Muyuw people for most of the present century. Some government person was always on Muyuw while I was there: among those that stand out and helped me the most are John Alesana and David Mitchell. Their support and friendship helped make this the study it is.

Simon Bickler and I have been together on things pertaining to Muyuw since we first met in 1991. We not only shared months on the island in 1995 and 1996, but have been engaged together over the island and its peoples on a weekly if not daily basis since then. Some of this entailed a three-way conversation with archaeologist Roger Green; some of it is represented in the maps and charts Bickler contributed to this volume. With many thanks the conversation continues.

The people of Muyuw island and I have now shared four years of our lives together. We are family. Nancy Coble Damon, my wife these long years, was with me for my original research on Muyuw and listened to much of the material presented here too frequently to tell. She brought her own rich life into my world of research, and, with our children, returned to the island with me in 1998. In 2009 my son, David, joined me for an important two months, effectively becoming a member of Wabunun—his observation that I was talking about the boat form that centers this work as if it was a mathematical expression defines a key contribution to this study.

This book sums a complex twenty-five year research period. Special thanks are due to Berghahn Books for taking on its challenge. In particular I thank Marian Berghahn, Duncan Ranslem, Jessica Murphy and Ryan Masteller. For their efforts this book is closer to the piece of literature I would like it to be. Ashley Carse also deserves thanks for a helpful reading of the Introduction and a useful comment about the book's title.

My immediate Muyuw hosts are now the grandchildren of the men and women who first tutored me in their culture. One of those elders, Aisi, blessed my research when I returned in 1995 to say I was going to study trees—he knew how important the topic was and that neither I nor any other Kula Ring anthropologist had pursued it. His sons, especially Dibolel, Ogis, and Amoen, are major sources for this book's contribution to Pacific studies. Aisi brother's children, first Milel then Leban Gisawa, and their wives, have been seeing me in and out of the country since 1991. We debate the wisdom, or lack thereof, of mining, timbering, oil palm plantations, and traditional versus modern practices. All of us have laughed and cried together with life's tragedies, and we shall continue to do so. Although I have never found it easy to leave my US home and to go to my home in Muyuw, whenever I leave there I marvel at the life these people have allowed me to share. Muyuw is the "basis" of my life. For this I pass profound thanks to its future.

PART I

Among the Scientists

New Perspectives on the Massim

"Like all savages, the natives are suspicious of strangers."
C. M. Woodford*

---

*C. M. Woodford, *A Naturalist among the Headhunters* London: Philip George & Son
P. 8. quoted in Damon (MS). Charles Morris Woodford was a Fellow of the Royal
Geographical Society; corresponding member of the Zoological Society; Fellow of
the Royal Geographical Society of Australia; and Fellow of the Linnæan Society of
New South Wales. He visited the Solomon Islands three times in the 1880s.

Introduction: Changes and Last Chapters

Sipum

This book is the culmination of research begun in 1991. I was interested in how people in and around Muyuw in the northeast corner of the Kula Ring located themselves in their environment. Returning to Muyuw after research there in the 1970s, I was informed about but not particularly interested in the vast changes—continuous gold exploration and timber and then oil palm plantation plans—that were once again making the island a proletarian speck in a renetworked world. By virtue of an 1895 gold rush, dreams about copra plantations, and the way steamships had apparently opened the Pacific to development, the island had been one such space for several decades at the turn of the twentieth century. But by my first experiences on the island in the 1970s its ties to the West had receded, villages were tentatively reconstituting themselves, and it seemed possible to investigate the culture at a remove from its European context. To this place I desired to return.

Yet from the very beginning, changing circumstances redefined this new study's purpose. One of these came during that initial return in 1991. The first half of this book documents the consequences of that encounter: I provide an account of the place of flora within the northeast sector of the Kula Ring. This book is an ethnography of flora. It describes how trees and other plants are understood and used to make and comprehend lifeways.

The second and third shifts came during and after my 1996 research. One is realized in the last half of this story. The investigation of flora led to an investigation into the structure and place of outrigger sailing craft in this cultural system. These forms create an argument about how the region relates to a major climate pattern in the southern world—, *El Niño Southern Oscillation,* from here on ENSO or *El Niño.* This phenomenon generates a problem of knowing, of knowledge structures. A general consideration of knowledge forms outlined in the first half of the book builds to their more focused pursuit in the second. Chapter 2, the beginning of Part II's ethnography of trees, begins with a bizarre event from my first research time, 1973–1975. One evening in 1974 one of my age-mate informants underwent a sudden altered state of consciousness reportedly because he saw a tree he was not supposed to see. Eventually we will see how that tree is a signpost for regional relationships, as is the conscience which enables the life I attempt to describe.

Suffusing the whole, the third unfolds around the ravages of time and history.

From 1991 I had planned to convert this research into a two-part comparison. The eastern half of Kula Ring is one part. China is the other: I visited Taiwan and the People's Republic of China for the first time just before returning to Muyuw in 1991. As I made repeated returns to Muyuw, so I continued visiting East Asia. Gradually a historical dimension merged the two. For I became increasingly familiar with the Kula Ring as a transformational moment in the Austronesian pulsation out of East Asia some four to six thousand years ago. Although footnotes and chapter 6's final synthesis allude to possible morphological conversions across these spaces,[1] this study synthesizes what I have learned of the northeastern Kula Ring.

One development from this synthesis became a profound sense of writing a last chapter in a Pacific story, undoubtedly one of many, nevertheless a last chapter. Among the events that heightened this realization was the tragic death in April 2009 of my friend, mentor, and jungle guide from 1995 through 2006–7, Amoen Aisi. Nicknamed "Sipum" because his ruddy face reminded people of a yellowish-orange flower identified with chicken pox, he was the son of Aisi, one my earliest instructors. In his late teens during my 1970s research, Sipum verged on an elder by my return. An ebullient man with aggressive intelligence, I treasured my association with him. But Sipum's death did not stand alone. His two sons predeceased him. Both drowned tragically, one in nearly inexplicable circumstances; some of its grief I shared with him and his wife in 2002.

Sipum became sick during a proverbially successful wild pig hunt he led one Wednesday afternoon in April. That hunt had followed a uniquely stunning fishing trip Tuesday evening; others went fishing too but caught nothing. Retrieved from the hunting grounds and transported immediately to the aid post located by the World War II airstrip at Guasopa, he was returned to his village, Wabunun, Thursday evening where he died in a sister's son's house early Friday morning. He was moved immediately next door to his own house where he was set up for visitors from near and far until mid-day. He was then buried. Following this, his daughter, his last child, came down with the same symptoms. Calling "Father I come to you," she died and was buried Saturday.

Sipum represented many of the tensions on the island. Many believe his wife killed him, their three children, one of her sisters, and her mother with witchcraft (*bwagau*); as I left in August 2009, the village was organizing action against to her. A picture my son took of us captures our anguish.

Sipum was a middle brother surrounded of two elder brothers with the extraordinary success in the *kula*, and younger brothers with phenomenal commercial success in the encroaching Western sphere. He became the village's expert on the bush, but not just that—he also made himself a sewing-machine mechanic and took great pride in fixing Singer machines across the eastern part of the island. He invented a special belt that allowed him to neatly insert elastic bands into the cloth skirts all women now wear. I learned these baffling but not surprising facts on the last day I saw him during 2007's brief return. His death, and his daughter's, however, brought to the fore another facet of his being, one closer to what attracted us to each other. Sipum *studied* traditional magic and the empirical world that gave it content—trees and winds and cloud formations—and was working on new material he had learned from people elsewhere in the region. He wrote magic down in a book. Fearing that unleashed powers from this practice led to Sipum's death as well as his daughter's, to say nothing of his sons, his subclan relatives insisted his books be buried in her grave.

A Last Voyage

I first settled on this project by trying to understand what people meant when, in 1991, they told me they used a tree called *gwed*—*gweda* in the Trobriands—to reproduce soil conditions for their horticultural fields. To pursue this issue I knew I had to make a general inquiry about the culture's flora. I thought I had finished collecting that data when I left Muyuw in late July 1996, exhausted and unknowingly malarial. I wound my way home, passing through Canberra's Australian National University and then Auckland, New Zealand, to see, respectively, Chris Gregory, an old friend and colleague, and Simon Bickler, a new friend and then a graduate student who had finished archeological research on Muyuw several months earlier. In Canberra I told Gregory that I unexpectedly learned a great deal about outrigger canoes, *anageg*, which Nancy Munn described in her classic account of Gawa (Munn 1986). He told me that I must meet Adrian Horridge, a legendary and founding figure in ANU's Research

School of Biological Sciences. Although retired, Horridge maintained an office and kept working in his main field, insect optics, and his intellectual avocation, Indo-Pacific sailing craft. We met for coffee in his building's outer public spaces. He quizzed me to see if I was worth his time, and apparently I was, as he invited me to his office. Among other things, I described springs people built into canoes and the peculiar cross grains of a tree species used to fashion keels. Although then a world expert on Pacific sailing craft, he claimed he had never heard what I told him.[2] Then as I talked about what puzzled me most about Muyuw sailing craft—those vibrating parts—he gently introduced me to his writings about Pacific sailing and invoked various physical principles—among others Grey's paradox—that might suggest lines of interpretation for my puzzles.

Surprised that the boats were still being made and sailed, Horridge urged me to return to learn more—I was conversant with the culture, unlike almost anyone else in the world, so who was going to study and describe this passing technology? For personal and physical reasons this was the last thing I wanted to hear. However, by the time I flew to Auckland several days later I started recalculating. Simon Bickler and I walked through Auckland's Maritime Museum, looking over its display of Pacific sailing craft. I was impressed by a tightly wound large craft from the Gilberts built by a European trying to replicate original forms; later I read that its twin, built by the same European, blew apart when it sailed in a stiff wind. We spoke with museum personnel about what I knew. Opening Haddon and Hornell's account (1936) to the Milne Bay sailing craft, they expressed that they were familiar with the literature, but they were intrigued by what I could report. Horridge was correct. My ethnobotanical project and familiarity with the language and people transcended my technical limitations.

So I succeeded in returning to Muyuw in 1998 to further investigate outriggers. I returned the following August, 1999, but had only several weeks in Alotau, Milne Bay Province's provincial capital. Fortunately I spent much of that time with an experienced sailor from Yemga, one of Muyuw's principal sailing villages, and Ogis, elder brother to Sipum, another old Wabunun friend who, since Sipum's death, has assumed that place as my principal tutor; much of the learning embedded in chapter 6 flows from Ogis.

While I had learned much through 1996, by the end of 1999 I had radically transformed the verbal and pictorial data base I had for these craft. However, my scant experience sailing these consisted of two simple trips in the 1970s when I paid these craft little attention and two short, riveting, trips in 1998. Therefore I arranged to return to Muyuw in 2002, hoping to spend time with Muyuw sailors on at least two *anageg*. Although my plans could not work out as imagined, I succeeded in sailing from the southeastern corner of the Kula Ring to my home away from home, Wabunun, in its northeast corner. Bad luck turned fortuitous for a developing thesis and my purposes.

The Voyage of 2002

In July 1998 I measured and drew an *anageg* from Nasikwabw that had sailed to Muyuw for ritual reasons. The boat's owner, called No. 2, was a son of a former Wabunun informant, and he consented to my inquiries. For my 2002 return I hoped to find my way to the island variously known as Koyagaugau, Gaboyin, or Dawson. I then reasoned that the Nasikwabw man would sail to Koyagaugau, carry me back to Nasikwabw, perhaps also to Yemga; I then anticipated hitching a final ride to Wabunun on another boat. Since fewer of these craft are being made and put into circulation, and given other issues, I was prepared for these plans not working out; and they did not.

Milel Gisawa, a youth in Muyuw's elementary school in the 1970s but by the 1990s a legendary announcer for Radio Milne Bay and organizer of the modernizing world for many Muyuw people (symptomatically, he died of diabetes in 2012), accompanied and helped me quickly pass from Alotau to Koyagaugau. We arrived there on June 14, 2002. Ending his campaign for the region's local representative to the Papua New Guinea Parliament (he lost), Milel went on to Nasikwabw and Wabunun, in Nasikwabw under instructions to tell No. 2 that I was waiting for him. On Koyagaugau I stayed with the nephew who had inherited the position of the great Kula man, Mwalubeyay (d. 1995).[3] Unexpectedly, on Koyagaugau Milel also found a young man, Onosimo, who had practically grown up in Wabunun. Milel instructed Onosimo, a polyglot like himself, to watch out for me. He did, and within a day we walked around the island and took a short canoe outing to the little island connected to Koyagaugau, Ole. Ole had several working *anageg*—called *kemurua* in this area—in various states of assembly and reassembly.

It became apparent that nothing was going to happen swiftly. First, mid-June is early in the time of the hard southeast wind, neither the best nor worst time to sail. However, a peculiar southern wind kept blowing, and this made it difficult to sail south, the general direction my craft for its return; and as I was to learn, it was especially hazardous to travel north, the direction we had to sail. Second, this region of the Kula Ring, Bwanabwana, was filling up with mwal, one of the two *kula* valuables. Hence preparations were underway for all of the remaining Muyuw sailing vessels to head there. Several people told me I should just wait for that; but I knew that sailing was unlikely to occur within the two-month period I had—more likely, it was a year or two away. Finally, intra- and interisland sports cycles are now a major factor in organizing travel among islands, so many youth were occupied by upcoming sports events.

These islands are connected by short-wave radio, so a week or so after my arrival on Koyagaugau I learned that No. 2 was not in Nasikwabw; he was in southwestern Muyuw at a volleyball and soccer tournament. Later, when I had arranged passage on the Ole *anageg Lavanay*, its owner had difficulty assembling a crew because many young men were practicing for a major soccer tournament, the Kula Open. New times bring new configurations of power and the body's relations to alienable objects: these times capture people's fancies now.

Worried that it may be months before No. 2 could come for me, I asked the small but regal looking owner of the *anageg Lavanay*, Duweyala, if he would take me to Muyuw, stopping in Panamut and Nasikwabw on the way. He was the younger brother of a Koyagaugau man named Gideon, also a polyglot,[4] who had become my tutor. Gideon had spent a lot of time on Muyuw's southwestern end. Both men, as was widely the case in these islands, spoke Muyuw as well as or better than me, and easily passed back and forth among the terms Bwanabwana and Muyuw use for the boats. On June 25 I moved from Koyagaugau to Ole to await our passage.

It was now apparent that I would not have several weeks on Nasik-wabw with its sailors and then more time in Yemga before finally heading to Wabunun. I needed to check what I was learning from people I knew best and who knew me well enough to correct my mistakes. This presented problems because I had no choice but to start learning boat details from Ole and Koyagaugau people, a significant issue because I was especially interested in the relationships between the boat's technical aspects and the way these redounded into the local culture. Bwanabwana culture, the ethnographer for which is my friend Martha Macintyre, is different from Muyuw culture. And while I learned much about the Koyagaugau and Ole variants of it, I by no means became nor wanted to become an expert in it or its language.

In any case, as Duweyala made steady progress outfitting his boat for our voyage, two matters posed challenges. The first was an infection on my left leg and a related or unrelated fever that was clearly becoming serious by the first week in July. At one point Gideon looked at my leg and told me it was rotting. Assuming the fever was malaria, I took an acute dose of lariam (Mefloquine) the day before our scheduled departure and told Onosimo, who had now agreed to sail with us, that if I became delirious and got out of control he was to knock me out rather than let me drown or delay the trip. I told nobody else about the fever for fear they'd refuse the trip, thereby voiding two years of planning. Fortunately by the morning of our departure the fever broke; however, the leg remained swollen.

The other problem was the wind.[5] By this time, and into August, Duweyala would look at the flowering "Beach *Calophyllum*" (*Calophyllum inophyllum*) and say that its flowers appear with the southeast wind. Although its direction wavered, the wind blew, unusually, from the south and remained very hard. It slackened enough that we were able to depart on July 4. As is usually the case, lots of people helped our boat into the water and worked to get us on our course before jumping off and swimming back to shore. I suspect they left with a mixture of regret and nostalgia. Throughout this region people are giving up *anageg* for the new faster class of outrigger called *selau,* fiberglass dinghies and their outboard motors, and diesel vessels. The latter ply these waters less frequently, but they carry much of the interisland loads of pigs, vegetables, clay pots, and other things for which *anageg* were designed. Yet the *anageg* remain carved into people's hearts and minds. As they helped us get underway, many men spoke to me about the joy and thrill I'd have on this voyage.

I carried a GPS with me to regularly record our position, direction, and speed throughout the trip. Map 0.1 and table 0.1 display the GPS information.

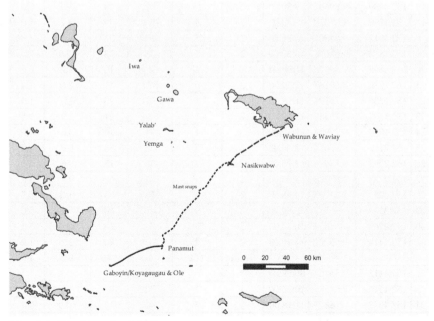

Map 0.1. The Voyage of 2002.

Table 0.1. The Voyage of 2002, July 4–15, 2002[6]

GPS READINGS
FROM OLE TO PANAMUT; PANAMUT TO NASIKWABW;
AND NASIKWABW TO WAVIAY
July 4–15, 2002

	Longitude	Latitude	time	Voyage	Distance (km)	Bearing	Time (mins)	speed (km/hr)	knots
1	151.4269	-10.3997	9:54:00	1		71			
2	151.51	-10.3611	10:48:00	1	10	65	54	11.1	6.0
3	151.5678	-10.3225	11:29:00	1	8	56	41	11.1	6.0
4	151.5147	-10.2836	12:18:00	1	10	64	49	12.0	6.5
5	151.7394	-10.2514	13:19:00	1	11	70	61	10.4	5.6
6	151.7908	-10.2392	13:54:00	1	6	76	35	9.9	5.3
7	151.8514	-10.2358	14:40:00	1	7	87	46	8.6	4.7
8	151.8633	-10.2372	15:00:00	1	1	97	20	3.9	2.1
				Total	52		306	10.1	5.5
9	151.8656	-10.2375	4:44:00						
10	151.8608	-10.2222	5:18:00	2	2	353	34	3.0	1.6
11	151.8831	-10.2958	5:43:00	2	4	40	25	9.1	4.9
12	151.8644	-10.1469	6:22:00	2	6	339	39	8.9	4.8
13	151.9553	-10.0881	7:08:00	2	12	57	46	15.5	8.4
14	151.9964	-10.0231	7:54:00	2	8	32	46	11.1	6.0
15	152.0372	-9.95667	8:42:00	2	9	31	48	10.7	5.8
16	152.0989	-9.89667	9:30:00	2	14	62	48	17.7	9.6
17	152.1158	-9.87778	9:59:00	2	4	298	29	9.2	5.0
18	152.1517	-9.84306	10:36:00	2	5	45	37	8.9	4.8
19	152.1828	-9.81056	10:59:00	2	5	43	23	12.9	7.0
20	152.1808	-9.80528	11:09:00	2	1	340	10	3.7	2.0
21	152.1786	-9.80167	11:20:00	2	0	329	11	2.6	1.4
22	152.1736	-9.79194	11:46:00	2	1	333	26	2.8	1.5
23	152.1694	-9.78389	12:07:00	2	1	333	21	2.9	1.5
24	152.1725	-9.77944	12:17:00	2	1	34	10	3.6	1.9
25	152.2111	-9.7625	12:50:00	2	5	66	33	8.4	4.5
26	152.2292	-9.75028	13:05:00	2	2	56	15	9.6	5.2
27	152.2783	-9.67778	14:05:00	2	10	34	60	9.7	5.2
28	152.325	-9.61167	14:58:00	2	9	35	53	10.1	5.5
29	152.3889	-9.55778	15:58:00	2	9	49	60	9.2	5.0
30	152.4269	-9.54694	16:24:00	2	4	74	26	10.0	5.4
31	152.4414	-9.55222	17:10:00	2	2	110	46	2.2	1.2
				Total	114		746	9.2	4.9
32	152.4417	-9.55167	09:57:00	3					
33	152.4519	-9.53389	10:35:00	3	2	30	38	3.6	1.9
34	152.4972	-9.4875	11:18:00	3	7	44	43	10.0	5.4
35	152.5814	-9.43889	12:22:00	3	11	60	64	10.0	5.4

36	152.6364	-9.40278	13:08:00	3	7	56	46	9.4	5.1
37	152.6947	-9.36611	14:03:00	3	8	58	55	8.3	4.5
38	152.7564	-9.33528	14:59:00	3	8	63	56	8.1	4.4
39	152.8061	-9.31222	15:43:00	3	6	65	44	8.2	4.4
40	152.8381	-9.3	16:13:00	3	4	69	30	7.5	4.0
41	152.875	-9.27722	16:45:00	3	5	58	32	8.9	4.8
42	152.8981	-9.26167	17:05:00	3	3	56	20	9.2	4.9
43	152.9014	-9.26056	17:12:00	3	0	71	7	3.3	1.8
				Total	60		435	8.3	4.5

I hoped to play whatever active role I could on our voyage. The trip's purpose was to begin to parlay the intellectual models I was constructing into experiential knowledge. But I was frustrated in these attempts right from the beginning. First, of course, we all knew I had little sailing experience, so I planned to spend the first leg of the voyage just watching (though maybe also bailing). But second, my companions quickly determined that because of my weight they should put me in the rear of the boat (actually the prow, which, because of our direction and the wind, had to follow for the whole trip). There I stayed until we finally drifted from Waviay to Wabunun twelve days later.

The trip from Ole to Panamut, a low island formerly occupied and now a source of food for stranded sailors, was under clear skies and mostly uneventful. We pushed off before 10:00 AM and saw Oleymat, a higher island near Panamut (from which Koyagaugau's sister island Ole gets its name), by about 10:15, long before Ole and Koyagaugau had sunk beneath the horizon; just before 3:00 PM we finished poling the boat into its shallow resting place for the evening. It's about twenty-two miles from Ole to Galakalaieliya, an islet on a reef extension north of Panamut, where we finally stopped.

Although the trip was uneventful, I noticed fairly quickly a discrepancy between how I was told the boat would be steered and how it actually was. These craft have to sail so that the outrigger float is into the wind. The force of the waves coming into the vessel is such that the steering oar (*kavavis*)—technically not a rudder, but rather something approaching a "lee board"—is supposed to be lifted up as a wave passes from the outrigger float to the keel. Otherwise the energy might snap the piece in two. Then as the wave passes underneath the boat, the *kavavis* is inserted back into the water to redirect and gently lower the boat as the swell and boat pass each other. A kind of rhythm can develop with this raising and lowering, and in 1998 I briefly experienced that rhythm in an *anageg*, when a number of crew starting calling out "lift" and "lower." Their words were not so much commands as voices of collective participation. I knew that men sometimes sang during their trips and wanted to experience a sense of rhythm potentially related to raising and lowering the rudder, among other vibrating boat parts. And so for this first leg I carefully watched for a rhythm, but I could never quite understand the pattern. The same thing happened on the next two legs of our journey, only then the wind was stronger and we sailed with two leeboards. I was told my companions would coordinate their movement, and there was sometimes talk between, usually, Duweyala at the back and whoever handled the forward leeboard a meter or so in front of him. But the pattern remained opaque to me.

Duweyala was also paying very special attention to the mast (*vayiel*), as it was made from the wrong kind of tree. And once I talked him into the voyage, one of his motives was to go to Sulog in south central Muyuw to obtain one of the two types appropriate for the part. I hardly noticed another item, but it was evident and I eventually understood the issue: While Duweyala did not hesitate to look all around him to gauge speed, direction and distance, he also kept watching the *bis*, the streamers, tied to many places on the sail's sides.

We spent July 5 at Panamut because I wanted to see it. I frequently heard about Panamut because it figures in origin travels of the clan to which I had been assigned and because most Muyuw people who have sailed back and forth between the islands talk about it and its food resources. According to the accounts I received in Ole and Koyagaugau, it was occupied and also the center for mediating north/south relations until about 1900. However, my crew chose not to violate prohibitions placed on the island—this was why we beached instead on the islet beyond it and why when we went there I was only allowed to explore a few hundred meters inland from its northern end; anything worth seeing—such as its potshards and a range of significant trees and their ages—would be at the southern end.

While several of us explored what we could, a crew member remained behind to rig up support for the mast. Duweyala had decided that the mast was not strong enough for the wind, and so he ordered the adaptation.

Although it was mostly cloudy and windy early in the morning of July 6, Duweyala ordered us up about 4:00 AM. We were loaded by 4:40 and began drifting to the northern end of the Panamut structure. By 5:10 AM the sail was up; we arrived at the Nasikwabw landing twelve hours later. Since the day varied from being very hazy to cloudy, setting and maintaining a course was difficult, and Duweyala kept looking back at Panamut—and its light beacon—until it completely disappeared. On a clear day the mountains of Misima to the southeast and Normanby to the southwest could be used as additional backsights for more of the voyage, but not that day. The voyage south from Muyuw or Nasikwabw is considered simple because you cannot miss islands stretching to the east and, to the west, the whole mass of New Guinea. By contrast, the trip north from Panamut is more challenging as it does not have similar features to help with orientation, so markers, including wave structures and the appearance, disappearance, and reappearance of bird life are watched carefully. The difficulty is noted by an adjectival term for this segment of the voyage: *nuwa veveyua.* My Muyuw informants say this is a Misima word phrase meaning "insides fly out," illustrating the nerves sailors have while traversing this distance.[7] In this long stretch of sea, islands to the south and north may be invisible for hours, and bird life, important for navigating and gauging distances, also disappears. It is not the place for a breakdown.

Duweyala navigated by studying the swells and the shadow cast by the sun passing over us; that shadow changed over the course of the day and was fixed to the time of the year we sailed, so it was much less constant than the star routes people prefer to use on these voyages. In any case, the wind remained strong the whole way; as we got closer to Nasikwabw, waves towered over us as they came almost from our rear. It was an eventful trip.

During this voyage the GPS compass was sensitive to the relatively wide arcs the boat traversed as it went up and down the waves, hence the readings in the chart. Yet many of the hand compass readings I took suggested we maintained a fairly steady course close to 40°; this was the same direction my watch guard, Onosimo, detailed the evening before when I asked him where we would sail the next day; a cognitive map of island whereabouts is well-formed among these people. However, about 9:30 Duweyala started yelling; the mast and sail was radically changed, and we seemed to head into the waves close to a 120° direction for about ten minutes. Just before 11:00 we experienced another episode like this, heading off at 60–70°. Right after we returned to our course the mast snapped, the sail blew into the water, and we started drifting. We were in an emergency situation so I had no choice but to stop asking questions, refrain from anything else that would seem impertinent (like taking pictures), and concentrate on taking more frequent GPS readings—the only way I could be of any help whatsoever—so, if need be, I could chart our movement on a map which remained rolled up in my bags.

Duweyala and the crew instantaneously swung into action. Their first act was to rescue the sail. Later I learned that none of them had ever had an experience like this, but one principle everyone knew is to never let the sail get into the water. Sometime later when the danger was over and the expressions on everybody's face had changed to obvious relief, Gideon looked at me and said, "We are seamen"; days later Duweyala told me he was never worried, only determined.

We drifted north for about an hour while the crew rigged up a new mast from the remains of the old. By 12:15 we were underway, our course adjusted more to the east to account for our drift. Soon birdlife appeared, and shortly after 1:00 PM one man sighted Nasikwabw.

It was well into August before I thought I understood what had happened to us. Duweyala was watching the telltales (*bis*) on the sail as well as other parts of the boat, keeping an eye out for eddies off the main wind current that might immediately capsize us. Two large eddies were the cause of the major course diversions. I do not know if another accounted for the snapped mast because it was obvious we were sailing near the limits of the tree used for it. Later Onosimo told me he had cautioned Duweyala about the wind's speed and encouraged him to lower the sail. But Duweyala was concerned about his "cargo"—me—and figured that moving faster to Nasikwabw was safer than moving slower.

We remained on Nasikwabw from July 7 to 15. During our stay Duweyala and Gideon prepared a new mast for the final leg of our voyage. Nasikwabw is full of the *Calophyllum kausilay*, the same species as the broken mast. Although this type of tree is good for a keel, it is inappropriate for a mast—yet it had to do. Shortly after we were underway on July 15, both Gideon and Duweyala expressed great pleasure in how the new mast was performing. Yet the hard southern wind continued to blow, and halfway into our voyage both men realized it was not going to work. As they had for much of the voyage from Panamut to Nasikwabw, the two men steered, each with a *kavavis*. Although the wind was strong, this leg of the journey was uneventful. The Sulog Mountains in south-central Muyuw are almost always visible from the shoreline of Nasikwabw, so we did not encounter any interesting navigational issues during our journey. But as we veered toward our destination, the wind came increasingly from our front. I had read Ben Finney's 1994 account of sailing in the mock-up Polynesian rig twice while I was waiting in Koyagaugau, and I began to think that we were sailing much closer to the wind than Finney thought Oceanic craft were capable. Knowing my lack of sailing experience and therefore fearing for my ability to record what I was seeing, I took a photo to try to capture a flag in the wind that seemed to be blowing into us. The arc of the sail roughly paralleled the fluttering flag as we approached our destination, and it may very well be that in this context the sail was functioning like an airplane foil, pulling rather than pushing us. Several years later, another friend and expert on Pacific sailing craft, Erik Pearthree, examined a model sail made for me and remarked about how the "back" of the sail was smooth compared to the rough "inside." The inside is the part designed to catch the wind; the backside is smooth perhaps to facilitate the wind's rapid movement over its surface.

I had envisioned a dramatic landing at Wabunun, my final destination. I had watched many an *anageg* sail to the island and be dragged up on its brilliant white sandy beach, but I never participated in the act from a boat crew's perspective. However, we sailed to Waviay instead, a small island on the outer lagoon just across from Wabunun. The next day we would sail across. I insisted that I be allowed to do something, bail the boat or handle the "rudder." However, I was not needed; in midmorning we merely hoisted a coconut frond and little more than drifted onto Wabunun, customarily drawing scores of people to help beach the canoe and greet the crew. For me *anageg* voyaging was over.

But not for Duweyala, his crew, and passengers. Originally he intended to stay in Wabunun just long enough to cut a new mast at Sulog. A fleet of Nasikwabw boats planned to follow us to Muyuw, and Duweyala intended to return to Nasikwabw with them in a couple of weeks, where he would then replace parts of his vessel, entirely retie it, and return home. However, that peculiar south wind kept dominating the weather. So he switched plans and reworked his boat in August in Wabunun, where more of the appropriate trees were available. Since it appeared for a while that no diesel vessel would be heading our way, partly because of the severe weather, I began plans to return to Ole with Duweyala, an idea most thought absurd; I was the wrong kind of person (a white man) on the wrong kind of boat (a "New Gin" wag). However, a ship finally came to the island and I left on it on August 15. Duweyala remained on Wabunun for several weeks, finally sailing back to Ole under a wind so undesirable that he veered to Normanby Island well to the west of Ole before doubling back.

Shortly after returning to Alotau I learned that an El Niño was brewing; this is what had led to the persistent south, rather than southeast, winds. By December a drought was building in the province, one, however, that did not compare to the 1997–98 ENSO. During my 2009 return I learned that our voyage and Duweyala's return was the last round trip an *anageg* with the traditional sail had made.

Practicalities (1): Original Intentions

I did not start this research to investigate outrigger canoes or trees; rather, I first returned to Muyuw in 1991 to focus on the island's "ethnogeology." About 1990 I determined to give a double twist to my research and teaching by taking up a new topic at my original research site and then turn my interests to China. When the project finally started in 1991 only one thing was certain: I wanted to learn about phenomena that were beyond my experience and training. In this I succeeded. And this is why this account combines, to the best of my ability, storytelling features—Dickens is my model—with soil and ecological sciences, recent concerns about climate and culture, and a theory about how the focus of my endeavors, Muyuw, fits into the structural history of peoples who occupy the Indo-Pacific.

Anthropology is a science about people in specific spaces and times conducted by people from specific spaces and times. I stress the personal aspect of this work because my discoveries have been so dependent upon the willingness of other people to explain so much that was beyond my understanding. If this begins with my close and distant friends from the Kula Ring, it now includes many other people from Australia, China, England, France, New Zealand, Taiwan, and of course the United States. Among these people I was an interloper fearful of their suspicions and dismissals.

The plan was to take up a new topic in an ethnographic context I already knew and then turn my redefined focus to East Asia. I desired to change my theoretical interests from theories about social systems emphasizing exchange, production, and ritual to questions oriented by environmental research. I conceived my project to be an exploration of the ways that "technical" knowledge related to the cosmological and social structures described in many previous publications. Although my regional focus on "China" changed over the years, when I started the project it could just as easily have been on India; I wanted to explore the continuities and discontinuities across these regions, which, in fact, have been interrelated for millennia. A foray to India in 1998 became a turning point in this story. In Buddhist origin mythology I found interest in flora that paralleled what I learned about Muyuw in 1995 and 1996 (see Damon 2007).

I was not, of course, sure how this deliberate maneuver—taking up a research agenda for which I was not prepared and moving from Melanesia to Asia—would turn out. Yet that was the point: I wanted a challenge that would be transformative.

Although sympathetic to the environmental movement that had been defining Western culture since the 1960s, I was professionally uninformed and, trained as a symbolic anthropologist in the Anglo-French tradition, largely unsympathetic to what I understood of ecological anthropology. Yet it was my intent to enter this new territory, and I wished to do it not just by learning from a subdiscipline of anthropology—I also wanted to initiate dialogues across the "arts" and "sciences." I did not, and still do not think divides between the branches of the academy are healthy for either understanding or making the world we inhabit.

I put the plan into action by the mid-1991 return to Muyuw, preceded by a week each in Taiwan and Yunnan Province, People's Republic of China. By 1995 I had planned on two additional returns; to date there have been seven, intermixed with enough trips to China and Taiwan to begin to transform that part of the design into a real one.

Lacking a deep natural science background, I selected the area about which I had the most interest and the greatest amount of positive knowledge, as I understood the island as a westerner and as my Muyuw friends presented it as a humanized landscape: so, geology as the area of interest for exploring the movement of water on the island. Had I been fully cognizant of subdisciplines in the environmental sciences, I would have added hydrology. My new research was going to focus on how islanders conceptualized their place as a physical entity, and how that knowledge related to the practices I already knew.

I had held a lingering interest in geology and always found geological writing about the southwest Pacific fascinating—the crashing of the Pacific and Austral-India plates that generated Melanesia. Geologists I had read or met on the island were engaging people clearly dealing with a thicket of complicated relations. The geomorphologist Cliff Ollier, who had also described many megalithic ruins across the northern side of the Kula Ring, wrote that the region was so complex that only a poet could describe the myriad of conflicting forces that had, and were, producing it. A result of larger encompassing forces, the island is a mixed classic coral atoll—or atolls—with a volcanic center circumscribed by a coral platform. The northern side of the platform rises as much as one hundred meters above the sea; the southern side seemingly slides beneath it, creating barrier reefs and extensive lagoons. The region is tectonically active.[8] The island's occasional movement is fully incorporated into the reigning cultural system. Elder instructors in the 1970s told me that earth tremors— *nikw*—result from the Creator moving around to look at something. The Creator supposedly holds the island on his—perhaps her—head, just as women carry loads on their heads. The strain of this load is represented in a Muyuw mortuary sequence, called lo'un (Damon 1989b). Designed to end the debt created by a father producing a child, in the ritual people witness selected women straining to hold and walk with a gigantic basket of yams and taro on their heads. Some people also tie tremors to the onset or end of droughts and exceptional wet periods. Arguably the conception is related to the ways in which Muyuw cultural forms are carved out of a dynamic that also produced Trobriand culture some hundred kilometers to the northwest; Trobriand chiefs are credited with controlling droughts and inundations there.

On hindsight my initial program was apt. These ideas were formulated before I was aware of the complicated ways in which concerns with water organized the vast cultural and technical systems of South and East Asian societies, before Lansing had published, or I was aware of his immensely stimulating work on Bali (Lansing 1991, 1995, 2006),[9] and before I was familiar with Kirch's wet and dry systems in Polynesia (1994). Such research would have lent credibility to questions I wanted to ask, though not necessarily to the phrasing of my hypothesis. Rather I was being guided by intriguing perspectives from a former University of Virginia graduate student, Christopher Taylor, who developed an understanding of Rwandan society based on the movement of liquids (Taylor 1992, 2001). From my earlier research and interests, I constructed a model that suggested the islanders had an organic understanding of their island's physical presence. Aspects of the island are named as if it is a body. It has a "back," and a "stomach," "head," and "feet"; current mortuary practices draw on an isomorphism between bodies and the island for placing the dead; burying Sipum's magic with his daughter followed from those ideas. A particularly prominent bay at the center of the island's "stomach" is called Kalopwan. Etymological analysis of this word would focus on *pwan*, which may be translated as "anus." The probability that such metaphorical references might suggest well-formed understandings of the island's physical structure and the movement of water on it seemed high. It would then be a matter of exploring the references and coming to a detailed understanding of the implicit and explicit knowledge they presumably contained. I established contact with a geologist in the University of Virginia's Environmental Sciences Department in preparation for future consultation then proceeded to the island in 1991 to explore those metaphors and to reorient myself to the language and people.

Twenty years later I do not think I was wrong about the cultural importance of geological settings, nor to peoples' models about them. In 1996 I watched Muyuw people dismiss Western scientists' ideas about how water moves under the island. The geological advisers were exploring conditions for a potential gold mine. Muyuw people quickly realized that the experts knew nothing about how underground water moves and how its movements change—considerably—according to rainfall variation, tidal forces, and shifts caused from the Creator looking about. By contrast Muyuw knowledge—more observational than modular—was profound and continuous. By 2000 I believed that whether one is looking at symbolism or human ecology, from East and South Asia to Australia and into the Pacific, populations built themselves out of their relations to water, and its obverse, fire. By the time I left Muyuw I felt I could argue that the Kula Ring, nearly the entire social system of eastern Papua New Guinea, had been organized to deal with its geological settings amid unpredictable circumstances of El Niño Southern Oscillation—thus the availability of water.

Nevertheless, my work would not focus on geology and underground water flows. During the weeks of my 1991 return I explained to my closest friends my interest in water and the island's physical forms. They laughed at me. I suspect my reaction to their laughter—dropping the topic—was hasty, yet it was conjoined with a startling positive encounter. During the course of those weeks an elderly woman casually mentioned that people used a tree, *gwed*, to reproduce soil fertility. Muyuw gardens are like Chinese temples, complicated constructions consciously encasing the swirl of relations that are—perhaps Shiva-like[10]—the culture's fundamental principles. I knew a lot about them, and had written about them in a major chapter in my first book. But I had never heard this story before, and I had no recollection of any other Kula Ring ethnographer writing anything similar. Moreover, everyone I talked to confirmed the woman's story; Sipum's father, who giggled at my water ideas, added that the tree was understood differently across the islands to the Trobriands, and by there, for him *Kilivil*, it was so important one would be killed if he cut it down. Hyperbole aside, he was telling me something important about something important. My focus was defined.

I have spent the last twenty years trying to understand what was entailed in what I was told those weeks in 1991. How the news of that tree led to the voyage of 2002 is a plotline of this book.

Practicalities (2): Getting Ready

When I returned to my home after the 1991 trip, I set out to engage the experts and gain the expertise I would need to carry out my intended research. My environmental sciences mentor, Hank Shugart, was pleased with the turn of events. A forest ecologist interested in modeling procedures, he had experience in Papua New Guinea and Southeast Asia and was anxious to learn more.[11] Among others, he quickly referred me to a new colleague in his department, Dr. Stephen Macko, a geochemist who uses stable isotopes to trace nutrient flows. Shugart immediately assimilated the story about the tree to the hypothesis that it was a legume and so probably fixing nitrogen; the analysis of stable isotopes should be able to determine that. It seemed that there might be a close fit between the Muyuw model and the kind of information we could gather by tracing stable isotopes from trees to soils to crops. But I needed to fit that tree into the larger body of knowledge that people had about surrounding flora. And for that I would need to learn as much as possible about plants across the culture. Shugart told me that he could not wait until I got attached to a systematist, the expert who would identify the floral voucher specimens I would collect.

After visits to herbaria in Hawaii, Washington, DC, and Harvard, I eventually became associated with systematist Peter F. Stevens, then a curator of Harvard's Arnold Arboretum and Gray Herbarium, later at the Missouri Botanical Garden and the University of Missouri—St. Louis. Stevens became the principal person who directed my collecting and identified my voucher specimens because he had an interest in the flora of what systematists call Malesia, more or less from Southeast Asia to Australia and the Pacific Islands. While the evidence of his passion for plants became clear, it took a while for me to realize the depth of his scholarship (e.g. 1994a, b). Known as a radical in the world of botany, Stevens told me that he didn't think botanists had a coherent idea of "species"; he has named many (Wikipedia lists them). But he exhibited precisely the kinds of characteristics Shugart was waiting for me to discover: deep skepticism, encyclopedic knowledge of flora, and, English eccentricity. Not incidental to these qualities was Stevens's fascination with issues pertinent to anthropological work: impressed by the limitations of his culture's analytical models, he did not look favorably on the anthropological use of our ideas on other regimes of knowing. In any case, I did not know then that our association would lead me to become fascinated by the genus for which he is perhaps the world's expert—*Calophyllum*.[12]

By the time we met, Stevens was deeply engaged in the problematics—and anthropology—of classification (e.g. Scott Atran and Brent Berlin and their associates) and it was about then, circa 1994, that I began to refamiliarize myself with questions about "classification" in anthropology. Brent Berlin's work (1992), which Stevens had reviewed, was my first turn. And Berlin kindly invited me to his department for a lecture after my original return in 1995. I told him that I intended only one more research trip—he smiled and said, "There will be more."

By this time I was teaching a course eventually be called Ecology and Society. Somehow that activity led me to discover William Balée and a 1994 conference he was holding at Tulane on a new turn in ecological anthropology, "historical ecology" (see Balée 1998); he let me attend the conference. Carole Crumley's pathbreaking collection, *Historical Ecology: Cultural Knowledge and Changing Landscapes*, had just appeared; I read it on the plane to Tulane.[13] To my good fortune, among others, Crumley, Joel Gunn, Darrell Posey, and Laura Rival (whose 1993 essay I incorporated into my teaching) attended the conference. In one short time I was introduced to many of the people who were then transforming what, for me, had once been an unconvincing ecological anthropology.[14] Although my biases in Anglo-French social anthropology continue, through these individuals I received new ways for thinking about the society/environment relationships I sought to investigate. Crumley's volume—its language of biomes, ecotones, and patches in particular—pervades this study.[15] Balée's work and friendship became useful and critical for both surveying and analyzing the forests I have now explored.

Although specific directions I first learned from Crumley's volume are employed throughout this study, most significant for me was a reorientation of the scope of the inquiry. I discuss what I mean with respect to Gunn's chapter "Global Climate and Regional Biocultural Diversity" (Gunn 1994b; see also Gunn 1994a) because, as my data base and understanding of the specific ethnographic materials became richer, Gunn's questions became more suggestive.

Gunn attaches a hypothesis about the concrete, instrumental knowledge contained in a society's collective representations to how societies position themselves in time and space. This is a time and space as defined by various macro relations, such as, among others, regional geology and regional effects of the sun/earth orbital dynamics, the latter of which partly determine the amount of solar energy the earth regularly and irregularly receives. The operant theory here suggests "culture" is knowledge, and that that knowledge is constructed with respect to important environmental dynamics. This idea varies little from the Malinowskian understanding of "charter," meaning that distilled cultural constructs impart models for social action. Here, however, the concern is with human action on a landscape rather than social relations per se.

Appearing at a time when it could still be claimed that climate dynamics follow from orbital cycles too large to be effected by humans,[16] and following a theory of solar variation found in the work of Reid Bryson, Gunn suggested that rituals, folklore, and myths might store, or "capture," information that would allow cultures to maintain climate-specific procedures across climate epochs. The relevant time periods here were inside the Holocene, in the last ten thousand years or more. The climate epochs that various scholars delimited were denoted by terms such as the Roman Climatic Optimum (ca. 300 BCE–300 CE), the Medieval Climactic Optimum (900–1300 BCE),[17] and the Little Ice Age (ca. 1300–1900). Time, or the processes that occur in what we call time, becomes important here because it may be cyclical, accumulative, or destructive. Social systems sooner or later must configure themselves with respect to this variation, some of which they may create. At one point during the Tulane conference, Gunn casually mentioned that most of the earth's soils and water have to be considered anthropogenic, i.e. created by human action, and not just recently, but for millennia. Although I was not wedded to a fixed nature/culture divide as anything more than a possible culturally significant contrast, Gunn's phrasing was startling and preparatory. The preparation was twofold: first, it helped me consider Posey's work in the Amazon as a possible model for analysis;[18] second, it helped me take more seriously what I was told in 1991—that by means of trees people were making their soils.

Crumley and Gunn were archaeologists, and it was partly in relation to their work that the initial comparison—Melanesia to East Asia—I conceived at the start of this work transformed from a typological to a time- (and space-) driven question. Yet I was not inspired by them alone. As I made my way home in 1991, I stopped in Auckland to meet an old friend, Geoffrey Irwin, and a student he was sending to the University of Virginia, Simon Bickler.[19] Bickler fulfilled a plan Irwin and I discussed when we initially met during the first Kula Conference in Cambridge, England, in 1978. I was then fascinated by the megalithic ruins I saw on Muyuw, and I conveyed enough about them for Irwin to envision follow-up studies (see Damon 1979).[20] In our plan, Irwin would tutor a student in Pacific archaeology and then send him to the University of Virginia where he would be exposed to our two Melanesianists, Roy Wagner and me, as well as to the expertise of the university's new archaeology program. Although our research projects were not formally linked, Bickler and I both went to the region about the same time in 1995, and again in 1996. Immediately upon our arrival in Wabunun in 1995, we took a boat to the Budibud islands southeast of Muyuw; coincidentally this was our first exposure to Sipum's magnetism and magnanimity. He accompanied us to Budibud and, by his diving wits and fishing magic, kept us well fed. I introduced Bickler to villagers all over the island. We became good friends. My sense of Pacific archaeology has expanded over the past twenty years along with his, especially as he teamed with the elder Polynesianist, Roger Green (d. 2009), with whom we regularly met to discuss the similarities and differences throughout this "region," a space that contracts or expands depending upon what kind of perspective, temporal among others, one gives it.

An early result from Bickler's syntheses (Bickler 1997, 1998; Bickler and Turner 2002; Bickler and Ivuyo 2002) showed an epoch of monument building from perhaps before 900 CE to roughly 1350–1400. By the latter time he could argue for a new mortuary system (close to the system throughout Milne Bay evident at contact in the late nineteenth century) and, perhaps, the beginning of the Kula. Since pot production and exchange dynamics also changed about this time this, transformation almost certainly meant a new regional setting; this is why I wanted to see the potshards on Panamout. This periodicity fits two of the temporal epochs to which Gunn oriented his work. In fact the divide between the Medieval Climatic Optimum and the Little Ice Age are critical centuries in Melanesia, Polynesia, and throughout Southeast Asia and perhaps the world.

Yet these facts alone were not all that served to expand my sense of the relevant spatial and temporal framework. Through my appearances at Australian National University I became increasingly alerted to James J. Fox's Austronesian project, first being drawn to Clifford Sather's work in Borneo (e.g. Sather 1993). Sather describes social systems in Borneo that, almost eerily, have feels to them reminiscent of the Kula region. By good fortune, Sather had learned things about tree symbolism in Borneo that played closely to the technical information I was learning. At one point I asked him how long the people he worked with paid attention to trees. He answered, "Four hundred years," a calculation made from the way people there organized tree knowledge, their method of tracking human generations. When experienced through the life cycle of forests, time takes on a new dimension.

However, I had much to learn before the ideas inscribed in this account played a concrete role in what I was about to investigate.

Practicalities (3): Confessions

This account is about the environmental research I conducted, centered on Muyuw, between 1991 and 2014. As it was intended at the time of its conception and as its goals shifted as I learned, its ambition was not designed to be about what was happening on the island during my time there. Nevertheless, a rough outline of the current situation is important to understand the island, for the last thirty years are probably epoch making.

Muyuw in particular and Milne Bay Province in general are at one of the extreme points of the modern world, and have been so since the late eighteenth century and probably throughout the last 2000 years if you consider the trading systems across the Pacific and Indian Oceans that gave birth to the West. Insofar as the modern world-system is concerned, what will happen on the island will largely be determined by what happens elsewhere in that world—so rising and falling gold prices, demand for various marine resources, timber, oil palm, and the ability of the centralizing state of Papua New Guinea to sink Western forms of education and wealth into the region's social fabric; some would add to this list Christian mission activity, whether externally or internally directed, and its ability to further inculcate its values. As I conclude this volume, it is unclear how these factors will affect this dot in contemporary geographies, but it is likely that the thirty-year interval between about 1980 and 2010 will amount to a great transformation. I did not research this but I watched and participated in it.

First, for almost all the time I have been on Muyuw, somebody has been surveying the island's mineral resources with the idea that they could pick up where the first mining phase, which ran from about 1894 to 1940, left off (see Nelson 1976). By my first return in 1982, the 1970s run-up in gold price sent some of Australia's larger firms to the island. Then the relative successes of mines on Misima, the Highlands of Papua New Guinea, and then Lihir put what the miners called Woodlark Island in the hands of smaller venture firms, though there was often a tie to larger companies and almost always a presumption that if deposits proved significant, a larger firm would move in. When I left the island in 1991, one such firm thought it was on the verge of an endeavor that might eventually lead to the removal of the Coral Sea.[21] When I left the island in 1996, the first leg of my journey being on a gold-mining company's plane (Auridiam Consolidated NL), the miners were pretty sure they had discovered enough gold to clear their expenses, though they thought they might have to ship the ore to the processing plants that remained from the then-closed Misima mine. These interests were bought and sold a number of times so that by 2005 a different set of people, still mostly Australian, were involved. These people—with a new name, BDI—had held a small interest in the Auridiam project, but by this time, having succeeded in creating a diamond mine in Indonesia, they felt that with their geological expertise, they could do a better job than Auridiam. I had lunch with the group's director, Lee Spencer, in Sydney in 2005. A fascinating and experienced man who knew and respected Tim Flannery's works,[22] Spencer talked about what the company would do for the island, especially concerning education and health. I wondered what was keeping the price of gold high enough to make a place as difficult as Woodlark appealing to such people; the first obvious answer was just that the place was there, a high mountain to climb.[23] But Chris Gregory explained how development in India was driving up the price of gold used for bangles in its marriage system. Thus, Muyuw "development" appears to be driven partly by India's marriage system. Asia as a sink for Western-generated gold is, in fact, a familiar pattern.

Other patterns are as old or older. By 2009 virtually the same company and personnel had been moved around the checkerboard of international mining capital so often that it was neither easy nor interesting to follow the shifts. The company on the island was called Woodlark Mining Ltd.; it was owned by something called Kula Gold Ltd., which in turn was held by Pacific Roads Capital Management and Rand Merchant Bank Australia. Kula Gold Ltd listed itself on the Australian stock market in late 2010.

At one point the firm was a fully owned subsidiary of the South African firm, Rand Merchant; Pacific Roads Capital Management is a Sydney-based equity firm that finances mining exploration and expansion. Almost all of PRCM's website pages show pictures (presumably projects it finances) demonstrating rugged landscapes transformed into roads, open pit mines, or other massively altered vistas. These images are power displays. In 2012 I was told American (pension) money funded much of the activity.

This power displays itself on the island. When I questioned a set of the miners[24] about their future success, given the island's long history of frustrated hopes, they asked me to consider the capital and effort I could see with my own eyes compared to other attempts I had seen since the early 1970s. Almost the whole igneous center had been leased for mineral exploration, and virtually all of it had been or was being surveyed by teams of Muyuw youth. Wherever initial survey paths looked promising, bulldozers smashed through the jungle and rolling hills so that one of two large drilling machines could extract cores to map the area's mineral potential. At least one drill worked day and night. Each drill set consumed about ten 200-liter drums of diesel fuel over each twelve-hour period. This was not an incidental use of power. Sipum's youngest brother noted a scale change. He correctly told me that the first miners came to the island with picks and shovels; now they had these massive machines. In his account of Australasia, Tim Flannery (1994) hypothesized that Europeans came out of an environment conducive to the exercise and use of raw power because, as did North America, the landscape seemed so rich. By contrast, the flora and fauna—including human—that became adapted to Australasia had to learn how to make a lot out of a little, to husband meager resources by seeing how far they could be extended rather than how quickly they could be extracted. "European power" now throws the dice on these islands.

One of the subtle signs of change was on display the evening of June 26, 2009. The mining company moved its television set, usually located in its mess hall, outside to accommodate a hundred or more workers, their friends, and families so they could watch a State of Origins rugby game beamed from Australia. The many Muyuw people were no less passionate about the game than were the attending Australians.

The success of this transformation is far from determined. The same capital jump the new mining exploration has instigated first appeared in the transition from the Neates' Kulumadau Enterprises to Milne Bay Logging, a company that started cutting and shipping out raw timber in 1982 (see Damon 1991, 1997). Ten thousand logs were cut and loaded for mostly East Asian markets within the first six months of operation. At that time, the company owners figured they had twenty years of cutting fifty thousand cubic meters per year. They were a relatively large operation for the country. By my return in 1991, owner Rolly Christensen had married into the island and cut roads all around it. Yet his operation had shrunk in relative terms because of the arrival of large Malaysian companies elsewhere in Papua New Guinea, and he realized that for the long term he needed to "go green." There was not as much wood as he thought, with large swaths of the island producing only five to fifteen cubic meters of wood per hectare. He told me that halving his output meant dropping to roughly twenty thousand cubic meters per year and creating the capacity to cut finished lumber for more local markets, with another five to eight thousand cubic meters. It turned out that much of the wood sized above the minimum standards—fifty centimeters in diameter at breast height—was in fact diseased; he was learning that the place was cursed. By 1996 he was a broken man with all but the occasional cutting of ebony prohibited, he thinks, by people he refused to bribe. He was effectively gone from the island by 2000, having moved back to Australia, though as of 2009 he occasionally returned to do work for the mining company and to plant yam gardens. Realizing how rapidly Christensen exhausted his machinery and now witnessing his failure, Muyuw elders on the island by 2005 or so began shifting their evaluations of Christensen compared to his predecessor, Don Neate (see Damon 1997). However, he left behind a relatively large pool of men who had experience with heavy machinery. Hence, by 2009 mining people were thankful to find what they considered a skilled and teachable labor force.

Muyuw elders were not surprised that the timber company did not work out. And while some middle-aged men living in the central part of the island who were firmly committed to "development" said they would now try mining, others recounted a myth from the early gold-mining days about a spirit who kept hiding the gold from whoever sought to take it away: it could be found but not removed. In a version of this story I heard in 2002, the narrative was tied to the 1891 hanging of a Muyuw man at the dawn of the colonial epoch. The hanging was a real event turned into a founding myth of the contemporary social order. It is also bound with original creation stories that account for the island being "cursed" by the Creator. Although permission was granted to go ahead with mining, the fall in gold prices by 2015 led to a decision to put the mine to sleep.*

One is tempted to take the curse seriously. By the latter half of the first decade of the twenty-first century, the island experienced the second rise and relative collapse of marine resource extraction since the 1980s. In the early 1980s the expanding Milne Bay Logging company attracted a flux of Trobriand people, mostly but not exclusively men, who worked in logging or got involved in the bêche-de-mer (Trepang) fisheries. But the reefs were quickly exhausted. By the latter half of the 1990s that activity started up again as East Asian wealth increased the number of people who could afford sea cucumber soup and transformed the means of reef exploitation. Traditional outrigger craft were replaced by fiberglass dinghies and twenty-five- to fifty-five-horsepower Yamaha outboard motors, which, of course, consume great amounts of petrol. Politicians running for the national Parliament distributed dinghy or outboard motor sets to influence voters. Young men could dive into reefs named but never before exploited. Unlike the reefs nearby, sharks were not used to people here—hence new risks for both sharks and men. But the new reefs too were soon all but exhausted, so much so that the Papua New Guinea Fisheries Department was forced to create a bêche-de-mer season.

This action brings me to the last transformative process outlined here, the long-term but visible attempts to make over the existing cultural template. At one level, the point of transition here follows from the fact that the person now making decisions about Milne Bay Province's marine fisheries is a middle-aged man born in Wabunun slightly before I arrived there in 1973, Leban Gisawa. He was a very small child when I was first visited the island. While his father was one of my more important informants, Leban was not in my vision, though some of his elder siblings were. He has a BS from the University of Papua New Guinea and was an experienced researcher on matters of Pacific fishing before he became an administrator; in addition to supplying me with fish and turtle identifications, we have engaged in a discussion on the degree to which traditional human activity might create nutrients for the tuna-rich Solomon Sea to Muyuw's north. He is like a number of other young men from the island who work in the national or provincial governments in Port Moresby or Alotau, some of whom have advanced degrees from Australian National University or other Australian universities, and some are lawyers and doctors. These people represent a contradictory but very thoughtful future, for both the country and the island. While thoroughly ensconced in Papua New Guinea's modern sector, they do not wish to see the base of that life—resource extraction—make all of the inroads that the present demands.[25] This contradiction is apparent on the island as well. One of my informants who provided details on the sailing craft that came to dominate my last decade on the island is an elementary school teacher. While he takes great pride in these boats and his knowledge of them, he was also pleased that one of his daughters was away in an Alotau high school where, he hoped, she was headed for computer programming.

If these young people are now a leading edge, they reflect, in combination with demographic expansion, on-going processes being led by a combination of the school system and the Christian church. Demographics first. When I was first on the island there were about two thousand people in all on what is considered "Woodlark Island" for national demographic purposes (all the islands between Budibud in the southeast and Iwa directly in the center of the northern side of the Kula Ring). Now there are closer to six thousand. Wabunun had slightly over one hundred people in 1975, counting a few from an apparent split between brothers just east of the main setting. As conceived by contemporary reckoning this place, called Topwelekel, is close to Wabunun's origin point. That point is a tree and associated relics that are never to be seen lest one run amok, matters taken up in Part II. In 2014 Wabunun looks quite similar to the way it did in 1973–75. However, through other splits between brothers or witchcraft accusations, three new daughter villages transform "Wabunun's" population to over 500. Other early 1970s villages have had similar transformations. In Wabunun's case the original split to the east, Topwelekel, has grown into a well-formed double-rowed village like the Wabunun I described as the island's standard. On another beautiful cove a hundred or so meters down the shoreline to the west another large assortment of houses has metamorphosed from a couple of houses isolated by witchcraft accusations into a full-blown community. It will be interesting to see if the bifurcating fractal forms evident elsewhere in the culture appear here as it expands over decades. Above the five- to ten-meter rise right behind and to the north of the main village is another assortment of houses. The logger Rolly Christiansen bulldozed the space between my departure in 1995 and arrival in 1996. Originally cleared to house a large United Church meeting in mid-1996, the place was deemed a good one for expansion, coming between the old village and a plot of land designated for a school, including a large field occupied every Saturday by soccer games and individuals marketing small items, baked rolls, sweet bananas, and various marine resources. The house spaces were set up to go in north-south rows, paralleling the directionality of a new church centered at the southern end of the desired village space. Both the new layout of the houses and the church building were partly conceived to go against Muyuw's traditional ordering, thereby locating a new order on the land. So far the pattern has not quite been accepted; the new alignments both to the east and west of the village garner more people than this location to the north.[26]

Wabunun's church was designed to be the central United Church building on the island, and a seminary/training center remains associated with it. However, while throughout the 1970s many people across southeastern Muyuw trekked to the village every Sunday for services, as well as late-afternoon kula talk, now, among other things, the increased size of so many other areas has allowed them to build and sustain their own church buildings. Wabunun no longer functions as the church center it once was.

Beyond demographics, another subtle change is coming about. Whereas in the past the 'smartest' young men were drawn into the church leaders, now that is not the case.[27]

But the United Church is no longer the only significant Christian voice on the island. A minuscule Catholic presence remains and is better known now that Italian interests are pushing sainthood for one of their brethren killed on the island in the 1850s. But the real new change was a vocal and active presence of the country's Christian Rival Church (CRC).

Although this is not a development I researched, and it probably has roots into the 1980s timber activity, the CRC rapidly gained momentum near the old and new mining and timber center called Kulumadau shortly after my 1991 visit to the island. I was told that by 1993–94 major splits developed all across the island. Sides were taken pitting the CRC against the United Church. Among other things, the CRC prohibited all Muyuw customs, including betel nut chewing, smoking, mortuary rituals, and the kula. Adherents of the United Church defended all of these practices. The situation was extremely tense in 1994. Yet by my arrival in June 1995 things had settled to a point that people in the southeast no longer felt threatened. They did tell me, though, that all of the people in the central part of the island, what is called Wamwan, had either joined the CRC or followed its practices and so refrained from the region's major traditional practices. This is an exaggerated truth, at least insofar as the kula is concerned. Yet there are many people in that region, often employed by mining interests, who do not participate in the practices I have learned and described. In 1995 the owner of Milne Bay Logging told me that he felt caught between the two movements. He had friends, employees, and affines in both groups. He tried to support each church. Few people in southeastern Muyuw seemed caught up in the conflict, but my elder informants in 1995 and 1996 rehearsed with me, defensively in my estimation, the virtues of the kula. They were speaking against injunctions coming from the CRC. I had the feeling that had I begun this research in 1994 rather than a year later I might have felt pulled into exploring the situation … and therefore engaged in a different literature[28] over the last fifteen years than the one that has captured my attention. And had I done this, without question I would be creating an account closer to the contemporary situation than is the intent of this book. On the one hand, even though it might be said that the United Church is retaining a dominant position on the island in spite of the CRC onslaught, there is a more evangelical cast to the religious tenor most people now carry compared to what I experienced in the 1970s.[29] On the other hand, regardless of their differences, Christian organizations work to undermine what was Muyuw culture. And they all have succeeded, in part, to make the present Muyuw dismiss, denigrate, and to some extent hate what they once were and are (see Sahlins 1992).

It is almost a universal belief across the island that the elders, some still alive, most dead, were stupid and simple, and until the present order—the aforementioned hanging fits into this conceptualization—there were no villages, gardens, boats, kula, or anything. Men ran around in the forest, leaving their wives and children, should the latter cry, for fear of being speared by other Muyuw men. More than once when we came across potshards at the foot of some giant tree, whether doing my tree surveys or following up Bickler's attempts to find old village sites, people would point to the shards as signs of their former existence living in the forest. And especially as I began to appreciate and understand the complexity purposefully built into the largest two classes of outrigger canoes, my instructors themselves would become surprised, and in a way delighted, by what I was learning. The experience also moved them from the "Christian" consciousness that dominates their present. Learning those other levels is what this study is about.

* * *

When I first decided to conduct this environmental research, I also decided I would not worry about the thesis I put forward in my book *From Muyuw to the Trobriands,* that there was a set of transformations that ran the length of the northern side of the Kula Ring. However, when I was told that the practices with respect to the gwed tree varied across the line of islands, I knew I had to pursue that part of the indigenous understanding. So I attempted to organize this research as a cooperative project in which I would handle the eastern side of the situation and Linus Digim'Rina would handle the western portion. Digim'Rina was then a young anthropologist from the Trobriand Islands finishing his PhD at Australian National University under Michael Young's tutelage. Although the research never got funded, Digim'Rina helped me start and we have maintained close contact as the project has matured. Many ideas presented here were first sketched in lectures he set up for me through his Department of Anthropology and Sociology at the University of Papua New Guinea.

Data acquisition for this work started in 1995 and remained driven by my initial methodologies, survey procedures. In the abstract there were two tiers to these practices. First, my University of Virginia environmental science advisers convinced me that analyzing carbon and nitrogen isotopes would likely tell us much of what we needed to know.[30] So my responsibility was to gather soil specimens and bits of tree and plant material across a wide spectrum of places. Dried by the sun and a storm lantern I kept in my Wabunun house, these were to be analyzed by a mass spectrometer during the fall of 1995 so that we could get an approximate idea of the situation for my return in January of 1996, at which point, the plan went, I could pursue our results in greater detail. I expected to have provisional results from the scientific point of view by then, as well as a much surer grasp of indigenous technical knowledge.

The second procedure concerned learning about trees. In general I followed two techniques: I conducted surveys of areas more or less formally determined—a hectare, for example; and I asked people what tree was used for what tool, boat, or house part and where the tree came from. Both of these methods were modified as I learned more over the course of the research. Since 1995 I have collected a total of 329 voucher specimens,[31] though some of these are repeats. Copies of these specimens have been deposited at the Papua New Guinea herbarium at Lae, some at the University of Papua New Guinea's herbarium; a few have been left at Canberra, Singapore, Sydney, and most at Harvard.

I have now walked through, visually examined, collected, and talked about plants from many of the regions across the northern side of the Kula Ring and down to Koyagaugau and Ole in the southeastern corner of the area. My examination of the Trobriands is restricted to Linus Digim'Rina's home village, Okaibom, but I have walked around all of Kitava, Iwa, through Gawa, and around the western end of Nasikwabw (the rest of the island is impenetrable) and the central populated islands in the Budibud arc. Muyuw, including Mwadau Island forming its western limit, is about sixty-five kilometers long and thirty kilometers wide. Of this area, socially and botanically, I know the southeast the best, but I have walked and studied along the island's northeastern end, surveyed a hectare close to one of the lakes in its southwestern quadrant, and walked and studied the Mwadau sector. I collected plants from the Sulog area three times, but there is much that I do not know about this igneous center where the flora differs from the limestone shelves found elsewhere.

With an exception or two I always walked and collected plants with one or more of the Muyuw people. I deal with the variations in what they know in chapter 2. Although some people found what I was doing insipid, others were anxious to tell me more than I could understand; this frequently occurred with Sipum. On our walks he'd often pick something out and tell me what it would cure. One time when we were walking through an early fallow garden area, he hacked the aerial root off a *Pandanus* tree, stripped off its outer covering, and told me the sap from the phallic-looking object cured venereal diseases. I made casual notes of these kinds of reports but decided ahead of time that I'd leave medicinal questions for somebody else.

But another episode with Sipum a day or so before I left the island in 1996—presuming it would be my last time—forced the realization of another limitation I imposed on myself. We were walking along a beach heading back to Wabunun when Sipum started expounding about the flora close to the water's edge, noting how interesting it was because it was in this liminal location. He practically gave me a lecture out of the writings of Tom Beidelman, Mary Douglas, Edmund Leach, and Victor Turner. The problem was that this was the anthropology I already knew; I had practically made a decision to ignore anything that led me to repeat that kind of analysis over again. That plan partly broke down but has remained an aspect of my intent to explore new paradigms, not prove the utility of older ones. The experience with Sipum, however, stands as a cautionary light as this project comes to completion yet impels another.

I shall end with another expressive episode near the end of my 1995 research. I had spent a long day of walking about the gardens and forests near Kaulay in north central Muyuw with Vekway, my chief contact and host during my earlier times in Kaulay. We hoped to resume our relationship. He remained an ambitious gardener and a good, if impatient, informant. Although he gave away all that he grew, he was especially expert about yams—in fact, he argued emphatically against something one of my best Wabunun informants had told me, that yams grew only at night. Vekwaya insisted they grew at both times, though water shortage during the day made it seem as if they didn't. However, time had not treated him well. He was probably arthritic and walked with a limp. He found my traipsing difficult and boring. In any case, on this particular day I was still struggling with initial overviews of everything botanical. Without exactly knowing how or why, I realized that human/forest relations in Kaulay were different than in southeastern Muyuw. As we sat down to rest, I looked out over a garden to see a mass of older trees just beyond it. A vista providing an exciting challenge beckoned in that tree line. Vekwaya, however, was rummaging around in the weeds where we were sitting, parting this one and that one with his fingers. He then bent over, looked down, and pulled up a plant about the size of his fingernail.

"This is a good one," he said. "It is for women."

I did not yet know those big trees, yet here was Vekway telling me I had to know things almost too small to see—making me realize that there were limits to my research.

Notes

* Just as this book is going to press I have received a message from Kevin Neate (September 3, 2016) reporting that Kula Gold has been taken over by GeoPacific, a company that intends to expand exploration activities beginning in December 2016.

1. Alemida 1990, Petitot 2009, and Turner 1990 reorient these speculations.
2. He mentioned Mekong River boats with sewn hulls, potentially significant given the Austronesian expansion out of the south and southeast China coast.
3. Throughout this region I am known as a man from Wabunun. These people had provided a lunch for my family on our voyage to Muyuw in 1998. Mwalubweyay married a Wabunun woman in 1973 shortly after my wife and I arrived and he became thoroughly engaged in the village as kula relationships and as his new wife's children matured. Although I compensated the Koyagaugau people, I participated in expected hospitality relations.
4. Although Gideon has spent years in Muyuw and was well-known in Wabunun, those people knew him as Talopet, from the English "interpreter," because of his language skills.
5. Bwanabwana categorize and name wind differently than do Muyuw.
6. Calculated by Simon H. Bickler. Bickler converted standard longitude and latitude positions into decimal form and calculated distances and speed. The original data for Table 1.1 can be found in the online accompaniment to this work, Chapter 1, Table 1.1, The Voyage of 2002. Subsequent references to positions and speed refer to my original recordings.
7. In Muyuw this would be *nuwa-* (+an obligatory possessive suffix for "inside") *veyo* (fly).
8. A spreading zone, the Woodlark rift, exists between the northern and southern halves of the Kula Ring, making the area geologically interesting (see Abers 2001 and Abers et al. 2002). Suzanne Baldwin from Syracuse University directs significant National Science Foundation-sponsored research on the area now (see Baldwin et al. 2012). Recent mining research on the island contravenes the prevailing geological model, which is one of continuous, gradual rise. Lee Spencer, a director in the current mining endeavors, believes catastrophic alteration defines it.
9. See also Schulte Nordholt (1996). My ideas about Bali and the pertinence of focusing on water were transformed by the predissertation research of a University of Virginia student, now Dr. Laura Bellows.
10. See Wheatley's 1983 account of symbolism and political forms of Southeast Asian regional systems.
11. Shugart had already had a significant interest in anthropology and uses modeling procedures identical Lansing's.
12. See http://www.umsl.edu/~biology/faculty/stevens.html.
13. Kirch (2000) is an archaeological response to this volume for the Remote Pacific.
14. By 2000 a new ecological anthropology has emerged that is far different than what it was in the 1960s and 1970s; however, Kurin (1983), drawing on symbolic constructs I was leaving aside, anticipates the new standard.
15. "Concepts in Historical Ecology" (1994) was my introduction to Winterhalder's work and some of the parallels between ecological and anthropological thought. The paper continues to stimulate even though my mentor in things ecological, Hank Shugart, finds the piece lacking, in history among other things.

16. By the summer of 1994, global warming was becoming a factor that a few experts discussed. Conversation outpaced understanding (it was suggested I buy land around my home, Charlottesville, Virginia, because it would soon become beachfront property); Michael Mann's hockey stick of recent global warming was four years into the future. Reid Bryson's work was probably the most interesting investigation of climate and culture. In 2001 I team-taught a course on climate and culture with Michael Mann and Bruce Hayden, a senior climatologist who had studied with Bryson. Generational knowledge conflicts between the two were palpable. The University of Virginia's collection of environmental scientists was shortly to produce Ruddiman's important thesis that humans started affecting the earth's climate six to eight thousand years ago. Tim Flannery's 1994 suggestion that Australian aborigines transformed Australia's climate thousands of years ago was a distant outlier on what then seemed possible. See Ruddiman 2003, 2010; Flannery 1994.

17. Or Medieval Climatic Anomaly—see Mann et al. 2009.

18. I gradually became familiar with the posthumous 2002 collection of essays (Posey and Plenderleith) and incorporated some of them into my teaching.

19. Bickler did computer programming that supported Irwin's important and pathbreaking study of sailing and discovery in Melanesia and Polynesia (see Irwin et al. 1990 and Irwin 1992).

20. In 1978 the Lapita Project was just underway, so Melanesian prehistory known today was barely conceptualized. Irwin guessed that Milne Bay Province was a likely route east by the makers of Lapita pottery some three to four thousand years ago. Investigations have not sustained that deduction, but recent work raises related possibilities (see Tochilin et al. 2012).

21. First the island's igneous core would be removed, then its sixty-five-kilo-meter-long coral shelf, then the Coral Sea, because gold was everywhere. Although the geology behind the plan was logical, the model is a millenarian fantasy. I've discussed it with André Iteanu—gold is the encompassing value of our times.

22. Through 2009 all of the Australian miners I met knew who Flannery was, yet they paid no attention to global warming and did not care to follow Flannery's move to that issue.

23. I discussed Alfred Gell's use of *"anticipatory joy"* (1992: 184. Italics in original) with an Auridiam official. He reported that most mining CEOs are well off; what fascinates them is the challenge, not the money. This challenge is the encompassed value of these forms, thus forming one of the dualisms of our times.

24. My son and I were invited to a dinner with many of Woodlark's executives at their headquarters, Bomagai, on August 1, 2009. Like virtually all of the Europeans I've met on the island, these are extremely interesting, interested, thoughtful, and experienced people. Its CFO, John Wadkin, was hired because he had successfully capitalized the Highlanders who "owned" the land on which the extraordinarily successful Porgera Gold Mine was located. Ideally, he would work the same magic for Muyuw. Their operation was not con-

ceived to be of the cut-and-run variety; they desired plans for the long term, their own future and what they understood to be that of the Islanders.

25. During the first decade of the twenty-first century they fought off repeated attempts to turn the island into an oil palm plantation. I participated in these efforts and helped forge ties to other persons—Tim Flannery, Christopher Norris (a member of a team from Oxford that studied the island's endemic cuscus in the 1980s), Glen Barry from Ecological Internet, Jeremy Hance from MongaBay.com and others. If a majority of people on and off the island sought to prohibit this activity, there was a split on and off the island. I have friends and close associates on each side of the divide.

26. The elders who did much to fund the new church had almost died out before it was realized with considerable ceremony in 1996. Since my original arrival on the island I have maintained a reserved and unsympathetic relationship to the church. When I explained that they could have maintained an east/west orientation and not only preserved Muyuw's fundamental axis but the axis of many Christian churches throughout the world, I received an unsympathetic scowl.

27. During my earliest fieldwork the brilliant Kula elder Molotaw worried about the church's ability to garner the region's brains. He thought they were needed and put to better use in the kula where they would achieve responsibilities for the *whole*. People in Wabunun alerted me to the intelligence shift by remarking how people now advancing in the church had not excelled at school.

28. Joel Robbins's work centers around this literature. See Robbins 2004, 2009 (and Damon 2009).

29. The effort lodged against Sipum's wife in August 2009 is part of this new stance to the world.

30. None of these people were experienced in tropical soils; one of their hopes was that they would learn from what I discovered. I received advice about gathering soils from experts at Australian National University, which was countermanded, correctly, by an experienced soil scientist with much tropical experience who was based in the University of Hawaii; this scientist's findings were later confirmed through Robin Hide's friendly commentary whenever I visited Canberra. Hide's contribution to this work is enormous.

31. Throughout this volume I specify scientific identifications with respect to the voucher specimens I collected. Damon 181 refers to the one hundred eighty-first specimen in the collections I made and deposited in various herbaria.

CHAPTER 1

Return to the Garden

Gwed, *Locating Intentions, and Interpretive Puzzles*

What was the *gwed* tree and how did knowledge about it and other trees relate to the ornate, temple-like garden structures I already understood? I returned to the northern side of the Kula Ring in June 1995 to begin answering these questions and follow wherever the answers took me. The plan was simple: collect voucher specimens of the *gwed* tree so that a scientific identification could be made, specify its ecological contexts, collect soil and plant samples across a range of environments so we could figure out if nutrients passed from the tree to the soil to the plants, and, finally, learn as much as possible about indigenous understandings of the tree and contexts associated with it. To what extent was there a well-formed social logic behind the reports that charged this research?

Everything started out perfectly. I arranged to meet Linus Digim'Rina in his home village of Okaibom, located in the Trobriands. The day after I arrived he took me to some fields and showed me how Okaibom people used the tree. People sometimes cut it down and sometimes burn it, but regardless of method they kill the tree in the early stages of garden preparation. They then crowd crops around its base. In the first garden it was easy to see that taro plants were shorter the further removed they were from the tree; later I saw similar differences in a yam garden. I took pictures of this pattern. I collected soil samples next to and several meters away from the tree, small portions of taro leaves similarly proportioned, and bits of the dead tree. These would be dried, returned to the University of Virginia, and run through a mass spectrometer to check nitrogen and carbon levels and types. I gathered similar materials from a range of other environments near Digim'Rina's village in our week together, and I followed the same procedures over the next two months across eastern and central Muyuw, and on Gawa and Iwa Islands. Statements people gave me throughout the process were almost identical; the practice seemed to span the northern side of the Kula Ring. The tree gives something to the soil.

In Iwa and Gawa, however, I noticed a minor difference. While everywhere else the tree is killed before crops are planted next to it, Iwa people often plant seedlings next to newly planted crops—fruit and nut trees, yams, and taro. One Iwa man told me they worked with the tree so much that it stained their clothes and skin. A variation on the plan I heard everywhere I traveled in 1995, people reported that as the tree grows it conveys its beneficial properties to the crops planted around it. The model is straightforward. *Gwed* put "sweet" (*simakein*) things into the soil, and this aids the targeted plant. While Trobriand and Muyuw people I learned from took advantage of what the tree deposited in the soil before gardening practices killed it, in the Iwa variation the tree and the crops worked simultaneously. Another Iwa man added a different perspective, saying they planted the tree next to the crop so they could grow trees while they were growing the crops—they fallowed as they gardened. Just noise in 1995, this idea developed into something else over time. Across the islands of the northern Kula Ring, the horticultural regime is slash and burn. Crops are regularly alternated with forest growth. And from Muyuw to the Trobriands there are three basic categories for fallow periods. So far as I knew in any given area one might find people using every type of fallow, and I had no reason to suspect variations were anything other than chance or individual preference. By the end of 1996 this synthesis radically transformed.

The three types of fallow periods, presented in the Muyuw language—with phonetic variation the terms are known and used across these islands—are *digadag* (the shortest fallow period at five to fifteen years), *oleybikw* (twenty to forty years), and *ulakay* (forty years and more).[1] From this point of view, the Iwa man's statement that *gwed* seedlings are grown next to crops expressed the mathematical limit of the interval between gardening. Since fallowing runs concurrently with gardening the limit is zero. In 1996 I learned of the other extreme, fallows so long that there are no gardens. That condition is turned into a joke about Sulog people in south-central Muyuw, people who, until recently, gardened very little.

Gwed: The Tree

When I returned to the University of Virginia after the 1991 planning trip, my environmental science advisors greeted report of *gwed* and switch of topics with enthusiasm. They presumed that the tree was a legume, so it was probably fixing nitrogen, one of the three primary plant nutrients (along with carbon and oxygen). In temperate agricultural systems at least, insufficient nitrogen is often a major "limiting" factor. The nitrogen atom that some plants can take out of the air—"fix"—is different than the nitrogen atom that is found in human or animal waste. If the *gwed* tree was adding nitrogen to its environment tracing the movement of nutrients from the tree to the crops could be accomplished by analyzing stable isotopes. While figuring this out should be easy in principle, a question that first interested me concerned the tree: what was it?

In 1994 Sipum sent me a leaflet from the tree. Steve Macko, the University of Virginia geochemist, ran a snippet of it through his mass spectrometer and got a reading that was consistent with the possibility that it was associated with fixed nitrogen. I sent another portion of the leaflet to Peter Stevens to identify. He could only suggest that it was in the Meliaceae (Mahogany) family. About this time, before my 1995 trip, Mike Bourke, a geographer from Australian National University, contacted me about an ANU-sponsored survey he had participated in and which had taken him across the islands of the northern Massim (see Hide et al. 1994). He too learned of *gwed*, and had it identified as *Rhus taitensis*, in the Anacardiaceae family. I noted the different identifications, Meliaceae and Anacardiaceae, to another systematist in the Harvard Herbarium when I visited Peter Stevens in 1994; he said that while it was sometimes difficult to distinguish the two, there was a simple test: their tastes. Anacardiaceae often taste sweet, Meliaceae bitter.

This seemed like a useful bit of information because in 1995 I quickly learned the indigenous model: the tree is sweet (*simakein*) and gives sweet, good things to the soil. Several Iwa and Muyuw people told me to taste the tree to experience how sweet it was. It has rather milky white, relatively profuse sap. I tasted it, often, because it tasted bitter.[2]

Max Kuduk of the Papua New Guinea Herbarium and Peter Stevens identified all *gwed* voucher specimens I collected as *Rhus* or *Rhus taitensis*, a tree in the Anacardiaceae family. According to the systematic literature, it is widely distributed across the Pacific. Although there are suggestions that some people use the tree, there is no record of it being employed anywhere like it is on the northern side of the Kula Ring.[3] The genus, however, is noteworthy for its sap's peculiar bonding characteristics. Most significant is the Japanese *kiurushi* tree (*Rhus vernicuflua* or *Toxicodendron vernicifluum*). For thousands of years the Chinese and Japanese have used this tree as the source of high-quality lacquer. The quality in the sap that provides for this lacquer, urushiol, is formally named after the Japanese tree; it runs through the genus.[4] For North American readers, many systematists put "poison ivy" in the same genus—urushiol in the vine's sap binds to and irritates skin (see http://poisonivy.aesir.com/view/fastfacts.html).

Iwa people noted this condition when they commented on the tree staining their skin and clothes. Wabunun informants in southeastern Muyuw confirmed the condition. They sometimes use *gwed* sap to glue together cracks that develop along the strakes of their outrigger canoes. Moisture is necessary for the substance to bond.

From Muyuw to the Trobriands the *gwed* tree is found in frequently disturbed areas, most obviously in the garden fallow class called *digadag*. Seedlings may sprout almost as soon as an area is burned. Groves of saplings quickly appear. The first time I knowingly walked through a thicket of *gwed* less than ten centimeters at breast height (cmb) in Muyuw in 1995, just after arriving from the Trobriands, my guide said the ground could already be used again because of all the *gwed*. It had only been a few years since the area was last used but it would produce food.

I have never seen *gwed* growing in a mature forest. They do not appear in gaps in contiguous high forest created by a downed tree or in the meadows that dot mature forests unless they are breeched by mining or timbering activities. Muyuw do not believe the seeds require heat from the burn phase of their slash-and-burn system.[5] In many *digadag* fallows dozens of these trees may be found. As an area gets older the trees become increasingly rare. Areas called *oleybikw* may exhibit a few relatively large ones, perhaps to thirty meters in height. But in the oldest class of fallow, *ulakay*, where the canopy top is about forty meters, none are found. These fallow categories imply both absolute time and qualities of growth. By definition an *ulakay* will not have *gwed*, while the areas where *gwed* are found are invariably *digadag*, though they may be young *oleybik*. The tree is a property of a category of land, the land in turn understood by its trees.

Gwed are easily killed by fire and quick to rot. Because of the former characteristic they are often not downed in the slashing stage of garden preparation. People say that sometimes they die if they are merely swiped with a burning coconut frond, the usual instrument for distributing fires. *Gwed* are spreading trees. *Digadag* gardens often have a distinctive appearance, that of a cleared field with only *gwed* standing in its unique, leafless, skeletal form. Once I was able to recognize the configuration, I found it was easy to enter a garden to see how people often crowded taro or yam crops right next to the trunks. And often, but not always, I observed the height gradient of plants close to or at a remove from the trunks. That the height dispersal was not always as I expected it led me to become skeptical of the model toward the end of my 1995 trip; more on the consequence of this skepticism later. Unfortunately, it was not until 2002 that I learned of another planting strategy that might explain the apparently inconsistent results. For taro the issue is this: Most people hold that, other conditions remaining the same, the major factor determining the size of a harvested taro corm is the size of the initial plant. Taro reproduces asexually, as buds off the "mother" plant. The buds, "children," are small, and it may take several seasons before they become respectably large. Consequently some people plant the smallest taro plants next to *gwed* to make them grow larger faster. When they mature, these plants will not usually be the largest in a planted field, but it is held that they will be larger than if planted next to some other tree. In eastern, but not central, Muyuw, sometimes large taro seeds are planted next to *gwed* not so much because they will get bigger but because the association changes the constitution of the corm. For ritual occasions, the prized food (*munawun*) is taro that has been boiled, mashed flat, then rolled up and re-cooked in a clay pot prepared with squeezed coconut oil. Taro varieties most suitable for this recipe are thought to adhere better in the mashing and rolling stage if they grow next to *gwed*. There are two important points here: First, people focus on more than one positive advantage that should stem from the tree—absolute size or relative size on the one hand and the quality of the food on the other. Second, these conditions focus on specific details of the productive process. These plants and their conditions draw people into them.[6]

Through my 1996 research period I presumed what I saw with my eyes, recorded with my camera, and established by means of stable isotope analyses would provide sufficient proof of the quantitative effects (or lack thereof) that *gwed* have on plants. So I did not actually harvest any plants to count the results. By 1998, difficulties in getting the stable isotope data convinced me that a brief test would be interesting. With permission from the garden's owners, I harvested ten *parawog* (*Dioscorea esculenta*), four immediately adjacent to a *gwed* tree (within two meters) and six at least ten meters from any of them. The results were startling. Altogether the four plants next to the tree produced fifty-two tubers considered sufficiently large enough to be eaten or saved for seed for the next season. The number of tubers per plant ranged from eleven to sixteen, averaging thirteen. The six plants at least ten meters from any *gwed* totaled thirty tubers of comparable size. The range was four to seven tubers per plant (including a very small tuber), averaging five. My Trobriand colleague Digim'Rina would have been appalled at the small size of *all* the tubers based on Trobriand standards. Yet while the results surprised me, Muyuw people to whom I reported them expected something like that to follow. Realizing my sample was very small, I tried a similar experiment in 2002. The man who helped me harvest the plants this time did not think the *gwed* worked, but there was an observable difference in the results we obtained (see the online photos for this chapter).

Gwed have significance beyond their effects on crops. *Gwed* wood is used as firewood, but is not highly valued because it will not hold a fire if removed from the set of burning pieces. It is very light and easy to carve, so it is occasionally appropriated for small outrigger canoe hulls. Such outriggers, however, would be expected to last for a year or two because the wood rots quickly. Muyuw *gwed* is rarely used in house construction. In a phenomenon once unique to central Muyuw, moderate-sized *gwed* are cut to be the first beams placed on foundation posts for yam houses because the wood is so light. In western Muyuw, yam houses are more permanent structures within village confines, so *gwed* are not used in those differently structured and valued buildings; the wood rots too quickly for the structure's purpose.

After a few years *gwed* start producing flowers, and at least some trees produce seeds.[7] The flowers cover the trees in brilliant white panicles, making the areas where they grow a sight to behold. The flowers turn into very dark blue or black berries with small seeds inside. In a grove, berries will practically cover the ground. It quickly occurred to me that whatever beneficial effects the tree might have could conceivably result from the dozens of fruit that cover the soil virtually every year.

In my experience the *gwed* on Kitava, Iwa, and Gawa were blooming from January on, before those in Muyuw. In Muyuw the rush seems to begin about the time the southeast wind can be expected to blow, thus from April. This becomes significant because the flowers' nectar is considered a favorite food for Muyuw's endemic cuscus, a species of marsupial (*kwadoy; Phalanger lullulae;* see Norris 1999). About the time these animals have gotten fat from the nectar, the hard southeast wind blows so much that people tend to avoid going out on the water. Cuscus meat then becomes a fish substitute. The relationships here are well-formed. With the apparent rise of serious timbering and gold mining on the island, the international biodiversity community feared for the animal's future, and Muyuw—known as Woodlark Island—was designated as a place to be preserved. So Oxford University biologists organized two research trips to investigate the situation (see Oxford University 1987, 1988). Although it had been presumed the animal would be found in the south-central mountainous region, the island's least inhabited zone, the biologists eventually found that it commonly lived around villages. Although animals in the *Phalanger* genus are known to be indiscriminate foragers, Muyuw people say *kwadoy*'s main food is *gwed* nectar and the nectar obtained from a vine frequently seen in early garden fallows. The nocturnal animal sleeps during the day amid the clumps of high forest scattered about garden areas. At night they descend to feed on *gwed* and other tender plants made available by human activity. Although it is largely the sport of youth, finding the animals is a practice most Muyuw males are skilled in. When asked, they readily list the trees in which *kwadoy* are frequently found—several strangler figs and a species of *Calophyllum* head the list. Altogether these facts suggest that *gwed* are an element of a regularly managed landscape similar to those Posey described for the Amazon (e.g. Posey 1985).

Through the end of my 1995 research I viewed these facts as if they were a part of a continuous tradition across the northern side of the Kula Ring. I had seen *gwed* in the Trobriands, Iwa, and eastern and central Muyuw, and I had the same basic storyline from these locations about how they were conceived. But back in 1995, although I was cognizant of transformations across these spaces, I did not yet know enough about the use of trees to fathom underlying systems. Moreover, while I knew that everywhere people gardened in *digadag*, empirical conditions and individual preferences in eastern Muyuw led me to think that the southeastern Muyuw preference was *ulakay*, the oldest class, and that there was going to be a gradual shift in garden arrangements between eastern Muyuw and the Trobriands, where *digadag* was overwhelmingly used. What I originally learned about Gawa and Iwa fit into that hypothesis. In order to get closer to actual relations and social ecology, I now review, and update, the garden structures that made me first realize I was being told something significant when I first heard about *gwed*.

Garden Forms and Geometrical Variation

My 1990 book hypothesized a transformation across the northern side of the Kula Ring. The transformations were most easily illustrated by comparing garden and village forms. I did not intend to review my original synthesis when this new work started; however, when the Wabunun elder excitedly replied to my first inquiry about the tree by describing differences extending to the Trobriands, I concluded that I should revisit that spatial transformation. If verified would they relate to the social ecology of trees? The results of this phase of the inquiry were surprising. This section and the next one lay out the first-order analyses that emerged relatively quickly. I begin with a summary of the Muyuw garden, a form that should be compared with the Atoni house of eastern Indonesia (Cunningham 1964).[8]

A standard Muyuw garden is a rich structure understood to reflect an order emanating from the Creator, Geliu, who arrived on the island by an outrigger canoe—an *anageg*. The garden's forms are meant to recapture powers that the Creator had at the beginning of time. Virtually everybody says, or used to say, that if the form is not made correctly, then crops will not grow. Without crops, no people: the form is a conceived condition of existence and inculcated into Muyuw consciousness from the time children can walk.[9] The shape is square to rectangular, consisting of at least four units defined by at least one east-west and one north-south path. The east-to-west path, from sunrise to sunset, forms the determinative axis for most of the culture providing primary garden, village, and house orientations. The east-to-west form is also likened to a tree, the "base" or "tip" form discussed in the next chapter.

The garden orientation is the culture's paramount form. Although all islands to the west have analogous orienting lines, their garden structures are not the exemplary form they are in Muyuw.

People shape garden paths by arranging logs left from cutting and burning the forest. The type of tree used to make this form is irrelevant—people use what is available. But where the logs intersect, the east-west log must be set first on a rock, and the north-south log placed on top of it. There are four such rock intersections at each meeting of paths. Called *ananun*, the rocks are likened to a breast's nipple (*sesun*). In much of Muyuw, four sticks are stuck upright at these corners and have a long limb set on them that extends down toward the ground and rests atop the log that it parallels: the whole structure forms a triangle, eight to each intersection. Although most people do not pay any attention to what saplings are used to form these uprights, in 1996 a Wabunun elder told me that one kind of tree—*akumal*—should be used for the two eastern uprights. Certain powers associated with this tree were then added to those associated with the structure; positioned to the east, the powers spread west through the garden.[10] A plant, unfortunately one I never collected, *tagasas*,[11] is left in the place of these intersections to grow through the fallow time so that a new garden follows the outlines of its predecessor.

Garden intersections are called "navel" (*pwason*). New units are added to the garden by increasing the number of paths, usually those going north to south, so most gardens are rectangles. If there are two or more "navels" created by the east-to-west extension, the easternmost becomes the focal point and is specified as the *omleyun*. The *omleyun* is opposed to the *omlakein*, which is a line beyond the western end of the garden defined by uncut trees. This inside/outside distinction makes the navel/*omleyun* the most contained part of the garden, and if garden magic is made it begins with the eastern side of the *omleyun*. It is first directed to the *omlakein* so that the magical powers extend out from the navel focal point.

In addition to invoking origin principles and the productive powers associated with them, garden navels model the most important and encompassing social relations in the culture, ideas about correct marriage relations among the culture's clans (*kum*). Although there are eight such units in the culture, four "original" clans are figured in the navel, and each is given a direction and place in this structure: east-Malas; south-Kwasis;

west-Kubay; and north Dawet/Kulabut. This form models the symmetrical exchange of people, prototypically women, but it can also be applied to men. That exchange also models the movement of pigs in mortuary rituals and, in principle, other exchanges fixed in the organization of male/female relations, all integuments of what marriage is (see Damon 1983c; 1990, chapter 4). This is a static model of social reproduction.

The idea of a "navel" as a critical place is also applied to the human navel—if it gets "sick" one's life is in jeopardy—and to the most important part of the highest class of outrigger canoe. If an *anageg's* navel is not properly trimmed, the craft malfunctions (see chapter 6).

Lines of stone or burnt wood running inside the garden's main sections are called *lapuiy* and create units called *venay*. Like the east-west path, these follow the direction of the sun's movement east to west. These units figure importantly in brother/sister exchanges, which in turn, by myth and social processes fundamental to the marriage system, are associated with subclans (*dal* or *wun*), although this social unit is not represented in a garden. That concept finds its place way beyond the *omlakein*, though not necessarily west of the garden navel. Separated brothers and sisters, sealed by the exchange of vegetable food defined as the content of one *venay*, identify the fundamental production relations in the culture, male and female differences conjoined in marriage. Following upon that, the primary crops planted in these small units, yams and taro, are first defined by a male/female contrast, and the two types should be planted next to one another. That association is likened to a marriage. Subsequent growth, new taro buds off the expanding new corm, and the yam tuber or tubers, replicate proximate and alternate generation relationships respectively. Whereas the imagery formed around the intersection of paths is static, that pertaining to these smaller units and their associated relationships exists in a changing time scale defined with respect to the maturation of crops and persons, the two conditioning each other.

A "beautiful garden" is one in which luxuriant growth becomes so mixed up that its initial design and the crops become indistinguishable, a mélange of forms—hence the form's Chinese temple or Shiva-like quality. Yet there is some reason for what gets planted where. In a garden devoted to yams, the innermost (called *utuwan*) and outermost *venay* should have either or both *gumawel* or *mangotan parawog* crop. The tubers of both of these plants tend to grow toward the surface of the ground, and the vine and leaf growth is luxuriant; furthermore, neither tends to taste particularly good. These properties are all deemed beneficial. First, if wild or domestic pigs approach the garden—hardly uncommon—they will see these plants first and eat them. In this regard pigs eat an inferior food first.[12] Loss is minimized. Second, the luxuriant growth not only makes the garden look good to people (and pigs) outside the garden, it also serves as a model for yams growing near them. The yams are said to become angry at that growth and try, therefore, to make their own growth equal to these two. In Wabunun, at least, there are no beliefs about taro quite like this.

Outrigger canoe imagery is superimposed over the forms and terms of the garden. This is an important usage to which I return in the last half of this book in part to answer the question of what is being said when a garden is talked about like a boat. Here I note that outrigger terms are usually the ones employed when gardens are produced. For example, people talk about erecting a fence for a garden as if they were going to cut the strakes (*budakay*) for an outrigger canoe.

In most of Muyuw, gardens and villages are ideally on two parallel planes. As gardens are oriented to the east-west axis, so are villages (and village houses). These forms begin to change in western Muyuw, and transform completely by the Trobriands. There the fundamental axis in a garden is defined by a village/bush distinction. Trobriand villages and gardens may lie along a single axis, a concentric form, whereas in Muyuw there are two parallel forms. From pictures in Malinowski's Trobriand ethnographies it may also be hypothesized that orientations in the gardens move from being primarily horizontal in Muyuw to increasingly vertical in the Trobriands: various pictures of Trobriand gardens show them suffused with staked forms, apparently primarily a matter of garden aesthetics.

Most of the above I knew before I began returning to the region in 1991. I now quickly review what I learned on my returns.

Although I was aware that gardens in the very southeastern part of Muyuw near Guasopa had none of the staked posts I learned about in Wabunun and saw everywhere else, I had never visited Kavatan, the easternmost Muyuw village. I incorporated Kavatan in my surveys to have access to the full reach of the island. And its gardens were less well-defined than those in the southeastern tip. Basic east-west orientations were evident, and I heard much of the familiar mythology and sociological modeling described above, but it was hard to see clear rectangles.

I learned nothing new about these forms in the rest of the southeast or central Muyuw. I was already aware that the basic form of western gardens shifted exactly ninety degrees. Hence the east-west axis elsewhere alters to north-south in western Muyuw. But I learned three new facts in 1996. First, one man described an elaborate vertical enhancement to his garden form that I had not experienced before, although it was consistent with a horizontal-to-vertical transition already noted. Second, internal divisions to *venay* create smaller units that western Muyuw people call *gub*, a term found for identical spaces in the Trobriands. As in the Trobriands, but unlike the rest of Muyuw, each of these units should contain a single crop variety, yam or taro. Whereas in the rest of Muyuw, for example, the two kinds of yams, *kuv* and *parawog,* should be planted in close proximity, in the Mwadau area of western Muyuw they should be in distinct *gub.* By Kitava and the Trobriands they should be in distinct gardens. Third, I presented village/garden relations the way they were rather than the way they were supposed to be. Mwadau village and its houses are oriented east to west like all other Muyuw villages should be. But in 1996 people told me this was by government intervention, and that western Muyuw houses should be scattered around like they are in Gawa and Iwa; the area of "Mwadau" was in fact taking that shape in 1996, though some of the form I reported from 1975 was still evident. Unlike in Gawa, however, western Muyuw houses are supposed to be oriented like they are elsewhere on the island. The sun "jumps" over them, which means their fronts or backs point north or south. Since many of the Mwadau houses now appear in dispersed hamlets strewn here and there, it was not readily apparent how they were oriented east to west. However, whenever I asked somebody to specify sunrise/sunset directions they easily did so, and all compass readings I took roughly corresponded with the conceived orientation. And yam houses are supposed to be, and are, behind human houses, rather than in gardens as elsewhere in Muyuw, or in front of houses from Gawa to the west.

All throughout this region people speak of their gardens being oriented to what I translate as a "principle." The word is from *kikun,* which is commonly used as a verb meaning "to follow." Most everyone in Muyuw knows that the principles for constructing gardens change to the west (reciprocally, the Iwa and Gawa people with whom I interacted knew practices in Muyuw were different than theirs). Western Muyuw people knew that from Gawa on there was a specific word for what is being "followed," *takulumwala.* Its pronunciation in Muyuw is *takunumwan.* In Muyuw the word refers to a place where dismembered corpses were deposited in secondary interment practices, usually at a remove from villages in the bush or ocean-side location. In Mwadau people were aware that the phonetic transformation between the Muyuw and Gawa-Trobriand pronunciations of the word went along with a shift in its overt meaning. The concept changes from scattered burial places in Muyuw to a principle of garden (but not village or house) orientation from Gawa toward the Trobriands.[13] Discussion of this matter with some people left me wondering if those to the west understand themselves surrounded by the work of the ancestors, landscapes well-defined by *takulumwala* keyed to stone markers or by actual lines of stone (as in the Trobriands), while Muyuw people see themselves surrounded by the bones of the ancestors from the unmarked presence of *takununwan.*

Although Gawa and Iwa each have a fundamental axis bisecting the island, southeast-northwest and northeast-southwest respectively,[14] the orienting *takulumwala* for gardens vary for each. In Gawa people spoke as if there were two *takulumwala,* one southeast-northwest, and the other northeast-southwest. Gardens should be defined with respect to these, although they do not have formal paths as in Muyuw, and they are not necessarily rectangular. In principle these orienting lines begin at rock structures Gawans called *vadayi,* which I was told are land border markers. While some Gawan megaliths I examined were considered examples of these *vadayi,* I came away with the impression that these structures existed more in principle than in fact. In any case, spaces inside a Gawan garden are divided into near squares. These replicate the terms and orientations found in Muyuw: they should be north-south and east-west. Combined, these two forms, the *takulumwala* and the east-west and north-south divisions, create a contradictory image, one based on the standards I experienced in Muyuw. There, internal garden boundaries exactly parallel the external ones; by contrast external boundaries in Gawa gardens are skewed about forty-five degrees from the internal ones. This generates odd shapes at garden peripheries.

Iwa gardens too are oriented to *takulumwala,* but the *takulumwala* themselves had no orientation that I could discover or one that any of my informants could enunciate. Most of these people were familiar with the Muyuw forms, so asserting that they did not correspond to some prescribed direction was experientially motivated. Having followed several lines and gotten nowhere fast, I drew a map of Iwa in my book at one point, crisscrossing many short lines within the map. Some were longer than others, but there was no orientation. Asked if my diagram accurately represented their understanding, people said yes. Named and "owned" garden plots were associated with the lines, but relative to the other islands the situation is confusing.[15]

On these small islands government policy consolidated people into villages, and rough conglomerations of houses remain evident on Gawa and Iwa. But people told me their traditional pattern was more scattered, and there was plenty of evidence of this. It is not meaningful to talk about relationships between villages and gardens in these little islands in the same way that you would in Muyuw and the Trobriands, where the patterns are visible and well-formed conceptually.[16]

During my June 1995 visit to the Trobriands, a guide pointed to a *takulumwala* as we walked through a garden. But the term did not mean anything to me. A map Linus Digim'Rina's brother drew for me illustrates, I believe, the situation. *Takulumwala* lines radiate from villages. However, I learned their significance only when I was on Muyuw in 1996, and then I

could only have recourse to Trobriand people married into Muyuw. Their
views were identical and unequivocal. "Ancestors" set *takulumwala* when
fighting stopped, order came, and fields and gardens were organized with
respect to them. They are the analogue of the Muyuw east-west orientation, defined by a state of time rather than direction in space. Their further
elucidation, however, is a matter for a different ethnographer.

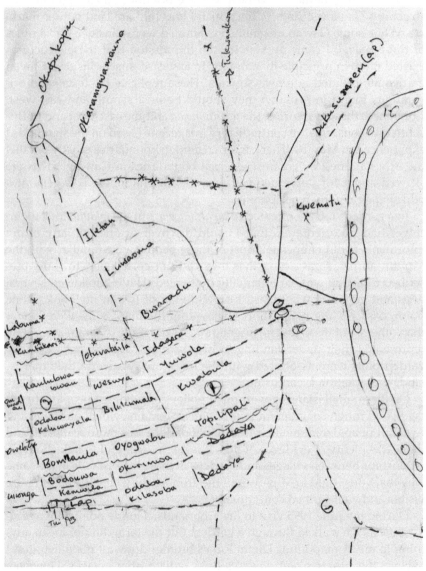

Map 1.1. Okaibom *takulumwala*.

Variation in the Intensity of Gardening Practices: From Kavatan to the Trobriands

Located on Muyuw's easternmost side, "Kavatan" is an old village name and by myth associated with one of the four old clans in Muyuw, the Malas clan. The clan association connects it to the Trobriands, and many people understand that linguistically and positionally there are family resemblances between the Muyuw "Kavatan" and the Trobriand "Kavataria," a village located along the shoreline of the central Trobriand lagoon.[17] Many Muyuw and Trobriand words differ by the absence of final vowels (Muyuw), and a Muyuw "n" is usually realized as "l" or "r" in the Trobriands.[18] *Tan* in Muyuw and *taria* in the Trobriands both refer to the high water mark caused by tidal action. Both places are considered fishing villages. One of Kavatan's responsibilities was to catch and transport fish to the peoples of north-central Muyuw, Wamwan; a well-worn coral path winds underneath a high forest canopy from Kavatan northwest to a landing called Suiyak, a point of departure for the voyage toward Wamwan.

Interestingly, the noun classifier for wooden things like rudders (*kavavis*) and paddles (*kalavis*), including in the latter the long paddle-like spoons that men use to stir sago dumplings (*alivin*), is *kave*. *Kavetan* is "one paddle-like thing"; *kavey* is two, *kavetoun* three. A working hypothesis might be that the name "Kavatan" implies that this place is the "rudder" of the island. In 2009 a number of people said that they could understand this hypothesis because rudders are prototypically at the base (*wowun*) of a boat, and Kavatan is the island's "base." Consequently Kavatan is where the island's New Year ceremony is supposed to begin.

In addition to conducting tree surveys in Kavatan, I wanted to explore how the village related to the model of Muyuw gardens and villages reviewed in the previous section. I have already discussed the gardens: they are poorly formed. I knew the houses were spread along a wide expanse of the beach. How did that relate to the model? Kavatan informants immediately agreed that Muyuw villages should be organized in two rows of east-west houses. And they said that before the government forced them to the shore, Kavatan was so organized. Signs of the old village sites could be seen in the interior. During our first visit there in June 1995, Simon Bickler indeed found potshards well inland from the beach. I saw those again in May 1996 when I returned. But more than those, by 1996 I knew forest regrowth cycles well enough to recognize that sometime in the recent past—two hundred years?—much larger areas inland had been used, for gardening if not villages. Yet this was not the pattern that eventually interested me; rather, what became significant was the enormous size of the resource base Kavatan exploited. Please note map 1.2. Although they do not claim it for themselves, Kavatan people regularly exploit the entire lagoon area that extends from their arcing shoreline out to Nubal Island. This is roughly thirty-six square kilometers. They also regularly fish along the reefs extending to the line marked Wamwan, a dividing line for the central part of the island's territory and Kaulay's; should they take a catch of fish to Kaulay, Kavatan people could use the reefs all the way to Kaulay's landing area.

Then there are their sago resources. Kavatan's list of claimed sago orchards begins at the headwaters of the Oscelio River, virtually the foot of Mount Kabat. That region overlaps with claims Kaulay people make. The total expanse for this one village encompasses nearly a third of the island. A list of sago types Kavatan men distinguish goes to twelve; by contrast, the number was about seven in the gardening villages along southeastern Muyuw. I'm doubtful that there are nearly twice the varieties in the Kavatan orchards, but it is not unreasonable to assume that the reciprocal of their inattention to gardening is a profusion of recognized differences in this substitute food source. Sago, however, is not just a substitute food source but, as detailed in chapter 3, is involved in various formal exchanges, especially with Budibud. Kavatan runs this exchange more frequently and more intensely than other villages. They do not use sago just for themselves.

The resource base of the tip of southeastern Muyuw is nearly as large as Kavatan's. Yet these are gardening villages, and inhabitants' excursions up the Oscelio and Sinkwalay Rivers for sago are more for ritual and

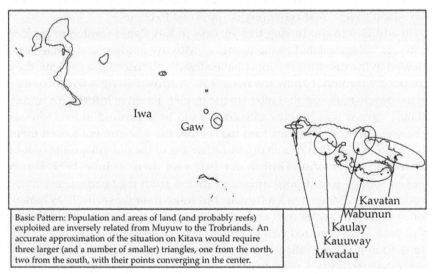

Basic Pattern: Population and areas of land (and probably reefs) exploited are inversely related from Muyuw to the Trobriands. An accurate approximation of the situation on Kitava would require three larger (and a number of smaller) triangles, one from the north, two from the south, with their points converging in the center.

Map 1.2. Extensive versus Intensive Resources Bases.

drought purposes than primary consumption. Moving toward Wayavat and Wabunun, the zone regularly exploited decreases. None of the people from these places habitually ventures much outside the lagoon framed by Sulog at the west, the southeastern tip at the east, and the reefs and islands running between these two points. Because some of the people from Wayavat descended from a line of villages that once laced the Kweyakwoya region across the top of southeastern Muyuw, some of their sago orchards extend up the Oscelio and Sinkwalay Rivers. But most of those people, like those from Wabunun and Unamatan, only go to the region marked on the map. Historical and archaeological remains suggest this southeastern area, probably unlike the Kavatan area, has tended to have a high population primarily devoted to horticultural pursuits.

The pattern: the further west one goes, the smaller the overt resource base becomes. Yet the population increases. While north-central Muyuw people exploit the sago orchards to their south, western Muyuw people claim no orchards; their need for sago for ritual purposes is facilitated through exchange with, primarily, Dikwayas and Kaulay. This demonstration, however crude it must be, intensifies when it crosses to Gawa and the islands leading to the Trobriands. There is a major jump up in the population of those islands; Iwa, not a mile square, has some eight hundred people.[19]

Several other factors correlate with narrowing resource bases. All concern a major intensification of productive activities. Some entail fishing methods, as none of these islands have barrier reefs and, therefore, lagoons to exploit. Iwa features a unique method for capturing *vakiya,* a bird of curious interest to this entire set of islands. Feathers included, these birds are about the size of a person's fist. Although my prime concern here is with the horticultural intensification, Iwa's bird-snaring system deserves a moment of reflection.

First the traps, *gabaku,* or *vakiya sikw*: *vakiya* refers to the bird caught, *sikw* the twine on the bent piece of wood used to spring the trap once the animal attempts to grab its bait. The traps are tapered rectangles about forty centimeters long and fifteen to twenty centimeters wide, made from a very light piece of wood. Inserted into one end of the form is a piece of wood that acts like a bent sapling, and in fact works on the same principle as traps used for pigs on larger islands. It arcs over the top of the trap toward the other end where there is a triangular concavity into which dirt, leaves, bait, and the mechanism for holding the bait down are put. The bait is a locust-like insect called *kakapla*. The insects are found on Iwa, especially in its oval basin along with all the gardens, houses, and hundreds of fruit and nut trees. The traps are placed on or near trees that ring and tower above the basin. Once set, the young men who pursue these birds hide a short distance away. When the trap is sprung, the arced stick snaps up, pulling a noose around the bird's head. It doesn't kill the bird, so the trap must be watched and attended the moment it is sprung.

Jerry Leach told me that Trobriand people believe the *vakiya* flies back and forth between Muyuw and the Trobriands. In the Trobriands its presence is a harbinger of a good harvest, and its image is an important symbol of chiefly hierarchy (Damon 1990: 1, 247n2). According to Leach, the bird is one of several kingfishers inhabiting these islands (*Halcyon torotora, H. cholris, H. sancta*). Iwa people, however, told me that what they call *vakiya* is what Muyuw call *kelkil*, the rainbow bird (*Merops ornatus*). Muyuw people believe it flies back and forth between the Trobriands and Muyuw, signaling the harvest when it appears. Iwa people reported that their elders said the bird comes from a place called Kinau, but they asked me if I knew where that was. Wherever it comes from, it is eaten in neither the Trobriands nor Muyuw, only in Kitava, Iwa, and Gawa. My Gawan informant also clearly distinguished the *kelkil* from the *vakiya* by, among other attributes, their songs: *kil, kil, kil, kil* for *kelkil*; *li, li, li, li* for *vakiya*. The first of these is related to harvesting activities (*-kil*) for yams. The second he related to *li*, from the verb, *kaleliu*, which refers to the initial cutting of small brush when outlining what is to be chopped, burned, and then planted. *Kelkil*, the Iwan *vakiya*, arrive among these islands in large numbers after the March equinox and pick off insects found in garden areas (they can also be found in meadows that dot Muyuw's heavily forested interior described in chapter 3). *Vakiya* as kingfishers are, in Muyuw, called *kioki*, adorn outrigger canoes, and are harbingers of arriving sailing craft, not yams.

The practices regarding *gwed* described at the beginning of this chapter are a dimension of the intensification process. From Gawa to Iwa, at least some people plant *gwed* seedlings directly into their gardens. This is supposed to have the same beneficial effect as elsewhere but by two different means. According to some accounts the living tree passes its values on to a growing crop; according to others the planted *gwed* seedling generates the reproductive conditions of fallowing while the same ground is also being used to produce food. These small islands are also known to use gardening areas as toilet regions *in order* to enhance their agricultural results.[20]

Iwa distinguishes between places that do—to the west, inclusive of Iwa—and do not—to the east—prune their yams. The practice is called *bid* or *bidabid*. Pruning enhances the size of the tubers. Two or three months before the tubers (*Dioscorea esculenta; teytu* from Gawa and Iwa to the Trobriands, *parawog* in Muyuw) are mature, their male owners—"women cannot do it"—go through every plant to loosen the soil around the tubers and extract smaller ones. According to their theory, the rest of the plant's energy goes into the remaining tubers. Something about the theory is correct. I never weighed harvested plants across this region as part of this study, but I have never seen large Muyuw *parawog* as big as the average ones in Iwa and the Trobriands. The fantastic piles of yams demonstrated by Malinowski's photos never really registered until I thought about their proportions compared to Muyuw's (see Damon 2000b). Another factor fits into this. Muyuw have one yam stake for every two or three planted tubers, and people strive to have *parawog* and *kuv* next to one another climbing the same stake. But to the west people become more assiduous in their staking; in Iwa and Kitava I saw individual plants with *three* stakes to them. It may be that staking has no certifiable benefit, but it requires constant attention, and that attention has an effect: every time diligent gardeners go into their gardens they remove unwanted weeds. Garden land I saw in Iwa and Kitava in 1996 looked like it had been swept clean. This is part of the intensification transformation.[21]

Compared to Muyuw people, those toward the west become fanatically interested in their gardening pursuits. And more than one Trobriand person became extremely intrigued by what I was trying to learn about *gwed*. One such person was a Trobriand schoolteacher I met in Gawa in January 1996. With great animation he relayed his own speculations about what the tree did; he thought it made the soil clumpy. By then I knew it wasn't a legume, so I no longer assumed it was fixing nitrogen; but another tree some people reportedly manipulated in the Trobriands, *nilg'* (or *liga* in the Trobriands, *Acacia*), was a candidate for fixing. I asked him if he used it, and he responded firmly that he did not. It made the soil too sandy and yam leaves tended to turn prematurely yellow if planted in association with it. He then talked about ferns he planted around his plants to enhance their growth.[22]

I close this discussion with a return to the comparatively lackadaisical practices in Muyuw. There is, first of all, a lot of variation in this. One Muyuw elder, a dedicated and passionate gardener, told me that he would bring this or that cut tree into his gardens at various stages of yam and taro growth. The properties of these trees would be deposited in the garden to enhance growth. This is the same principle that operates in Muyuw garden magic. Such practices are now disparaged, and few youth follow them. But this also means that those very same youth go into their planted gardens less frequently and pay less attention to the growing plants. Practices that in themselves may have been meaningless were devices that resulted in people attending to small differences.

Beyond these reflections, different practices across these islands are principled. Many Muyuw know that people from Iwa prune their yams, thereby obtaining larger tubers. But Muyuw also think those products are inferior. Relative to Muyuw's preferred "smooth" (*mukumuk*) yams, those resulting from pruning are "stringy" (*vatulin*). In place of intensified horticulture to the west, Muyuw plant many more seeds; while I was astonished by the large size of Iwa yams, I was also fascinated by their minuscule gardens.[23]

The cosmological/geometrical transformations across this line of islands accompanies a major contrast in the intensity of the exploitation of resource bases. More extensive productive activities toward the east become more intensive toward the west or, as in the case of sago making, cease altogether.

Gwed and *Atulab*, and Other Significant Trees

Information I had about *gwed* was confirmed and amplified in Wabunun almost as soon as I arrived there from the Trobriands in 1995. When I explained to a prime informant, Dibolel, what I had seen near Okaibom, he excitedly stated that Wabunun people did the same thing. Then he said there was also a "bad tree," called *atulab* (*Apocynaceae Alstonia brassii*), which, although it grows in the same contexts as *gwed*, presents opposite characteristics. *Gwed* spread out whereas *atulab* tend to be straighter, narrower, and eventually taller. The former have thin bark and soft, light wood easily chopped or carved whereas *atulab* have thick bark and very dense, heavy wood. People do not like cutting it because the bark and wood dull their machetes and axes.[24] And because the bark is so thick, burning doesn't readily kill it; it should be fired three or more times to make sure it is really dead and not just stunned. *Gwed* have profuse whitish sap that is supposed to taste sweat whereas *atulab* sap does indeed taste bitter, or "salty" (*yayan*). It is said to be "very salty," and it is said that the tree puts salty stuff into the soil, which is very bad for the plants; it is also bad for head lice (*kut*), so it is used to combat them. These two trees are not just different; they are opposed but paired. And by February 1996, after my trek from the Trobriands through Kitava, Iwa, Gawa, and then into eastern Muyuw I confirmed that across this expanse it was widely understood that *gwed* are good and should be sought after when planting crops, while *atulab* are bad and should be avoided. Forming the extremes of a continuum, other trees could be situated between these two.

Along with *gwed*, *atulab* became a focus of my investigations. Well known in this part of the world, *atulab* appears in disturbed landscapes, often finding niches in very meager conditions. But unlike its opposite, it is rarely used in construction. However, it is slow to rot, and as such it is cut for fence stakes while large ones are useful to support betel pepper vines. Large dead *atulab* may stand for three years or more before they fall, so people who plant pepper vines for chewing betel nut start a few next to these trees. They trade length of growing for diminished growth.

By the end of August 1995 I had gathered much indigenous exegesis and materials for biochemical analyses on both *gwed* and *atulab,* and I had been told about a number of other trees that were supposed to be good or bad for crops. In the latter category is an early fallow tree called *asibwad* (*Timonius timon*) that is said to be like *atulab,* especially in its root structure. Other bad trees included several strangler figs (*akisi* and *bwibuiy*; *Ficus,* but never identified beyond that). However, another person claimed another fig (*agigaway, Ficus tinctoria*), was good for crops, and he sought out their dead trunks whenever he gardened by them. Then reports gradually started coming my way about another good tree, *tabnayiyuw*.[25] By this time I was becoming dubious about many of these claims, for it seemed clear some of them just invoked principles of like begets like, common in Muyuw magic. So crops planted next to large trees grow larger than those grown next to small ones. A statement such as this, frequently expressed, also asserts that old fallows produce different and larger food than new fallows, and may be commenting on properties that are indeed physical. Nevertheless, such beliefs were different than specific assertions, positive or negative, about specific trees. So I expressed my skepticism to Sipum and another son of one of my good informants from the 1970s. They then marched me out to an *ulakay* garden full of killed *tabnayiyuw* and systematically showed me various crops planted at the same time, some near these trees, some meters away. The demonstration was astonishing; I made repeated photographs to verify the differences. While most of the examples were of a crop Muyuw call *vesop* (*Alocasia sp.*), yams were also planted next to this tree. The vines were dead, the stakes that had been used to support them were two or three times higher than Muyuw customarily employ, but the superior growth of the vines was easy to see), as were those of the betel pepper planted in this garden. Later I learned that yam seeds are always washed with water enhanced by *tabnayiyuw* sap (a similar practice uses liquid from the tree called *akuluiy*). Toward a garden edge was an old *atulab,* and close to the *tabnayiyuw* was a tree called *akewal* (Dysoxylum *arnolddiam*). The pepper had been planted on each of these at the same time, and the plant was just beginning to climb the *atulab* and *akewal* and was long from being harvestable. By contrast, the plant on the *tabnayiyuw* stretched ten or more meters up the tree, completely encircled it, and was very thick. This case, by the way, illustrates a principle implied in the discussion of *atulab* above. *Tabnayiyuw* rot quickly, so while the growth on this tree was rich and fast, within a year or so the tree would collapse; the *atulab* would stand for a long time. People sort differences.

This demonstration had two effects. First, it considerably widened my framework of research from its original focus on *gwed*. This was a completely different tree from a completely different fallow phase. Second, while I never doubted that much of the elaborate Muyuw garden structure referred to things that were real in the world—including a social system that located people in a regional setting—I remained skeptical about the reports of growth enhancement, especially because I was treating them as if they referred to biochemical, not social, differences. On this day that ambivalence came to a full stop.

The consequences of that realization began to play out in 1996 during research excursions in central and western Muyuw. In central Muyuw's Kaulay village I received the same information about *gwed* and *atulab* as I had elsewhere. But during the course of my 1996 return, I began to realize that the people there paid the ideas no attention whatsoever. They talked about a different tree, another fig (*kubwag*, never identified beyond the *Ficus* level). Wabunun people did not recognize it as one they knew. The best gardener I knew in western Muyuw had no knowledge of it, a fact a Kaulay source found incredulous. *Kubwag* supposedly makes the soil black and very rich. Although never as common as *gwed*, it was often found in the middle-aged fallow type called *oleybik*. Spreading rather than particularly ascendant, it was often the largest tree to be found. When this came into focus, I collected biochemical data for it like I did for the trees I had earlier singled out. And people did with it what people did with other trees elsewhere: They crowded plants around its base, and if the *kubwag* was downed, they were also purposely planted along the fallen trunk.

Yet this was just the beginning. When I toured western Muyuw, a region called Nayem, the shoreline from which Gawa is visible on a clear day and between which there is frequent intercourse among the islanders, I discovered that not only did people pay no attention to *gwed* but also that none of my informants, mostly elder men who were dedicated and distinguished gardeners,[26] could repeat what I had presumed was basic knowledge everywhere. *Gwed* grew in their early fallows like elsewhere, but the people ignored them. I also recorded finding *tabnayiyuw* in some of their older fallow areas, but, more or less like central Muyuw people, they ignored the tree. However, some people pointed to another tree, as it turns out another *Dysoxylum*, called *silamuyuw*. Unlike Kaulay's *kubwag*, which by name and leaf shape was not known elsewhere, this tree is known everywhere. But while eastern Muyuw would only say that it is "pretty good" (as they say of all but one of the Meliaceae trees), they do not follow practices that western Muyuw people do.

One thing was becoming clear by this time: a lot of attention was paid to minute differences. Moreover I could never predict when I would suddenly learn that this tree or that was good or bad. In 2002 I set up a contrast of soils in a garden between plants next to *tabnayiyuw* and a tree called *ukw* (*Sterculia sp.*), another old-fallow tree significant for reasons I return to later but was then, to my knowledge, irrelevant for gardening. Yet when I returned to the village one man said that it was sometimes "pretty good." And there seemed to be another level of classification that facilitated the focus on this tree or that, or caused people to ignore certain trees altogether. It is to a consideration of this level of organization that we next turn.

Intentions and Temporal Aspects of Fallow Regimes

I have repeatedly noted and contrasted the fallow types *digadag, oleybikw,* and *ulakay*. Although I had seen the terms' approximations in the Trobriand corpus, it was only when Linus Digim'Rina showed me around his village of Okaibom that it became clear they were also used there. In 1995 and early in 1996 I confirmed their relevance in Kitava, Iwa, and Gawa, although I was told the small islands did not have *ulakay* anymore. But I also gradually learned of new categories, and I began to realize that however constant their usage was across these spaces, there were fundamental differences in conceived and practiced horticultural regimes. In preparation for the biochemical analyses that follow, I discuss these now. This focus, however, verges onto the slightly different question of ecological zones and their floral composition, a topic for succeeding chapters.

The first new category I learned was *tawan*. I never heard it in southeastern Muyuw nor could I get it explained there once it emerged as a descriptive category. But it refers to the usually very large and old trees growing on steep rises that form the northern side of Muyuw and that encircle Gawa, Kweywata, Iwa, and Kitava.[27] Although these islands, and especially Iwa, show traces of the north-south slope that defines both Muyuw and the Trobriands, they may be envisioned as columns shooting up out of the ocean. On the small islands, people traverse these cliffs by steep ladder-like trails to the top platforms where they live and do most of their productive activities. One of the north-side Iwa trails, rarely used except during a portion of Iwa's New Year ceremony, is so steep and dangerous that escaping death while descending or ascending it is part of its significance. On all these islands, one to three relatively narrow terraces often break the inclines. And on these terraces people use the *digadag-ulakay* fallow terms. The physical rises themselves are called *papap*, a term accurately translated as "cliff." Although the tree growth and content is identical to that found in many areas classed as *ulakay*, the tree areas on the steep inclines are called *tawan*. These areas never seem to be completely cut for gardening purposes, but on some islands individual crops are crammed into convenient, relatively open places.

Unlike every other used landscape, *tawan* are not intentionally burned. Places that have been burned are called *sigob*, based on the verb for burn—*gob*.[28] Although often used as if it suggests a recently used garden area, *sigob* refers to any area that has been intentionally burned for gardening purposes. Sometimes it specifies a predominant or remarkable use of a certain area. During my first research in Muyuw from 1973 to 1975, my wife and I maintained our own gardens along with everyone else. Throughout this last research stint everyone in Wabunun knew where we had gardened and referred to the places as "Fred's *sigob*," even though in most cases others had gardened there subsequently. In the 1990s, if one flew over the sector of Muyuw south of the Sinkwalay River and east from Unamatan village, it would look like most of it is hardly touched by humans. But one knowledgeable informant once waved over it, saying it was all a *sigob*. Periodic burning, however long the periods, seems to be the first condition for the definition and divisions of fallow forests. *Tawan* excepting, firing underlies most horticultural practices across these islands; firing also forms an essential component of sago orchard management, although the practices there are on a very different scale. Firing, however, is not just a horticultural practice: infants and young mothers are also smoked, pigs burned before being sectioned into distributable pieces, and even outrigger canoe floats singed to look banded.

Kadidulel and *idal* were additional new terms. The first is commonly used in eastern and central Muyuw and was not known by my western sources, but when I described the condition to them they understood and said their term was *idal*. When I reported this new term to my authority back in eastern Muyuw, he distinguished the two. Both generally refer to poor soils, but in the east they are distinguished this way: *Kadidulel* soils are those in which trees do not get big, except for maybe an occasional *atulab*. *Idal* soils are those in which the taro or bananas do not get big. Whatever the perceived cause of the *idal* condition, *kadidulel* is said to come about from continuous gardening. Some people oppose *kididulel* to the term *kaleybikw*. This latter term refers to soil of sufficient strength, so that after the soil is used fairly quickly the tree classification reaches the *oleybikw* stage rather than *digadag*. Clumps of hectare-sized soils around Wabunun and Kaulay villages easily fit into these *kadidulel* classes.[29] These patches, relatively insignificant given the total Muyuw landmass, have rarely changed in their appearance over the forty years that I have experienced them in the Kaulay and Wabunun regions. Muyuw people who have been to Iwa, Kitava, or much of the central Trobriand gardening area claim that those areas are all *kadidulel*. At least a few Muyuw people use that fact to explain the intensity of the garden practices there: the soil is so bad that if it is not worked hard, it produces little. This observation implies that continuous action has led to a phase change resulting in a new regime of action.

These new categories were coming into focus as I began to realize that however widespread the three standard fallow categories were, there were major differences in the ideal, and often actual, way people gardened. Although it was evident that practically everywhere people passed among all three classes, it is also clear that ideal models exist, and these models, once understood, could be readily verified. People followed them, and if they did not they had specific reasons for not doing so.

I will have to skip Kavatan in this discussion because this order of modeling came into focus weeks after I last visited there. So I begin with Wabunun, representative of the island's southeast sector.

Wabunun is where I first experienced the fallow categories in everyday use, eventually over a nearly forty-year period. I know preferences and actions at an individual level and, for many of the village's members, the rather dramatic consequences of introduced coconut plantations. Although these were first established in the 1920s, the 1950s and 1960s saw major expansion. The people becoming the elders in the 1990s, the youth of the 1940s–1960s, reported to me that the shoreline areas were so dark with high forests that they feared walking in them; occasionally "small people," *to's*, would be seen there. Now the land is all *digadag* amid coconut plantations in various states of ruin.[30] Muyuw origin mythology shapes a transition from dark to light, a contrast reinforced by Christian ideology. For many of the elders I first knew in the early 1970s and in the 1990s, that contrast is palpable. In the 1970s the ambitious elders I knew thoroughly enjoyed knocking down *ulakay* forests for their new gardens, and were very pleased with the enormous sizes of taro and yam; the taro also has a different, softer texture. Many others were cutting *oleybikw*, areas that were last cut before or just after US soldiers were on the island during World War II.

Given all that, it was not until well into 1996 that I discovered that the village's ideal fallow type was *digadag*. The taro characteristic—not very big and hard—generated by this class of garden around Wabunun is widely known and admired by many; I first learned about it in Monivey-ova village in far northwestern Muyuw in 1975 when an elder spoke about Wabunun's great taro. But I did not realize that he was speaking about a registered differential practice, one of many. Once I learned the Wabunun paradigm, everyone agreed to its existence. One Wabunun man used the fact that "we are *digadag* people" to explain why he, representative of a type, did not know much about another region's flora.

Although this characterization is common in the area, not everyone sits well with its consequences. One couple has very different preferences: the woman prefers the food generated by the *digadag* norm, whereas her husband prefers food coming from older fallow classes. For reasons I'll review later, she usually prevails in decisions about the gardens they make by themselves, while sometimes he does when they garden with other people.

In north central Muyuw, gardens should be *oleybik*. There are exceptions: for example, my guide throughout from 1996, Talibonas, used *digadag* instead. When I learned the model and then asked him why this was the case, he said that his father had told him to use one *digadag* place; so he stayed there, widening the cycle of land so defined by cutting into surrounding *oleybikw*. By then, my former guide for the area, Vekwaya, Talibonas's elder brother, was having increased difficulty getting around, so *digadag* would have been easier for him. Yet he continued to use *oleybik*. The trees Vekwaya was cutting often had diameters of twenty to thirty centimeters in diameter at breast height, roughly twice the size of those found in *digadag* his younger brother was cutting. The favored firewood for this area comes from *oleybik* gardens, a topic I return to later. But after I learned the model, one of the ways I checked its usage was by walking through the areas where people were getting their firewood. It was easy to determine that many Kaulay people were following the practice much of the time. Among other things, their model systematically generated their primary cooking fuel, harvested from downed boles for years after an *oleybik* region is cut and used as a garden. Firewood from a *digadag* would be used up or rotted to nothing within a couple of years.

The western Muyuw ideal is *ulakay*. Neither I nor my informants knew how this was done before steel tools were introduced. But today the expression is very clear. Some people said they regularly went back and forth between *ulakay* and *digadag,* the former to build up their number of seeds, the latter to make garden preparation easier. Yet their model is why they paid *gwed* no attention. In the *ulakay* ideal there are no *gwed*. We shall return to these dynamics, but the important point here is that ideal fallows types help govern knowledge of flora.

The *tawan* (usually pronounced *tawala,* to the west) and its *ulakay*-like characteristics aside, the small islands between Muyuw and the Trobriands effectively plant in *digadag* only. And by the time you reach Iwa and Kitava, the garden areas become almost completely dominated by kunai grass. However, the realities here are complex and not what they might first appear to be; their reasons remained opaque for several years.

For my 1995 trip to Iwa, my Wabunun friends told me that Iwa "had no trees." Indeed, they saved their yam stakes from one year to the next, a strategy Muyuw people find ludicrous. During my longer stay in Iwa in 1996 their garden fences looked so flimsy I measured the diameter of the stakes: they were on the order of one to two centimeters. By contrast, in Kaulay they are three- to six-centimeters in diameter, roughly the same size as those in Wabunun. However, my first and continuing impression of Iwa, and only to a slightly lesser extent Gawa, was not that it had so few trees but that it had so many. The discrepancy in perception partly comes from, as noted earlier, the fact that gardens and villages and treed areas are all mixed up. Even more to the point, the place is packed with nut or fruit-bearing trees. A complete survey of trees on Iwa would be possible because it is so small. Several times I walked between Iwa's two main constellations of houses counting the breadfruit trees (*kum*) easily visible. However I got different counts each time and it always depended on how carefully I looked and in which direction: there are dozens. In one *garden area*, roughly thirty by seventy-five meters, there were seven breadfruit trees, two coconuts, six saido (*T. catappa*), three mango (*kwani*), one *kaboum* (a planted tree very important for outrigger canoes), one *kebwiba* (a species of *Pandanus*), and a *ganom* (an unidentified tree, but one that reportedly produces nuts like the one Iwa people call *tawakw*). *Tawakw* (*Terminalia megalocarpa*) is found by the hundreds on the island. Where these trees were absent, the area was planted in sweet potatoes and tapioca. In another random survey of about seventy-five meters along a path through gardening areas I counted every tree over ten centimeters in diameter at breast height and within five meters of the path. There were fifty-six trees, all but a few planted, and all for identifiable purposes the least auspicious for building houses or shade. In the list were twenty-five coconuts; seven of the island's all-important *saido*; four *tawakw*; four mango; two *natu* (probably *Burkella obovata*), a tree that produces an apple- or pear-like fruit as well as wood useful for outrigger strakes; a couple of other minor berry-producing trees; and one said to be planted for shade. Iwa is not first and foremost a horticultural area; it is devoted to arboriculture. Gawa plays the same theme, only in a slightly different key. One of the signs of these intentions I first found quite odd: While I was doing a survey of yam and taro names on Iwa in 1996, several people told me they did not know the names of their plants, and if I wanted to know the names I should go to Muyuw. This was just noise in 1996, but the statement finally made sense by the time of my brief return to the area in 1998.[31]

This data creates an image of varied intentions across these spaces, the intentions related systematically to knowledge of and expectations about flora.

The Biochemical Evidence

Beginning with the 1991 initial knowledge of *gwed*, coming across reports of Posey investigating biochemical transformations that the Kayapó might have generated in their landscapes (e.g. Hecht and Posey 1989), and wanting to extend my own analytical skills and empirical interests, I sought to build a biochemical component into my research strategy. Although my ecological advisers at the University of Virginia had little experience with tropical conditions, they too were interested in the adventure and were increasingly using stable isotope analyses to trace nutrients through ecosystems and between continents (e.g. Swap et al. 1996). In this experiment we hoped to combine our analytical skills and interests. And as noted earlier, first with the presumption that *gwed* was a legume and second that it, or something associated with it, was fixing nitrogen (the former is *not* a condition of the latter), focusing on stable isotopes seemed reasonable.[32] This section of the chapter outlines the developing results from this portion of my research. Complete data sets are provided in the photographic accompaniment to this book.

This research sought to combine two forms of knowledge, that of my Kula Ring teachers and a scientific tradition in the West. I thought I could intelligently combine them because people spoke of something passing from a tree to the ground to their crops; when I returned to investigate this idea further my recollections proved correct. The native model is of interaction across plant type boundaries and environments. Both the indigenous and the ecological models specified systemic relations.

In 1995 I gathered soil samples and pieces of tree and plant leaves from across a broad swath of these islands. These materials would be analyzed in the fall of 1995 during which time I expected to learn more about the biochemistry of plant growth. I would then return to the islands in 1996 so that I could make more focused collections and, critically, talk in greater detail about plant biology with local people; I did not underestimate what they might know, though how they knew and represented such information was an empirical question. However, the mass spectrometer broke, so I returned to Papua New Guinea in 1996 without the analysis of variation we hoped to create from the 1995 collection. I made many new collections and learned a great deal. But the additional soil and plant samples were gathered following intuition and intelligent hunches that had not been transformed by another order of positive knowledge.

In late 1998 I established contact with Dr. Shaw Reid of the Cornell Nutrient Analysis Laboratories at Cornell University. Reid helped organize analyses of many of the soil samples I collected in 1995 and 1996, a much smaller set from 2002, and a final one from 2009. And he became an important instructor for interpreting the data. These more conventional analyses do not trace nutrients through an ecosystem, which is what I had hoped for from the stable isotope analysis; however, given the limits of my data, they suggest significant differences in the nutritional backgrounds humans generate in these landscapes.

The materials presented are not sufficient for drawing firm conclusions. Yet the data are dramatic. At the time these data were gathered, anthropologists had collected limited materials to support the kind of research I was taking on as a novice. Posey and Hecht's work was interesting and inspirational only. Now the situation is much different. Hecht has more firmly established the case of Kayapó modifications of landscapes along biochemical parameters (2004), and modifications of the kind I shall be describing below are increasingly found in the literature. The Cornel soil scientist Reid, for example, had heard many accounts of non-Western horticulturalists recognizing and taking advantage of effects that trees generated in soils they customarily farmed; my stories were familiar to him. Although the specific data sets I have are very limited, they are now consistent with materials presented around the world, including the Pacific (Vitousek et al. 2004). The data presented here enable important hypotheses that can be stated relatively convincingly but which need further study.

I concentrate on results that appear the most interesting. A small note on methods first. Since neither I nor my University of Virginia environmental science advisers were experts in analyzing data from tropical soils, my method for analyzing it had to be one of gradually seeing patterns, repeating those patterns to anyone I could find who was more experienced than me, and learning from their reaction. This is the procedure Gladwin followed in learning about Puluwat navigational principles (Gladwin 1970). For me it is continuing. I came to an interesting set of conclusions in 2004, repeated them to my University of Virginia consultants, and they confirmed them. Realizing the limitations of this situation, in 2005 I repeated the material to an Australian soil scientist[33] who quickly found fault with the analysis, though in so doing he returned me to an original point of departure. In the account that follows, I follow this storyline, beginning with a set of stable isotope readings then moving to the soil analyses.

The stable isotope tests determined the amount of carbon (C) and nitrogen (N) in each sample and the relative proportion of each of the two C (ΔC) and N (ΔN) isotopes. Carbon and nitrogen are the two most important building blocks in organic life. Which of the two isotopes of each, ^{12}C and ^{13}C and ^{14}N and ^{15}N, a plant uses for its growth is immaterial, yet their relative proportions often indicate interesting aspects of ecosystem processes. I had hoped that by examining the amounts of C and N, the relative amounts of the two isotopes, and the ratio between these two

(C/N), we would learn something interesting about how the societies of the northern side of the Kula Ring managed their environment. I shall begin by describing one little test I conducted to see if there might be a close fit between what I assumed was an aspect of Muyuw mythology and a hypothetical biological process in yams.

In 1991 Sipum's father told me that yams only grew at night. Yams are thought to be the Creator's "string"; their growth is a sign of strength. I knew how this belief fit into creation mythology, but I began to wonder if it had another objective basis. For one of the first things I learned as I began to delve into plant biology was the difference between C4 and C3 plants. Details aside, these terms refer to different biochemical processes. While most plants are C3, many grasses, especially those coming out of the tropics, are C4. This allows them to effectively shut down when under stress of the sun and then start back up again when it is darker, thus being able to operate without the normal photosynthesis processes. As it turns out, different carbon isotopes characterize these processes, for example the C4 plants are more enriched in ^{13}C. With the way results were reported to me, this means that ^{13}C-dominated samples are going to show a number on the order of −14 in contrast to ^{12}C samples that are closer to −25. Since a plant running the C4 process is likely to do so only when under stress, I gathered yam leaves at sunup and then at noon from the same plants on a sunny day. Sunny days, however, were difficult to find in 1995 since much of Papua New Guinea was experiencing the La Niña stage of an irregularly shaped El Niño that began in 1991 or 1992. The hypothesis was that if the plant switched into the C4 mode at night, the sunrise reading would be on the order of −14, and if stressed it might be near −25 at midday. In fact both were about the same, −29.18 and −28.95, respectively. We concluded that the plants were thus C3, and in fact all C isotope readings I took of plants and soil, with one exception, were in the C3 range.[34] The one exception was a soil reading of −14.22 from a meadow covered in kunai grass in the south-central part of the island. Presumably kunai grass is a C4 plant.[35]

Interesting results are evident from the *gwed* samples. Please note table 1.1. For the ΔN reading, numbers 1–6 are all negative or just above zero. This means they are enriched in ^{14}N, which implies that their source of (fixed) nitrogen is more likely to be from the air than from human or animal wastes. These samples are from southeastern Muyuw (1–3 and 5), the Trobriands (4), and western Muyuw (6). They contrast dramatically with the next three items from Iwa, which show that *gwed* growth is influenced by human wastes deposited in the landscape. One leaf also has much more nitrogen (7; 3.56), but since this is not repeated in the other leaf (8) there is hardly data to suggest a pattern (roots [9] have less nitrogen than other parts of a tree). Although the ΔN numbers for soils show that they

Table 1.1. 1995–96: *Gwed* [*Rhus taitensis*] Samples Sorted by Ascending ΔN

Sample Number, item and location	%C	%N	ΔC	ΔN	C/N
1.) 26 Gwed fruit From ground SeMuyw1 digadag	52.096	1.004	−27.59	−2.32	51.89
2.) 25 Gwed lf 'Male' tree-i.e. no fruit SeMuyw1 digadag	51.109	1.551	−31.52	−1.9	32.95
3.) 24 Gwed lf 'Female' tree-i.e. fruit SeMuyw1 digadag	48.824	2.062	−30.86	−1.8	23.68
4.) 21 Gwd lf Gumakaraina kwabila; poor soil. CTrb5	50.307	1.78	−30.07	−1.09	28.26
5.) 146Gwed SeMuyuw Sibulobul DigTawtun 2	49.260	1.338	−29.92	−1.09	36.82
6.) 152Gwedleaff WMyw Mwadau Idal soil	46.732	1.619	−30.8	.55	28.86
7.) 84 Gweda leaf **Iwa** 1 Digadag soil26	48.854	3.356	−29.98	**3.37**	14.56
8.) 90gweda leaf young tree next to 89 **Iwa** Digadag	49.525	1.509	−30.59	**3.56**	32.82
9.) 85 Gweda root **Iwa** 1 Digadag soil26	45.785	0.645	−29.32	**4.78**	70.98
10.) Average of **Iwa** *digadag* soils (N=4)				**9.15**	10.3
11.) Average Trobriand soil s (N=7)				6.36	14.1
12.) Average Muyuw soils (N=41)				5.25	

are more enriched in ^{15}N than the plants, there remains an interesting difference between the Iwa (10; 9.15) and the Trobriand samples (11; 6.36). Although the Muyuw data comes from a much wider range of environments, the average is on the same order as the Trobriand average (12; 5.25). Tree isotope readings follow those of the soils.

Table 1.2 deals with a contrast discussed earlier from 1998. I compared harvest results from yams planted close to and away from *gwed*. The yams were the same variety of *parawog, Tatalawiu*. The plants next to the *gwed* tree averaged thirteen tubers per seed, whereas those ten to twelve meters

Table 1.2. 1998 *Gwed* [*Rhus*] Harvest Experiment

	sample number, item and location	%C	%N	ΔC	ΔN	C/N
13	183 next to *gwed*	41.9	1.1	−26.4	1.72	38.8
14	184 ten meters from *gwed*	41.6	1.2	−23.2	3.92	35.4
15	185 twelve meters from *gwed*	43.5	1.1	−25	3.9	39.4

away averaged five. The *gwed* were long dead, so there is no record of nitrogen contained in their leaves. Presumably they were of the same order as those listed in table 1.1. The yams were not quite dead so I sampled their leaves—the results here seem equally dramatic. Although the number for sample 13 is positive (1.72), it is considerably lower than the other two, 14 and 15 (3.92, 3.9), probably meaning that a greater proportion of nitrogen in this plant is from a fixing agent than in the other two. Hypothetically the source of the fixed nitrogen is associated with the *gwed*. Although the Trobriand and Muyuw data suggest *gwed* are associated with fixed nitrogen, I concluded in 2004 that nitrogen is probably not the significant contributing factor that accounts for the observed differences in productivity. This followed because the source of nitrogen makes no difference to the plant, and nitrogen amounts do not differ significantly in the samples. This is the logic that the Australian scientist faulted in 2005: I was taking as my unit of analysis a single leaf instead of the total plant. Since foliage is roughly proportional to tuber output, much more nitrogen must have been in the plants that produced nearly twice the number of tubers. And the governing hypothesis here would be that the *gwed* proximate to these plants was not only contributing fixed nitrogen, but more of it. But what of the Iwa which suggest less fixing—if *gwed* might fix nitrogen, do they do something else as well?

With respect to other trees mentioned to this point, similar results appear from *oleybik* and *ulakay* fallows. Table 1.3 assembles data on betel pepper, a variety of taro, and *vesop* (*Alocasia sp.*) from two different gardens growing with respect to four different trees. All samples were selected to test trees against one particularly singled out as good, *Tabnayiyuw* (*D. papuanum*). Two different gardens are at issue. Numbers 13, 14, 18, and 19 are from one garden, 15–17 and 20–23 from another. Samples 13–17 concern betel pepper growing up five different trees; 18–23 are of taro or *vesop* (a variety called *tankubay*) growing next to three trees or without any particular association. Two of these trees are considered bad, *asibwad* (*Timonius timon*) and *atulab*. *Aysasous* is not considered bad but is not avidly sought out. I selected it because I found plants nestled against it when I was first convinced that indeed *tabnayiyuw* did something quite extraordinary. The ΔN difference between numbers 16 and 17 suggests that the pepper plant there is associated with fixed nitrogen. This pattern is repeated between numbers 24 and 26, with 24 close to a *tabnayiyuw* tree and appearing meaningfully more enriched in ^{14}N than 26; however both have the same percentages of nitrogen.

This data suggest that many trees and other plants contain fixed nitrogen. This is especially true of those plants I tested from *kadidulel* sectors, but hardly those alone. *Atulab*, a tree that should be avoided and clearly

Table 1.3. *Tabnayiyuw* (*"Tab.,"* D. *Papuanum*) Comparisons

	ID No.	Item	%C	%N	ΔC	ΔN	C/N
16	80	betel pepper on *asibwad* tree 8/16/95	44.4	4	−31.9	2.51	11.1
17	81	betel pepper on *tab.* tree 8/16/95	46.5	4	−31.I	0.81	11.6
18	99	pepper climbing *tab.* meters from soil 12	45.1	4.6	−28.4	2.1	9.8
19	104	betel pepper on *akewal* meters from soil 12	44.8	4.5	−29.1	1.44	10
20	105	betel pepper on *atulab* closer to soil 12	44.5	5.4	−28.9	1.01	8.2
21	82	taro *asibwad* tree 8/16/95	45.1	3.3	−29.4	0.5	13.7
22	83	taro *tab.* 8/16/95	45.3	3.3	−28.6	0.42	13.7
23	95	*vesop aysasous* base meters from soil 12	43.8	4.3	−26.7	3.25	10.2
24	98	*vesop tab.* base some meters from soil 12	43.4	3.8	−26.7	3.45	11.4
25	100	*vesop* base of down and burned *tab.* meters from soil 12	45.3	3.8	−27.1	2.49	11.9
26	103	*vesop* not near *tab.* meters from soil 12	44.6	3.5	−26.9	5.2	5.2

an early fallow or pioneer species like *gwed,* also suggests association with nitrogen fixing (e.g. −1.51). Two samples of another important tree, known as *kausilay* in eastern Muyuw (*Calophyllum leleanii* P. F. Stevens), both in *digadag* but one a mature spreading tree and the other a sapling, suggest association with fixed nitrogen (−1.6, −1.74). This tree is discussed in detail in chapter 4. *Kubwag,* leaves and seed, from the *Ficus* that is also the target tree in central Muyuw *oleybik* fallows, suggests ^{15}N fixing (0.01, −1.37). A *tabnayiyuw* leaf from an *ulakay* area showed low numbers typical of higher amounts of [15]N (−0.48).

While the stable isotope numbers suggest something about ecological processes underlying many of the species tested to date, they fail to discriminate among factors that locals notice and use. But they do one thing very clearly, and that is show differences between surface collections of soils (top ten centimeters) and soil collected fifteen to thirty centimeters below the surface. Table 1.4 illustrates this for four sets of samples: two from *kadidulel* areas, soils classed as particularly bad; one *digadag* in which the soil is classed as good; and one from an area not gardened in many,

Table 1.4. Surface/Subsurface C&N readings

	Location	%C	%N	ΔC	ΔN	C/N
27	43b e10 SeMuyuwWasimoum Kadidulel1 top10	8.7	0.6	−27.8	5.3	15. 8
28	44b e11 SeMuyuwWasimoum Kadidulel1 U 30	4.6	0.3	−26.6	7.2	14.5
29	45b e12 SeMuyuwUkalel Kadidulel2 top10	7.4	0.5	−27.3	4.2	14.1
30	46b f3 SeMuyuwUkalel Kadidulel2 u25	3.8	0.3	−26.3	5.8	14.5
31	47b f4 SeMuyuwBweybwayet DigTawt1. Top10	12.2	0.9	−27.8	3.7	13.1
32	48b f5 SeMuyuwBweybwayet DigTawt1. U25	5.7	0.5	−26.6	5.7	12.2
33	13 SEMuyuw Galokio Top 10cm	4.6	0.37	−23.4	3.5	12.3
34	14 SEMuyuw Galokio Below 30	1.3	0.08	−26.8	6.3	15.7

many years. In each case the odd number is the surface figure, the even the subsurface number. Numbers 27–30 are the *kadidulel* poor soils, numbers 31 and 32 the richer soils, and numbers 33 and 34 are from the area effectively not gardened. With the exception of numbers 33 and 34 where the difference is near four, the surface soils roughly double the amount of carbon and nitrogen found below twenty-five centimeters; and the subsurface soils are considerably more enriched in ^{15}N than the surface soils. Findings like these are especially common with tropical soils where most of the nutrients are on the surface. These differences were run through the tests that the Cornell labs provided as well. I shall not, therefore, discuss them further in this account.

The Cornell Laboratory soil tests show the availability of primary, secondary, and micronutrients needed for plant growth, as well as pH, which is a standard gauge of acidic or alkaline soil levels. The data will be presented in table or graph form categorized first by the fallow or forest regimes already discussed, although given the single sample of the *idal* class of poor soils from western Muyuw, I do not always show values for it. Table 1.5 supplies the codes for the fallow types.

Table 1.6 shows pH relative to fallow types, where acidic soil register below 7 pH and alkaline soil registers above 7 pH. Most western food-producing plants do best in slightly acidic soils, 5.8–6.8. Below these levels soils become infertile, partly because the condition allows for the solubility of aluminum and manganese, which become toxic for some plants. Among the most acidic soils recorded are those from the meadows, and

Table 1.5. Code Key for Soil Types

Code	Indigenous Category	Translation
1	Idal	West Muyuw term for weak/exhausted soil
2	Kadidulel	East & Central Muyuw term for weak/exhausted soil
3	Digadag	Early stage fallow: 0–15$^{+/-}$ years
4	Oleybikw	Middle stage fallow: 15–30$^{+/-}$ years
5	Ulakay	Oldest stage fallow: 30–60$^{+/-}$ years
6	No gardens, forest	
7	Sinasop	Meadow

Table 1.6. pH and Fallow Forest Types

Kadidulel **2**	Max of PH_H2O	7.79
	Min of PH_H2O2	6.31
	Average of PH_H2O3	**7.11**
Digadag **3**	Max of PH_H2O	7.61
	Min of PH_H2O2	5.97
	Average of PH_H2O3	**6.89**
Oleybikw **4**	Max of PH_H2O	7.49
	Min of PH_H2O2	5.17
	Average of PH_H2O3	**6.09**
Ulakay **5**	Max of PH_H2O	7.26
	Min of PH_H2O2	5.18
	Average of PH_H2O3	**6.26**
Not gardened forest **6**	Max of PH_H2O	5.76
	Min of PH_H2O2	5.27
	Average of PH_H2O3	**5.50**
Sinasop **7**	Max of PH_H2O	5.84
	Min of PH_H2O2	4.45
	Average of PH_H2O3	**5.15**

these have levels of aluminum that, according to Dr. Reid of Cornell Laboratories, are nearly toxic (Bungalau [484 parts per million] and Sulog [169 parts per million]). Iron (36.5 and 7.1 parts per million) is also much higher for these two; one other location, also with a low pH reading (5.5) approaches this level (6.6) whereas in all other cases, values are less than 3 and usually less than 1. We shall have occasion to return to the meadows. When I first started becoming aware of them in the high forests I presumed they resulted from peculiar soils. At first sight these data are consistent with that guess.

Graph 1.1 suggests that gardening raises pH levels from too acidic to slightly too alkaline for the *kadidulel* soils. This is important for a number or reasons. Nitrogen, phosphorous (P), and potassium (K) availability is partly conditioned by soil pH. It is commonly stated that pH needs to be above 5.5 for nitrogen to be readily available and phosphorous becomes most available to plants with pH values around 6.5–7. These pH figures suggest that human activity—burning the forests—raises pH making the soils better for crops.

Graph 1.2 plots phosphorous (P) values against the regime types. A pattern is very clear: gardening activity raises the availability of P. First, ten samples registered no available P or less than two parts per million, extremely low. Of these samples, four are from soils below twenty-five centimeters. Of the remaining six, one is from a meadow and three are classed as not being gardened on or from the oldest garden class; one of these is from the Sulog area and was being gardened for perhaps the first time ever (neither kind of yam was growing well). Another is from a *digadag* from Gawa probably used a year or two before the sample was collected. One of the under twenty-five-centimeter samples in this set is a *digadag* from Iwa. Its surface sample is also less than two; Iwa people told

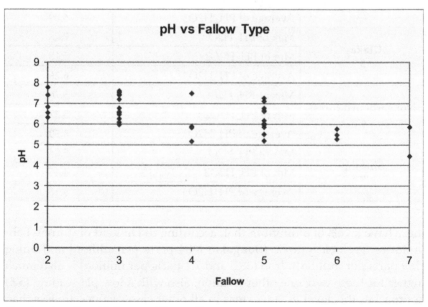

Graph 1.1. pH vs. Fallow and Forest Type

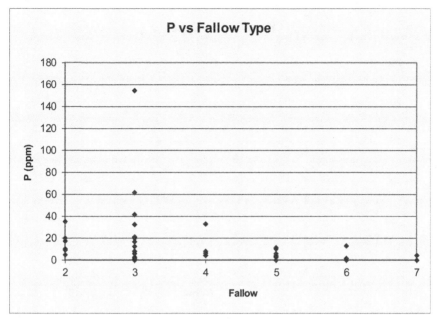

Graph 1.2. P vs. Fallow and Forest Type

me that this particular sector was very poor. Interestingly, the one *oleybikw* that stands out with a high P value is from a sample taken next to the target tree in central Muyuw, the *kubwag*. Its value (32.5 parts per million) contrasts dramatically with a sample taken about seventy-five meters distant (3.9 parts per million).[36]

In Western horticultural understandings, values of four parts per million and above are sufficient for most crops. Given this as a guide, most land has sufficient P, and it is hard to make a case that, for example, *gwed* make a major difference. In one Trobriand *gwed* test, the soil next to the tree had sufficient P (19.5 parts per million), but two samples 6.5 and 7 meters away had more (61.6 and 41.5). In a large plot of land near Wabunun called Wasimoum, in which there are many gardens and fallows running the gamut from *kadidulel* to *ulakay*, I took four different samples in 1995 and 1996. One of these was amid a small grove of *gwed* noted earlier in this chapter; I was told it would produce food already but it would be better to wait before using it again. All show a fair bit of P, but the *gwed* grove showed only the third most of the four, and a *kadidulel* area the second most (21.1; 17; 12.1; 7.8); the one showing the least was an area passing from *oleybikw* to *ulakay* status until it was cut, which was also right next to a dead *atulab* tree.

Graph 1.3 reviews potassium (K) plotted with respect to regime types. While not a structural element, potassium is significant because it controls

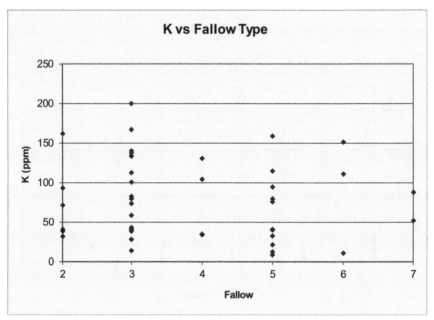

Graph 1.3. K vs. Fallow and Forest Type

enzymes that influence many metabolic processes. These include root development and the opening and closing of stomata. The latter controls water retention, a critical process in tropical climates.

From the point of view of Western soil science, less than 100 parts per million of potassium creates deficiencies in plants sufficient to affect growth. Of the forty-four soil samples, thirty are below this number, ten of those from below twenty to twenty-five centimeters. All of the *kadidulel* soils fall under this rough scale (at 93, 71, 40, 13, 8). Of the fourteen samples over 100, eight were classed as *digadag*. Digim'Rina classed another from the Trobriands as an *ulakay* (*ulaka*), though by Muyuw standards it would have been called a *digadag* because its soil was taken about twenty centimeters from a *gwed*. Of the remaining six, two are in areas not gardened. One of these had a very low K value (1.5) and a below average %N (0.37%);[37] three are classed as *ulakay*, one of which is the Wasimoum garden area noted earlier with the least amount of P. There is one *oleybikw* in the group, from the central Muyuw garden next to the *kubwag* tree. Its K value (104 parts per million) contrasts dramatically with the sample purposely taken about seventy-five meters away in the same named ground and the same garden (34 parts per million). The two highest samples (199.97 and 166.67 parts per million) are from the Trobriands and Iwa, and both were right next to *gwed* trees. As noted earlier, I took samples 6.5 and 7 meters from this first Trobriand *gwed*. Both have respectable values, but half that proximate to the *gwed*; the closest is 100.02, and the furthest is just under that mark at 79.12.

The data assembled here are not sufficient to present strong conclusions, but they suggest some hypotheses. And the first among these is that repeated gardening on these soils enhances some of the prime nutrients needed for better plant growth. This apparent fact might be related to the earlier observation about the presence of the ancestors, i.e. time, in the Trobriand landscape, a presence that becomes coherent by Iwa. That presence, repeated working, apparently enhances soil fertility. In my earlier research period (1973–75) I was told that land had to go through a sequence of (six) repeated gardening before the "badness" of the soil was worked out of it (Damon 1990: 159n26). Then magic would no longer be needed on the land. I received a similar story in 1996 when sampling *ulakay* soil near the recently reconstructed northwestern village called Lidau. The woman who helped me gather the soil said it wasn't very good because it had not been worked enough times. The data presented here are the biochemical evidence for which these indigenous comments are an accurate reflex.

The second major hypothesis is that targeted trees, especially *gwed* and *kubwag*, are significant. Although it now seems obvious that ("fixed") nitrogen may also be an issue, it is not yet clear that nitrogen availability is a problem. But these trees augment the amount of available P and K. Potassium is also the one that tips into the relatively deficient set of numbers in the *kadidulel* soils. While *kadidulel* P values seem fine, K becomes deficient.

Finally there is the issue of the quality of the crops. Although older Muyuw fallows produce larger crops than the younger ones, there are strong individual and ritual preferences for the qualities generated by *digadag* soils and especially the soils close to *gwed*. These are root crops of course. And although the mechanisms are not known, the tree enhances K. Potassium is supposed to be especially vital for root development and it is the nutrient that seems to stand out here. Its absence or presence seems to generate the differences these peoples have worked into a formal model based on the *gwed* tree.

Further analysis of this material and a more complete description and analysis of the stable isotope materials will be presented elsewhere. Moreover, the horticultural focus of the discussion necessarily neglected those interesting meadows. I return to their biochemical properties in the next chapters devoted to a broader picture of the use and management of landscapes. That discussion will bring us to the case of western Muyuw. As noted earlier, western Muyuw is the only place where I did not hear the mantra that *gwed* and *atulab* put good and bitter stuff into the soil and therefore should be sought or avoided.

From Sweet and Bitter to Modeling

I draw this chapter to a close by pulling together data on what seemed to be a growing contradiction as my studies continued through 1995 and 1996. The contradiction was this: western Muyuw aside, although everyone elegantly stated the sweet and bitter model, the most dedicated gardeners contradicted it. They would sometimes do this right after restating the model. The major issue is the sweet/bitter dichotomy. These people would say *gwed* does not put sweet stuff in the soil; in fact "we" do not know what it does—but it does something. Knowledge that it did something is one reason why some people were interested in my work. I might answer a question clearly implied by two sets of ideas they simultaneously maintained: one, that *gwed* put sweet things in the soil they did not believe; the other, that *gwed* did something positive they experienced and acted upon. But the beliefs associated with *atulab* are more interesting. Some people easily moved from the model—it puts bitter, bad stuff in the soil, so avoid it—to a different description: that the difficulty in killing the tree is compounded by its peculiarly dense and fine root structure. Rather than large, long-running roots extending hither and yon, a dense ball of very fine roots surrounds the base of the tree. If the tree is just stunned, it comes back to life. Then roots surround anything planted near it and more effectively pull nutrients out of the ground. Since phosphorous is absorbed from root hairs and this tree has more fine roots than other trees, this is a problem. It is interesting that of the four Wasimoum soil tests I took, the one with the least amount of phosphorous was right next to an *atulab* tree.

I eventually became convinced that discriminations people were making with regard to trees and plants were about the objective world. It is now reasonably clear that this objective world is one of their making. The data suggest that burning the forests increases pH, thus enhancing the availability of major nutrients. And the appearance of specific trees, clearly conditioned by human action, in varying fallow regimes enhances specific elements otherwise found wanting in this environment. Social action along the northern side of the Kula Ring differentially transforms the conditions of existence. But long before I had achieved a satisfactory hypothesis about what these transformations might entail, a different realization presented itself: the paramount importance of formalized models in governing social action.

First born of the discrepancy between the *gwed/atulab* contrast and what some people readily enunciated, upon reflection a second question arose: how to understand variation across these islands. Although it first seemed that ideas about *gwed* were constants, in fact they were not. They were as discontinuous as the garden forms that crystalize the differences across this region. With our problem now located in the question of "modeling," we can focus on the place of trees in the construction of cultural practices.

Notes

1. Based on my limited Trobriand experience and how Dr. Digim'Rina used these categories, their relative relations are constant but the absolute ages vary. An Iwa man told me that Iwa had no *ulakay*.
2. In 1995 and 1996 my guides tasted trees to help identify them as we traipsed through the forest; I tried this once only to be reprimanded because they knew that I knew nothing about what might be poisonous.
3. In 2002 I learned that while the species grew on Koyagaugau and Ole, people did not share the practices and knowledge known to the north.
4. For information on this genus I am indebted to a brilliant PhD qualifying paper by a graduate student at the University of Virginia, Sharon Teeler (2001).
5. There is some indication that the species R. *typhina* needs heat to sprout http://www.fs.fed.us/database/feis/plants/tree/rhutyp/fire_ecology.html
6. This conclusion follows a remark Robin Hide made to a presentation at a seminar at Australian National University in 1999. I believe these conditions may be generalized throughout the Indo-Pacific (see Damon 2012).
7. Unlike the case in the Trobriands where *gweda* are considered "female," Muyuw do not regularly class trees as male and female. But they do recognize that some trees have male and female counterparts. When I began to realize that not all flowering *gwed* produced fruit, I asked people if there were male and female trees. My best and oldest informant had noticed the same pattern and thought the trees were gendered. Neither the systematic literature nor Peter Stevens settled this issue; the genus is not noted for such a characteristic.
8. See also Fox (1993) and Waterson (1997).
9. In 2009 I encountered two mature Muyuw women who did not know the model. Surprised, I started a modest survey to see if the model was a standard form. It is.
10. That elder died in 1996. In 2006–7 I tried to learn more about this tree. None of the elders at that time knew the tree nor practiced anything special with these uprights. One man finally remembered seeing something like that "long ago." These garden patterns now function as relics, not principles generating new elaborations and experiments. See Robbins 2009.
11. The plant is similar to *atawas* (*Rutaceae Euodia hortensis* J. R. and G. Forst), a small herb with a powerful smell associated with death and used to cover up the stink of corpses in funeral practices.
12. This is the food I was usually served!
13. I cannot find any discussion of house or village orientation in Scodditti's work on Kitava. Kitavan villages are found all around the extremities of its

flat top. The largest ones tend to be in parallel lines running roughly perpendicular to one another. Okabulula runs roughly northeast-southwest (78 degrees/258 degrees), Kumwageya southeast-northwest (160 degrees/340 degrees). Houses in these villages did not follow an evident pattern. The smaller hamlets were more or less circular in orientation.

14. I was told Kweywata's orientation was like Gawa's.

15. Damon 2005 interprets this confusion as symptomatic of a phase change between the very distinct Muyuw and Trobriand patterns. That publication supplies ethnography not reproduced here.

16. Kitava needs to be researched. At one level, the three largest Kitavan villages have all kinds of fenced-in gardens inside or running inside the villages. In Okulubula with Linus Digim'Rina in January 1996 I mentioned the striking pattern; he agreed. On another level, however, the interior plateau of Kitava—dominated by kunai grass (*le*)—is one massive garden area separated from the villages.

17. One of Wabunun's best Western-educated men noted that there was a similar correspondence between the southwestern Muyuw village Kauwuway and a similarly named village in the Trobriands; like the Kavatan/Kavataria couple, these two are associated with the Malas clan.

18. Muyuw are highly conscious of this difference, referring to a final vowel as a word's "wife" (*nakwav*; lit. "its wife"). When they hear the pattern (increasingly common toward western Muyuw), they understand it as misspoken Muyuw. The "vcv" and "vc" pattern is probably a product of stress differences spanning the region.

19. According Australian National University's Michael Bourke, Iwa's population density is the highest in Papua New Guinea. This population variation is evident from Seligman (1910) and is a recognized problem for the government, for Iwa, and for Muyuw. Small numbers of Iwa people periodically relocated to Muyuw to live by the 1980s and 1990s. Now southeastern Muyuw sustains an entire Iwa village.

20. I realized people were relieving themselves in their garden areas when I was in Iwa but Wabunun people who told me the reason; they denigrated the practice. According to Robin Hide in 1999, this is the first recorded case of its kind in Papua New Guinea.

21. A Trobriand man briefly married into Wabunun reported that Trobriand gardens were treated like children: you tend them very carefully. By contrast, he said, Muyuw people obtain poorer results because they pay their gardens less attention.

22. I sought stable isotope readings of the ferns (from Muyuw) this man listed. They are associated with fixed nitrogen. Muyuw do not, however, plant them next to crops, for they consider them a sign of poor soil—when they see them they avoid planting near them. One is called *kaydigadag* (*Micrsosorum cf.cromwellii*, an unresolved therefore unofficial name according to Peter Stevens, personal communication, 5/27/2016). Its nitrogen number is −2.65. The other fern, *signavan* (Stevens just identified this as a "fern"), has a reading of −4.35. Any number near zero or below signals fixed nitrogen.

23. A Wabunun man who has been teaching school in various locales across the northern side of the Kula Ring has similar views of garden sizes and the intensity of Trobriand production methods.

24. My experience with this tree grew as Simon Bickler's archaeological investigations continued in 1995 and 1996. When I realized people did not like chopping *atulab* because it dulled their steel tools, I wondered how they managed with their old blunt stone tools. But that may have been their advantage, for the blunter edges might have been very good at mashing *atulab*.

25. Repeated collections of this tree, easily classed as a *Dysoxylum,* were not confidently identified at the species level until David Mabberly, in the process of reworking the genus, reconfirmed Stevens's assessment—*Dysoxylum papuanum* (Merr. and Perry) (Mabberly, personal communication October 15, 2002). Although another tree in the Meliaceae family is famous for its beneficial properties (the neem tree, *Azadirachta indica* A. Juss.), my report was the first Mabberly had heard for any *Dysoxylum.*

26. Including the son of the island's most distinguished gardener, Meskolos, then deceased (Damon 1990: 140, 153–54).

27. In central Muyuw the term pairs with the oldest fallow term, *ulakay,* to specify directionality rather than fallow type. So *"tawan"* can take on the meaning of "seaward" or "north," whereas *"ulakay"* can mean "inland" or "south" even when the specific area referred to is, strictly speaking, an *oleybikw* or *digadag.*

28. The Trobriand term appears to be *ligob.*

29. Eastern Muyuw's southeastern tip has undergone massive change, first when Europeans planted coconut plantations in the early twentieth century and then when US soldiers built a fighter airstrip during World War II. Extra runways, roads, and courses for hiding planes turned much of it into compacted coral strips that act like *kadidulel.*

30. And ruins. Potshards, shellfish refuse, and stone tool materials litter the area. The entire shoreline area has been in and out of formal occupation and abandonment for hundreds of years.

31. I reviewed these facts with Australian National University's Michael Bourke in 1998 and 1999. According to him, smaller islands in Melanesia tend to be densely planted in nut- or fruit-bearing trees. Muyuw villages closer to these small islands have more trees than those that are more central. Wabunun, Kaulay, and Dikwayas (my standards of reference until deep into this study) hardly have any while Kavatan and Mwadau, eastern and western villages respectively, have many. Budibud to Muyuw's southeast is also densely planted in fruit or nut trees. However, Bourke's generalization is not always the case. Nasikwabw has relatively few compared to Iwa and Gawa, and while Yemga has more than Nasikwabw, it also has few compared to Iwa. For both Nasikwabw and Yemga the issue is their easy relation to horticultural-based villages, those of eastern Muyuw for Nasikwabw, Yalab for Yemga.

32. After my 1996 return, Dr. Debora Lawrence, who has research experience in Southeast Asia, joined the University of Virginia Environmental Science Department. The procedures of this study would have been much different had she been an early advisor. As others have now told me, particularly Dr. Brian

Walker (CSIRO Wildlife and Ecology, Canberra, Australia), P (phosphorous) and K (potassium) are often more critical for tropical crops than nitrogen.

33. John Armour, senior soil scientist from the Centre for Tropical Agriculture in the Queensland Government Division of Natural Resources and Mines; I met with him on July 18, 2005.

34. As noted in the Introduction, Kaulay's Vekwaya disputed the assertion that yams grew at night. He insisted my source was confusing growth with the relative amount of water in the vines. The plants go limp during the day because they lose water faster than they can absorb it. He insisted they still grow even with the water loses; further that yams are like people so they rest at night, growing mostly during the day.

35. Soil samples from Iwa are also slightly higher than those elsewhere, probably from greater amounts of kunai grass.

36. These samples were taken in 1996 after I learned about the tree. In 1995 I took a sample from near these two. Its value is 6.2 ppm. I did not know about *kubwag* then, and I do not know where this sample was located with respect to the other two or the *kubwag* tree.

37. Altogether 53 soil samples were run through the Mass Spectrometer. The average sample had .46%, the range from .958 to .03.

PART II

Toward an Ethnography of Trees

The special distinguishing characteristic of Malinowski's field technique lies ... secondly in the *theoretical* assumption that the total field of data under the observation of the field-worker must somehow fit together and make sense. (Leach 1957: 120)

A patch is an ecologically distinct locality in the landscape; it is problem- and organism-defined, relative to the behavior, size, mobility, habits and perceptive capabilities of the population.... In general, patches are localized discontinuities in the landscape which affect behavior." (Winterhalder 1994: 33)

CHAPTER 2

The Trees

Classificatory Forms, Landscape Beacons, and Basic Categories

> It is important to describe folk taxonomy, but it is equally important to doc-
> ument how folk taxa are used in everyday thought and interaction with the
> world. (Coley, Medin, Proffitt, Lynch, and Atran 1999: 228)

One Evening in 1974 and That *Teili* Tree

I paid little attention to trees until I began this study, but I experienced their importance during my first fieldwork, receiving one particularly dramatic message during an unexpected event in 1974. Central to the ep-isode was a nut-bearing tree Muyuw call *teili* (*Terminalia catappa*), known as *saido* from Gawa to the Trobriands. It was located behind and east of Wabunun on the little peninsula that gives the place its name—Wabunun translates to "inside" the "point." The tree is conceived to mark the vil-lage's founding (*unumug*). According to the story, Iwa men paddled to the area long ago, planted the tree, and said, "Do not forget us; start your village here." Near the tree they embedded three stones to be a permanent hearth and planted some *vesop* (*Alocasia sp.*), all to ensure the village's pro-ductivity. These items, including the landmark tree, are *boman*, tabooed, to remain out of sight, lest one "run amok" (*kabalawein*). Although everyone in Wabunun knew where the tree was and now and again cast his or her eyes on it, I never found it until 2002 when I was walking on a path that went right by it. Yet except for the outline of that tree, what makes the village's presence possible should be invisible.

This is what happened: Ogis, one of Sipum's elder brothers, and a now longtime friend and teacher, came across the tree and unumug returning from a day's gardening. With a gait deeply etched into my mind he men-acingly stumbled into the village near dusk waving an axe and machete. Elders, settled in groups behind their houses rehearsing the day's events while eating or waiting for this or that evening meal to be served, ignored him. However, village youth and one other adult immediately recognized his abnormal state. The youths taunted and chased him until he chased them back. The lone adult, a large, powerful hulk of a man and the village's most notorious deviant, kept maneuvering around Ogis trying to subdue him. He failed, sometimes because of Ogis' erratic charges the children generated. Ogis would swing his axe and machete at them— and occasionally at a mother as well—and they would run shrieking and laughing away. Now and again he 'gardened' in the village by chopping at house posts, village coconut trees or some other piece of wood, say a canoe. Sometimes he commanded people to go out to real gardens with him. His unmistakable mode simultaneously mimicked typecast Europeans and Trobriand chiefs. Since it was dark people refused to follow him. He had times and places upside down, the confusion emphasized by his speaking in stereotyped Australian-accented English; in fact Ogis did not speak English. Years later I was told that slipping into another language, even if the speaker does not consciously know it, frequently happens in this state.

Eventually the elders were convinced that Ogis had truly run amok. "He really did it well!," one of the mothers he attacked later said. So the village went into redressive action by making itself disappear. As we were ushered into our house everyone else hid inside theirs, extinguishing lamps and plastering doors and windows shut with woven coconut fronds. After some time Ogis sauntered up to his brother's house. When he banged on its front door the crowd of men inside pushed it over on him. The aforementioned hulk pinned him to the ground. One of his father's "hit him" with magic and he regained some consciousness. Several men then hauled him into his brother's house and left him slumped against a wall. Lights went back on, the evening resumed with new stories to tell and Ogis slept well into the next day. When he finally awoke with a searing headache he was oblivious to what had happened.

This was one of the more stunning moments I experienced on the island, and I often talked about it on subsequent returns.[1] In 1982, while trying to find the tree, I came across graves of elders I'd known, broke into tears, and feared running amok. That experience cooled subsequent attempts to search for the tree until I was told about it again in 2002, at which point I was beginning to piece together how it made sense. I found it and photographed it. With a breast height diameter of several meters, it is enormous, presumably quite old. A spreading rather than ascending tree, it does not now appear as a village landmark. This is because coconut trees now line beaches near most villages, including Wabunun, and obscure the tree. These stem from 1920s responses to government coconut plantation ordinances from the prior decade. Along with the *casuarinas* trees (*yay: Casuarina littorale*) that once, reportedly, lined many shorelines and that still mark Wabunun's eastern end, this one origin tree would have stood out as firmly to the sight as it does in the minds of Wabunun people today.

Since I had never before explored the ways flora organizes Muyuw and related cultures, the 1974 experience with Ogis running amok remained a senseless oddity. This tree, from another place, anchors appropriate cultural consciousness by means of its story and its taboo status. What is there in the organized knowledge of flora that appears to be determinative, formative, for this region? This and the next chapter provide contextualized answers to this question. Along with the formal model of sweet and bitter trees and the recognized gap between that model and what some people know as a reality, this tree sets the stage for a more considered exploration of trees, forests, and forms spun from them. This chapter moves from individual trees to a model of the whole that they enable. The next chapter begins with holistic images of forests but ends with how tree types demarcate intelligible parts. With modeling capacities clearly delineated, I return briefly to the *teili* tree. Chapter 4 then details one set of trees as it begins to lay out practices that organize the totality of the eastern side of the Kula Ring. This shows how ideas internal to the understandings of trees articulate mutual relations between wholes and parts, a significant theme in many Austronesian societies (Sather 1993).

Questions of Method and Limits

Independent of the measures that led me obtain the data in chapter 1, I followed two formal procedures in my research. In the first method, I conducted formal surveys of areas in which I sought names and uses of plants, collecting voucher specimens when suitable. These places included selections of forest, from hectare-sized areas to strips of fallowing land, and smaller plots of high-forested areas nestled in gardening regions. These latter regions are called *tasim*. A focus of the next chapter, *tasim* are "patches," as specified by the Winterhalder quote in the epigraph of Part 2 of this book. Their existence and formal creation were unanticipated findings during this study, although it was long before they came into focus. So, for example, I learned about them first in Wabunun and organized my surveys of them there. In Kaulay in central Muyuw I surveyed the same kind of unit. Only from more extensive and unfocused walking did I learn that Kaulay *tasim* took a different shape. This fact parallels the discontinuities in garden forms and knowledge about trees described in chapter 1.

Included in the plot surveys were formal attempts to figure out what kinds of trees were used for fence stakes and yam poles across the islands. Surveys like these latter ones mediated the second main procedure I followed in my research, where I started by studying objects people were using—firewood, tool handles of many types, houses and yam houses, and eventually boats—and moved to questions about the trees. Because I knew so little about sailing when I started this adventure, my initial inquiry into boats was perfunctory. Yet this investigation became all-encompassing. One of the ethnographic reasons for this relates to the category *tasim*.

The information in this chapter has been selected from the experiences these methods generated. For example, I collected a great deal of data about which kinds of trees are used in building houses. Although these structures become transparent repositories of technical information about trees and a record of the environments from which they come, I shall save for a later time a full exposition of this information. Where the topic fits into this chapter's argument, house wood adds little. This is undoubtedly because across these islands, houses are not a major repository of sociologically significant information, although the content and form of the materials used to make them keeps pace with contemporary social trends; mechanically fashioned lumber replaces more traditional materials.

What I learned following these procedures led to their modification so that I eventually pursued more narrowly defined topics. Especially during the earlier months of 1995 and 1996 these strategies entailed searching for particular trees, or sets of related trees. These inquiries resulted in much time devoted to walking through the island's forests. As I learned, of course, several pursuits became more focused; the fruits of those inquiries dominate the remaining chapters of this work.

Although I move toward presenting what an ideal member of this region knows, there are differences in how as well as what is known. Discounting phonetic transformations that cross this region, many important trees carry the same name. But a few trees carry different names, and some people were quick to recognize such differences. I sometimes showed voucher specimens collected elsewhere to Wabunun instructors. They readily picked out plants they did not know, those they knew with different names, or noted common names referring to different plants. A name *abuluk*, for example, is used with one kind of tree (Damon 181, Celastraceae *Bhesa*) in the far eastern Muyuw village of Kavatan, and a different one (Damon 200, Myrtaceae *Syzygium*) in Wabunun (and Gawa).[2] *Akuyak* is the name of a large tree common in Nasikwabw; my Wabunun informants said they did not have that tree there, yet they used the same name for another. Sipum stared carefully at a tree my Kavatan informants called *aydidawi*.[3] Eventually he agreed that the name was correct, yet the kind around Wabunun was slightly different. The tree *atilotal* (Damon 210, Euphorbiaceae *Omalanthus sp.*), considered important (for attracting women) by a central Muyuw man, went completely unrecognized by Wabunun informants. A very common tree in early fallow systems is called *kwakwis* in eastern Muyuw and *wageo* in western Muyuw (Euphorbiaceae *Macaranga sp.*). This tree is very well known, its wood used for non-weight-bearing uprights in the houses and yam house interiors, and its soft velvety leaves are appropriated for personal hygiene. Although I do not have sufficient information to analyze the structures of these differences, as this and the next chapter proceed, the ways some of them are used to structure the socioecology of the region will be apparent. Variation structures this system. Alternative ways of naming the same phenomenon are undoubtedly part of that form.

Although not uncommon, discriminations of the kind Sipum made were special. Some guides were indifferent to tree names and bored by my attempts to learn them; other guides found the pursuit fascinating, especially when we came across plants they did not know. These "patches" of knowledge are interesting in their own right. Chapter 1 noted how Iwa people who didn't know yam and taro names of plants they were using told me to go to Muyuw to find them. "Personality" differences accompany regional differentiation across these islands. For me, Muyuw in general and Wabunun in particular is home; everywhere else is a strain. Whenever I was told to leave a place, I was in a mood to do that as soon as possible. Yet why do Iwa people not know names of crops they plant? Does this "deficiency" relate to the *teili* tree origin myth?

Knowledge of flora in this area is highly differentiated and known to be so. Concluding a portion of her subtle analysis of design patterns on Vakutan canoes Campbell writes, "looking for difference amongst a sea of similarity is a recurring theme in Vakutan thinking"(Campbell 2001: 86). Although the differences in this area are greater than a novice could be expected to understand, the essential thrust of Campbell's statement should be generalized beyond Vakuta's borders. The fundamental pattern of life in this region is heterarchical, to use the term Crumley (1994) brought to our attention. The region is defined by varying value systems.

In addition to idiosyncratic factors, age, place, and gender determine who knows what. I return to these differences in various contexts throughout this chapter, for these variables have important bearings on the nature of the classificatory principles evident in the region. If plant names at that place are the business of those people, and not us, then the social system is not likely to generate totalizing forms of knowledge we associate with Linnaeus and his predecessors and successors.

Although I spent most of my time with men, I had no qualms about seeking information from women, and eventually I found it nonproblematic when women corrected or disagreed with elder male informants whom I knew to be among the most knowledgeable people. Although Muyuw say that women do not know trees because they do not make, and rarely sail in, boats,[4] many women know a great deal about flora. This follows from conditions of production. It is said that men produce gardens because they chop down the forest and plant the crops, while women just tend and distribute a garden's produce. On the face of it this should mean that men know more than women; however, Muyuw women gauge the qualitative and quantitative success of gardens, and everybody understands that success follows from the forested land out from which the produce grows. Although there are men who pay much attention to such matters, judging this success is women's responsibility, and they often tell men to keep gardening in the same area because the trees are good or tell them to move on because they are bad. Muyuw gender responsibilities are woven together; what appears to have subordinate position sometimes has a determinative value. In a particular fence-stake survey, one woman was more confident about identifications than two (knowledgeable) men.

I believe the gender content of knowledge changes as one moved toward the Trobriands, because responsibilities and obligations for gardening activities shift. The major dividing point is probably Iwa, for it is there that the practice of yam-pruning begins. Men assume the analysis of and responsibility for deciding how plants are growing, which tubers to prune and which to let grow. Iwa informants emphatically told me that women could not perform that task. The existing ethnography west of Muyuw and my observations in the Trobriands, Kitava, Iwa, and Gawa all suggest interestingly different dynamics in the labor processes there. Although my surveys in the islands to the west were too sparse to demonstrate the point, I suspect that suitably different distributions of knowledge follow varying labor processes. Perhaps related to such differences is one fact that did emerge: people on Gawa and Iwa had a greater tendency to understand trees as male or female. In Muyuw there are male and female trees, and various conditions for this usage will be described throughout this work. But that dichotomy is not a major classificatory principle. In Iwa and Gawa, people easily listed sets of male and female trees, a form of categorization that did not exist in Muyuw; "We do not say that," I was told when repeating to informants a paradigm I had learned to the west. Although discontinuous relations across the northern Kula Ring are a topic in this book, and I have recorded names of the same plants that differ substantially beyond minor phonetic shifts (for the latter *gwed/gweda*), an analysis of varying floral names is beyond the task here.

The Trees: The Problem(s) of Classification and Categorization

A distinction designed with heuristic value only organizes this account. On the one hand I pursue what I call "classification;" on the other I focus on "categories." By classification I mean how terms of reference are part of a naming system, and by this I draw from the work associated with Berlin and Atran. By contrast, categorization deals with how what is known is used in or for some other domain of activity or knowledge, for me exemplified by the work of James J. Fox (1971; see also Fernandez 1998). These might respectively be entitled the nature of classification versus the social use of classified forms. I mean for the consideration of the former to turn into the latter. Together they effect a passage from trees to forests, from information to meaning.

The Nature of Muyuw and Northeast Kula Ring Classification

In 1996 when I was well along in my study of trees in the genus *Calophyllum*, chapter 4's subject, I performed a test to see if men who I regularly worked with could correctly identify the island's six named *Calophyllum* from individual leaves, leaves I had taken from voucher specimens collected with their help. I could not yet pass the test, but it would not be difficult to become so accomplished with this set. While two of the three men got all of them correct, the other missed some. But what was interesting was Sipum's reaction when had him take the test: he swore at me. Once he examined the leaves he ordered them quickly. Aside from size and shape differences, the angles of the genus's peculiar leaf venation stood out for him once he realized he had to focus on minutiae. But he lectured me on how they distinguish trees, by location, shape, size, appearance of the trunk and tree, as well as other characteristics such as the trees' spatial distribution; what happens to it when it is cut down; what it does in seawater; how soft, hard, dense or heavy (*vat*), or light (*aweyan*) it is; how it chops,[5] burns, and rots; what happens to it when it is bent;[6] how it smells and in some cases tastes; and how deeply branches intrude into a tree's center. Once when we were moving through a forested area verging on the mountainous rise near the island's south-central core, we passed by a *Calophyllum*, and Sipum called out its name, slapped it with his machete, then said that it was as big as that kind got—considerably smaller than three of the other kinds. He considered it ludicrous asking for identification by a single leaf, and told me so.

At the simplest level of classification these matters are important for an analysis of Muyuw classificatory forms. Over the course of this research

I concluded that the names used to class trees rest atop, uncomfortably and incompletely, a rather vast knowledge of properties. At one point in 2002 I drew up a list of features similar to those in linguistic studies (e.g. Chomsky and Halle 1968). More than once people explicitly pointed me to trees or vines that were particularly good for certain uses, but when I asked for a name I was told, "I don't know." However, rather than beginning at this vast array of perceptual differences, I shall begin at what some would consider the top of the Muyuw system of classification.

Classing Life Forms

Among specialists (e.g. Atran 1999b; Berlin 1992) there is a consensus that taxonomies tend to show commonalities of rank and exclusivity. In Atran's version of this consensus there are 1) analogues of kingdoms—plant versus animal; 2) life forms—fish, bird, tree, herb, among others; 3) distinctions that combine or separate the Western sense of genus and species; and 4) the lowest, varietal forms. With the exception of the lowest level, names that peg these differences tend to be short, unanalyzable word-forms. Varietal terms are often binomial and usually composites of meaningful relations. In Muyuw, for example, taro (*lawuw*; *Colocasia esculenta*) is considered a "tree" (*kay/ke*) largely because its stalk is trunk-like and also because its form differs from yams, which are considered "vines." Although not all taro varieties had names I or my informants could understand, many do, and a good example is one called *kaynikoya*. It grows taller than other taro, and the name connotes this. While the *ni* morpheme is probably a locative, informants were certain of the *kay* as a "tree" designation, while the form *koy* translates to "mountain" or "hill" in Muyuw: "tallest plant (taro)" is an appropriate literal translation. Facts like these could be said to substantiate the fact that this variety is in the tree class, distinguished into the type *lawuw*, "taro," then distinguished among them by its meaningful name, a property of the organism, understood as mountain-like, i.e. explicitly contrastive.

The class designators for life forms may, however, be used with the content of other forms. A "vine" called *ulibutobot* (Damon 95, Vitaceae *Parthenocissus sp.*) is called a *kaykal*, literally "tree fence." It is one of a class of plants that can be used like a "fence" (*kal*) around the human body because of the form's complex leaf structure. It has five-leaflets that tend to circle around a space, a capacity sought for medicinal or social purposes. The leaves are rubbed between one's hands with water or seawater added to them. Messaged over the body, the concoction causes a slight skin rash or irritation, the sign of its protection.

Table 2.1. Muyuw Life Forms

Gamag "persons," including "spirits" like *yeluw*, *aluw*, and *tokway*	*In* fish, dugong, lobsters	*Bwaloug* clam-like things	*Man*[7] birds insects mammals snakes	*Ke/kay* trees	*Auyow* weeds	*Yamwik* Epiphytes	*Vatul* vines
Some generic-like terms e.g. "turtle" (*gonam*), "shark" "tuna"								
Hundreds of species-like terms (some with no names); a few are understood to be in defined "groups."								
Some variety-like terms e.g. taro, yam, and other crop names, and male and female of some organisms								

Table 2.1 shows the eight life-form distinctions. These forms easily divide many aspects of the Muyuw universe and what may be suggested as distinguishable levels of rank beneath them. I illustrate this form to provide some examples and then discuss naming forms. I did not explore how these categories might relate to phenomena that are not alive, like rocks. But this might be worthwhile. Muyuw consider the many coral as kinds of "rock" (*dakul*), and they know that they grow. One person told me that rocks in the sea grow, and those on land do not. Perhaps this form mediates primary classes described above.

There are at least eight major life forms.[8] Noun classifiers used in counting the classes are one way people distinguish these basic life forms. *Yamwik*, a category I gloss as epiphyte, basically means "big leaf." And leaves take their own counting/classifier system, *moi*. So "that leaf" is *moinatan*. "Two leaves" is *moiney*, three is *moinatoun*, four is *moinevas*, and five is *moinanim*. Part of a noun classifier may also be employed like a relative pronoun. For example, there is a relative pronoun based on the "*man*" category that can be used with birds, insects, pigs, and women. *Manawen* translates to "that one" in English. I specifically asked women informants if this implied women were equated with pigs. Without denying that *manawen* may reference a woman, they were insulted and emphatically refuted the equation, proving the point to me by illustrating how you counted women, the same way as men, so *teitan* (one), *tei* (two), *teitoun* (three) and so on, while pigs have two different classifiers.

More or less unique classifiers hold for many of these life forms. Yet things inside these classes have their own more specific classifiers. "Pig," for example, may be counted with the classifier used for other items in its class, hence *manatan* (one), *manay* (two), *manatoun* (three); but more often one hears *bulatan* (one), *buley* (two), *buleytoun* (three). For "tree" (*kay* or *ke*) and "vine" (*vatul*) the classifiers are *kay* and *ul* respectively, the former part of a construction with deep Austronesian language roots.[9] Species-like names among these two life forms tend to begin with all or part of this classifier. An inordinate number of "tree" names begin with /a/ or /ay/, probably a contraction of the classifier *kay*. Earlier I noted the plant *kaydigadag* (with some speakers *aydigadag*), which, however, is classed as an *auyow*, a weed, not a tree. Another is *amwakot*, a tree in the Mahogany family (Meliaceae). Three vines are *ulbunibwan* (Damon 62, Linaceae *Hugonia jenkinsii* F. Mueller), *ulagubaguba* (Damon 248, Linaceae *Hugonia jenkinsii* F. Mueller), and *ulakaykay* (Damon 108, Melastomtaceae *Memecylon*), whose names have specifiable meanings (roughly "eagle-claw vine," "hand-over-hand vine," and "vine-as-big-as-a-tree").

While in their unmarked usages these life forms are quite distinct, in a marked sense one can be derived from another. When I asked, for example, if a specific vine was a *kay*/tree, I was told no, it was *vatul*, a vine-like thing. However, at least five different good kinds of tying material derive from five different trees. The pictures for this chapter illustrate the extraction of one *vatul*, *im*, from the shoreline *Pandanus* tree called *loud*.

Some items are classed in a genus-like fashion. The most prominent examples concern "fish." There are well-formed ideas and simple names concerning the categories "turtle" (*gonam*), shark (*tagligal* or *kwaw*), and ray (*lepeyay*). Inside these designations are species-like entities. There are four "turtles" (*panowan*, *wedal*, *tadiyay*, *poun*). These are avidly sought for the "strength" that comes from eating them, contrasting to pigs which are sought for their fat. Turtles are understood according to their morphology (used in boat-part design), taste (related to their smell), habitat and temporality, and egg-laying preferences. I've collected up to seven names for sharks, six for rays. I have three names for tuna, one of which works as a generic-like term that, when marked by two others, also serves to distinguish a species-like form.[10] The form we understand as "eel" is named—*buliwad*. Differences are recognized among them—the largest, for example, is black—but my informants could not name them. The same holds for mackerel. Muyuw call them *didayas*, and some people told me there are two kinds, not distinguished by name. Whether or not they are distinguished by name, many fish are also grouped in units called *bod*. Fish serve as clan (and subclan) totems, and seem to be the most prominent such entities. But while people will usually name one fish as their totem, that fish serves as the exemplar for a particular feature, say long noses, big lips, or thick skin. These are the features that distinguish "groups." This set of facts underlines the impression related above with respect to trees, that in fact what really stands out are differentiating features, names resting somewhat chaotically on them. In the case of fish the relevant feature for totemic uses is generalized up rather than specified down. *Kuduwal* (probably *Agrioposphyraena barracuda*) as the totem for the Malas clan concretizes a differentiating "principle"—long bodies.

While genus-like designations seem relatively common with the lifeform class "fish," nearly the opposite holds for trees. There, a level closer to our understanding of species prevails. Although there is a tendency to create unnamed groups, rarely are trees bundled into larger named units.

Atran (1999: 124) describes a level he calls "generic species," a slot of knowledge that specifies a recognizable form even though people might not have a name for a particular case. Below I discuss a tree that came to be called *ayogal* while I was on the island that fits a slot like this. This tree, but not its name, was well known to most Muyuw. I came across a number of striking instances like this. One was an enormous tree growing next to a frequently used path that tied Kavatan village to the landing and bay along the northern side of Muyuw. None of my best Kavatan informants had a name for the well-known tree. This was in May 1996. A short time later I came across the tree in a high-forested location amid Wabunun's garden areas. I brought some of its distinctive seeds to informants to see if they had a name for it. To my astonishment several people had detailed understandings of the tree's bark, wood, weight and density, and leaves, but like the Kavatan people, had no name for it. Since it is an ascending emergent tree, I was never in a position to acquire a voucher specimen for it; so I have no name either.

I have less of an excuse for two vines. These vines are a principle reason why Muyuw rarely carry water when moving across the island or in the bush. They are easily cut and water readily flows from the slice. One does not irritate skin while the other generates mild to severe itching—it is kept centimeters above one's mouth while drinking from it. But like the tree, neither of these forms carried names that my best informants, including the village's wild pig hunters, bothered to know. While these facts illustrate an order of reality that often gets named, they should make clear that understanding and knowledge about life forms do not have to be fixed to named phenomenon. Names class patches of information, but they are not a condition of intelligibility.

These two sets of examples contrast life forms that are well-known if useless and nameless with the vines that are useful but also well-known and nameless. We can close this exercise with a form that is well-known, useless, but named. *Gilpwapoi* is the name given to a tree frequently found in sandy shoreline environments. My voucher specimen (Damon 239) was hesitatingly classed only in the Anacardiaceae family; I doubt this is correct, for informants compared it to a tree they call *nelau* (Damon 45, Sapotaceae *Burckella obovata* [Forst.] Pierre), saying the only difference is that its fruit, similar to the *nelau*, is eaten by pigs but not people. While knowing that Europeans like and use the tree for timber, Muyuw dislike it and never use it because it rots too quickly.

All these forms have to do with a level of knowledge at which contrastive distinctions, and sometimes names, are easily stipulated. I shall argue that this is the prime level of organization, whatever the degree to which a kind of rank-ordered system can be deduced. This can be illustrated by a review of those trees that were, in fact, grouped together but never named as such. For throughout my surveys and specific collecting procedures I realized that people understood commonalities among a number of trees. Somebody would say, "There are three of those," or, "I think they are a group (*bod*)." Before going into these, however, a slightly more thorough examination of names is in order.

Although there are many exceptions, most species-like tree names have no identifiable "meanings." I know of nothing *gwed*-like or *tabnayiyuw*-like for two trees discussed in chapter 1. Examples of plants having names with specifiable meanings vary from function to material quality to appearance. One tree, often called *asimwalgayas* (Damon 70 and 104, Euphorbiaceae *Glochidioin sp.*), takes its name from one of its functions. It is frequently found in early succession areas that are repeatedly gardened. While the wood is hard and suitable for houses and posts, people derive the name from the way its leaves are sometimes used: to wrap (*simwal*) leaves from a tree called *gayas* (usually but not always Genetaceae *Gnetum*) collected for cooking. Another instance is the tree called *lagay* (Damon 190, Sapotaceae), found in the island's higher mountainous region. It has exceptionally hard wood and is often used to make axe or adze handles; in fact the term "*lagay*" means axe or adze handle. So, presuming the indigenous interpretation is correct, these trees are named for the way people use their attributes. Another plant (Damon 82, Polypodiaceae *microsorum cf cromwellii*), is "*aydigadag.*" Literally this means "tree of digadag" (even though the plant is classed as "weed" and not a "tree") and connotes land that has been gardened so often that it is getting weak. That is where the plant is often found. *Anakay mwatet* (Damon 270, Mysinaceae *Maesa sp.*), literally meaning "the snake (*mwatet*) tree," is a case of appearance seemingly determining the name. I collected it in central Muyuw. However, while that tree grew near the village of southeastern friends, they did not name it and used "snake tree" for a different tree. Another tree, of which I never collected a voucher specimen, is often found growing on the coral-limestone rocks right next to the sea. It is called *kaydibwadeb. Dibwadeb* is the name given to shoreline regions dominated by compressed and eroded coral limestone; hence, the name of this tree approximates "tree of the limestone," again suggesting the fact of a property-defined member of a type. Finally, there is a tree that nobody could identify. After a few hours, Sipum's father, the reigning and knowledgeable Wabunun elder, came to tell me he remembered his uncle and father-in-law had once told him it was called "crunchy" (*amwatat* [Damon 133, Sapindaceae]). When I gave him some of the leaves to look at he crunched them in his hand trying to figure out what it was. Usually green leaves do not make much of a noise. These did. This made him remember a story from years ago about hunting cuscus, the occasion for which was his learning the name. I could adduce more examples like this.[11] And later I shall return to these facts to give them a slightly different interpretation. Here, however, we can use these exceptions to the generalization that species-like names tend not to be meaningful in order to nevertheless suggest another generalization: paradigmatic difference takes priority over inclusive ranking.

Groups of Related Plants

This section reviews sets of trees that Muyuw told me they class in groups. It provides a rich overview of tree names, knowledge associated with them, and an empirical introduction to questions about names, knowledge, and place. The information moves the argument to the second of my heuristic slots, the question of categories. One might use this data as evidence of folkbiological (covert) ranking. In almost all the cases, members of the sets are in the same genus, given Western identifications. However, I suggest instead that this phenomenon underlines a concern with specifying fields of difference, relatively unique discontinuities in the presence of evident similarity. These discontinuities signal a, if not the, fundamental mode of understanding, organized discrete patches.

Almost the day I arrived in Wabunun in 1995 I learned that people understood some trees to be in groups. The trees I was first told about were those with many (visible) roots, for the most part strangler figs. Birds deposit seeds from these trees atop other trees and their roots gradually grow down, over time killing the host. I was first told there were two, shortly thereafter three such trees: *akisi* with "small leaves" (Damon 40); *bwibuiy* "with larger leaves" (Damon 41); and *agigaway* with larger and longer leaves yet (Damon 13, *Ficus tinctoria*). All are *Ficus* (Moraceae), only the last determined by the Lae Herbarium to the species level. After noting the similarities, my sources quickly distinguished the trees by the sizes and shapes of their leaves. All three are common. The first is the most dramatic, approaching bo tree proportions (*F. religiosa*), because its aerial roots—soon becoming trunk-like—spread the furthest. It is the one most likely to kill its host. Sometimes it leaves behind a hollow tube from the host's bole (which provides homes for cuscus, [*kwadoy*], or snakes), and it is the one whose behavior becomes grist for other domains of activity. On the one hand it is said to be very bad for any kind of garden crop (although one informant stated that *agigaway* was very good for crops, and used it like *gwed* and *tabnayiyuw*). This partially derives from the fact that it kills its host and partly from the profuse amounts of white sap it contains, the production of which is thought to suck needed moisture from the land (and so the crops). On the other hand, its use in love magic fulfills one of the culture's fundamental aspirations: that fancies of youthful love endure to the end. Employing the tree's leaves in love magic is supposed to keep a couple married until one of them dies, the idea drawn from the organism's behavior: it wraps itself around and is dependent upon another tree until the latter dies. It is interesting, however, that there is no attempt to equate these trees with another that exhibits similar behavior, a tree called *bwit* (Damon 117, Loganiaceae *Fagraea sp.*). This is one of the few trees found—and is known to be—in most environments. In frequently gardened areas it may be discovered growing with its trunk firmly planted in the ground so that one would hardly suspect it could, or would have to be, a strangler. Hard, dense, and rot-resistant, it is desired and often appropriated as a house post; for this reason, small ones may be left alive in cultivated fields with the hope that before another cycle or two the tree trunk will be large enough to use. From Gawa to Iwa these trees are employed as ceremonial posts centering ritual spaces, *dibedeb*.[12] Yet its distinctive orange flowers frequently appear in swamps, high forests, and even in the mountainous areas because, it turns out, it is growing from some canopy-top tree.

The term I translate as "group" is *bod*. It is not synonymous with the various kinship terms that nicely translate into the anthropological categories of clan (*kum*), and subclan (*dal/wun*). But one may say the people in those associations constitute a "group," *bod*. Nevertheless, I asked people if trees had clan- or subclan-like relations, which may, with care, be understood to be in container/contained forms. So is this tree in that group like you are in the Kavatan subclan of the Sinawiy clan? People denied that kind of analogy.[13] Some of the tree associations seemed informal and may in fact have been stimulated by my own inquiries: one guide and informant said with respect to a number of trees, "I think they form a *bod*." These groups were never named as such. In several cases I review there are only two members to the group. These tend to take the same name while being differentiated by contrasts such as male/female or a spatial positioning—for example, toward the shore versus inland.

The strangler set aside, the first group I learned about and actively sought to collect consisted of trees called *ababuyav, buyek, akunukun* and *ayayak*. These trees are thought to be similar because of their red sap, an indication seen in the name *ababuyav*. Informants derive the name from one of the words for blood, *buyav*. "Red blood tree" is how my north-central informants glossed the specimen I collected among them (Damon 275). Another tree, called *gagig*, is said to be like these because it also has red sap even though, I was told, its fruit is different: round rather than oval in shape. I include *gagig* here even though it was more of assertion of sharing a common attribute than suggestion that the tree was in the same group as the other four. Five of these trees were collected in southeast Muyuw. From western Muyuw I collected a tree called *kabibuyau*, which might be literally translated as "cause to bleed." I presumed it was the same as *ababuyav*. When I showed it to one of the southeastern men who helped me find the original set, he said it was the same as the southeastern Muyuw *ayayak*. By the time I collected the north-central tree (Damon 275), I recognized the name but thought it looked different. My southeastern informants agreed. Although they said a fair number of those trees could be found in a wet-soil area to their west called Poiyo, they did not recognize this tree as the same as theirs. Further, they said that while the leaves on their tree were larger than this one, their own kind of *ababuyav* died once its girth attained the size of a coconut tree while this one became a large and significant tree in its environment.

With the exception of the western Muyuw *kabibuyau,* and the north-central *ababuyav,* I collected all these trees on the same day in the same general area. However, they were in very different ecological settings,[14]

Table 2.2. Grouped Trees and Their Western Determinations

Indigenous name	Lae Id.	Harvard Id.
ababuyav (Damon 36, se)	Myristicaceae *Horsfieldia spicata* (Rosb.)	
ababuyav (Damon 275, nc)		Myristicaceae *Myristica*
akununun (Damon 30)	Myristicaceae *Horsfieldia spicata* (Rosb.)	
ayayak (Damon 29)	Myristicaceae *Myristica schleintzii* Engl.	
buyek (Damon 31)	Myristicaceae *Horsfieldia sylvestris* Warb	
kabibuyau (Damon 253)		Myristicaceae *Myristica*
gagig (Damon 32)	Flacourtiaceae *Flacourtia zippleliana*	

and my informants easily differentiated the plants, and, following them, so could I. *Ayayak* is usually a small tree found on a coral limestone base often very close to the shore, and when it was learned we were in pursuit of it we were directed to a little island right off the shore, a bump of compressed coral rock raised a couple of meters above the high tide mark in a sheltered cove. *Akununun* is a significantly sized tree found in a number of environments in higher forests, where it spreads toward the canopy top. *Buyek* is not found along the shore, and while it has a straight bole, it has longish branches with long narrow leaves. The branch/leaf arrangement generates the (incorrect) impression that the leaf form is pinnate rather than simple. *Ababuyav* is found in older forests, but it is a shorter understory tree, and, unlike any of the others, insects molest its leaves and fruit. While reasonably well known, these trees are not very useful. *Akunukun* saplings will be taken for fence posts, but that is about it. *Gagig* sap, however, is used to cure pinkeye-like infections.

Table 2.2 shows the determinations I have for these trees. In our terms they are closely related. With the exception of *gagig*, they are members of two genera in a single family. Note the correction my southeastern informant made concerning the western *kabibuyau*. Based on the names, I presumed it was equivalent to the southeastern Muyuw *ababuyav*, but he said that it was the same as the southeastern *ayayak*. Effectively I crossed trees in two genera; he kept them in *Myristica*.

Chance led to another set of trees. I often collected voucher specimens whenever I could find flowers or fruit, knowing they would make it easier to identify. So I collected one small tree, clearly a kind of *Ficus*, whose fruit came out of the trunk, sometimes on or near the ground. Immediately informants told me there were a number of these. So I collected all I could find.

Virtually all of these trees are found in highly disturbed areas. Exceptions are, however, *lalakay*, which I was told might be found in older forests

Table 2.3. Grouped Trees and Their Western Determinations

Indigenous name	Lae Id.	Harvard Id.
gipilapal (Damon 74)		Moraceae *Ficus septica* Burm. F.
adawab (Damon 75)		Moraceae *Ficus hispidioides* S. Moore
lalakay (Damon 93)		Moraceae *Ficus wassa*
no name (Damon 188)		Moraceae *Ficus sp.*
lalakaya (Damon 226)		Moraceae *Ficus sp.*
kapwala (not collected)		
kakupa (Damon 175)		Myrtaceae Syzygium sp.

as well, and the uncollected *kapwala*, which is said to be an old-forest— *ulakay*—tree. *Adawab* provides an exception: in addition to roadsides and frequently gardened areas (*digadag*), it is also found in forest gaps in interior areas. Although all of these trees are considered to produce good firewood, they are of little use other than as fencing filler. *Adawab* leaves, however, are thought to cure elephantiasis; scrotums are wrapped with them for relief. And *lalakay* fruit is boiled and eaten, having a sweet, pleasant taste. The one for which I have no name, Damon 188, is interesting because the absence of a name did not prevent informants from supplying a diacritical account of it. It is softer than the others so it could be used for nothing other than fence posts, and its fruit is never found on the trunk, only on the branches. I collected *lalakaya* in north-central Muyuw because it looked like others in the set and of course shares its name with one of them; my southeastern Muyuw teachers said it was the same as theirs (Damon 93). In general these trees were said to be similar because of the peculiar way in which the fruit appears along the trunk or the branches, and by virtue of their profuse white sap. *Kakupa* is then an interesting exception. It was collected from Kavatan in far eastern Muyuw after most of the others, but I was told about it in early 1996. Sipum's father, the man who had probably been my best informant for the longest time, was the one who said it was in the same group. However, Sipum disagreed. While it has fruit like the others—which was why I collected it the moment I came across it near Kavatan[15]—Sipum disputed the similarity because its sap was very different and its very hard wood contrasts with the others' soft wood. In fact, while the sap from the tree I collected smelled pleasantly sweet, there was very little of it. Informants also reported that the wood was so good you could use it for making keels for a class of outrigger canoes, and that it burned so well that it was good for smoking things, especially women and infants. All of these characteristics distinguish *kakupa* from others in this group. And it appears to be a totally unrelated tree, by Western criteria. (For the properties that Sipum specified, it is like other *Syzygium*.)

The trees in table 2.4 stand out clearly by virtue of common properties of the wood and the diversity of environments in which they are found. These trees are similar, so classed from their wood and very thick and sometimes peeling bark; all burn extremely well. Yet while known throughout much of the island, they are found in quite distinct environments. *Mamina* (the north-central name; there is no final vowel in the southeastern Muyuw word for the same tree) is found on wetter soils and becomes very common in middle-aged fallows (*oleybikw*), especially in north-central Muyuw where it appears in grove-like conditions; the trees are occasionally found in wet soils north and west of the southeastern rise, but it would appear that the successional cycles there are generally

Table 2.4. Grouped Trees and Their Western Determinations

Indigenous name	Lae Id.	Harvard Id.
mamina (Damon 18, 212)	Myrtaceae *Syzygium*	
alabuyo (Damon 113, 178)		Myrtaceae *Syzygium*
ameleyu (Damon 119)		Myrtaceae *Syzygium*
kikiyau		

so advanced that the tree is rare. It is an important tree, and I return to it in the next chapter. *Alabuyo* is probably found in wet soils, but I came across it most often in the drier limestone soils of southeastern and eastern Muyuw. Standing out by its whitish bark, it is a towering tree forming part of what Muyuw consider the suite of trees found in their oldest garden fallow (*ulakay*). *Ameleyu* are even larger and higher trees, found in the clayey soils toward the central and mountainous part of the island. They poke through the canopy and so are lightening targets, one case of which was my introduction to the tree. Although it is smaller, *kikiyau* looks very similar to all of these, but is found in swampy freshwater areas close to the shoreline. I did not make a voucher specimen for it, but would be shocked if it was not also in the same genus.

Ayeliv, alidad, and *atwaleu* form another set. I began to notice these trees because of the distinctive red or pinkish and thin papery quality of their new leaves; later I was told their flowers and fruit are similar in appearance, the fruit of the *alidad* large enough—four centimeters in diameter—for the exo- and endocarp to be eaten. *Alidad* and *atwaleu* are understory trees found in high forests but differentiated by the fact that the former is found near the shoreline, usually on compressed coral, whereas the latter is located further inland. These trees are not just noticeable; Muyuw pay attention to them because of their exceptionally hard, interlocked wood. Those qualities make them useful for handles for various implements. However, the hardest of them all, *atwaleu,* has a more restricted use because it rots quickly. The other two are appropriated for posts for homes or yam houses and are also special because the structure of their wood allows them to be used as crossbeams in outriggers connecting the main part of the boat to the outrigger float. Employed as such on outriggers, it is said the piece might crack, but it will rarely break in half because of the cross-grained structure of the wood. *Ayeliv* might be used for small and middle-sized craft, whereas *alidad* is the standard for the larger *anageg;* I return to *alidad* and its uses in chapter 6. These two trees, by the way, are among a number that are easily contrasted by their leaf shapes. *Alidad* leaves are said to be *kapakop,* broad or fat, whereas *ayeliv* leaves are *igiligil,* skinny or narrow.

Table 2.5. Grouped Trees and Their Western Determinations

Indigenous name	Lae Id.	Harvard Id.
ayelev (Damon 171)		Myrtaceae *Syzygium*
alidad (Damon 278)		*Syzygium stipulare* (Blume) Craven & Hartley[16]
atwaleu (Damon 285)		Myrtaceae *Syzygium*

From the determinations in table 2.5 it can be seen that this set is also classed in *Syzygium* genus. So far as I know, Muyuw make no attempt to assimilate the large number of trees that fall into that the Western classification, and nobody ever suggested these last two sets might be combined.[17]

The next group I review differs from the others because the trees tend to be found in the same environment, the oldest fallows (*ulakay*). Muyuw people do not have a unique term for older regions; however, it seemed reasonably clear that while these trees are found in areas that have not been gardened in a long time, with the exception of *atwabeba*, they also seemed to be rare in regions that were devoid of evidence of having been gardened in the last hundred years or so. This set of trees is of interest across the island because they are generally held to be beneficial to crops. And while many of these trees rot exceptionally quickly, they have extremely cross-grained wood that is difficult to chop on the one hand, but on the other, especially in the case of *silamuyuw*, very good for making canoe paddles. This type of wood does not easily snap, a significant attribute given the short, quick strokes customary for Muyuw paddling.[18]

In western terms these trees have compound leaves: a single leaf has some number of leaflets. And perhaps with the exception of *atwabeba*, all

Table 2.6. Grouped Trees and their Western Determinations

Indigenous name	Lae Id.	Harvard Id.
akewal (Damon 14)	Meliaceae *Dysoxylum arnoldianum*	
atwabeba (Damon 230)		Meliaceae *Chischeton*
atwabeba (Damon 288)		Meliaceae *Aglaia*
ayogal (Damon 301)		Meliaceae *Dysoxylum*
manib (Damon 170)		Meliaceae *Dysoxylum*
manbukoy (Damon 173)		Meliaceae *Dysoxylum*
silamuyuw (Damon 300)		Meliaceae *Dysoxylum*[19]
tabnayiyuw (Damon 15, 43, 302, 319)		Meliaceae *Dysoxylum papuanum*

of these trees also become large with significant boles and crowns stretching to the canopy top. There are, however, evident differences. In 1996 people from Gawa brought me to Muyuw. One evening before they left, the subject of these trees came up when Sipum asked me again for the name for the tree I designate *ayogal*. Although a prominent tree in the oldest fallows around Wabunun, people did not have a name for it. I received the name given here from a man in another village and told it to people from Wabunun, and so a name for the tree has become known (some people refer to it as "Kalatab's tree," after the man who gave me the name). It is similar to these others except that it does not branch out until its top. Rather than the spreading shape typical of the others, it has a long ascending bole. My friends wanted to know if Gawa people knew it. They did, and called it by the same name (I was not convinced of this).[20] I then asked the Gawa about *tabnayiyuw*, which was described as the same as *manib* except that it did not smell. They did not know *tabnayiyuw* but did know *manib*, so the analogy worked. *Manib* is one of two similar trees Muyuw people, and Western systematists, put in this group while recognizing their unique differences—they have a pronounced onion-like smell; very occasionally Muyuw use their leaves as a seasoning for boiled crops.[21]

Now to the *atwabeba*. Voucher specimen 230 was collected in Central Muyuw and identified by my instructors there. It was interesting, however, that when I showed it to my guides in Wabunun, they were confused. They smelled the seeds and were reminded of the *manib/manbukoy* set. However, to them, the leaves looked like they were from *akewal*. They eventually consented that it was the tree specified by my north-central instructors but said it was slightly different than the one they knew. The tree I collected as Damon 288 was a four-meter-high tree found in an area long removed from human activity. I had been shown *atwabeba* before that were canopy-top trees, but it was difficult to see them much less cut branches with flowers from them because they were so tall. When I saw this tree with a flower within reach I asked what it was and took it. It is interesting that these two specimens are probably from a different genus in the family. It would appear that people were recognizing a similarity and so grouping them. This similarity runs across a number of dimensions having to do with, among other things, size, shape, leaf appearance, wood quality, and smell. Nevertheless these similarities did not interfere with my informants recognizing and naming differences. Among Western systematists Meliaceae is known as a difficult family to sort.[22]

Another set may be briefly noted that includes the tree headlining this chapter, *teili*. It and at least two others are recognized as alike, and all three are determined to the genus level—Combretaceae *Terminalia*. The other two are *gaum* (Damon 107) and *sidagaum* (Damon 137). The nuts are similar in shape and all may be eaten, but those of *gaum* and *sidagaum* are very small, and on Muyuw nobody takes them seriously. The names *sidagaum* and *gaum* are understood to mark a similarity; however, the prime difference between the two is that *gaum* is found inland and tends to be a canopy-top tree in mature forests, while *sidagaum* is a shoreline tree spreading and overhanging the water. These trees tend toward being deciduous; their leaves turn red and fall off at the transition points. The appearance of nuts on the *sidagaum* is a "sign," *advutus* or *tavutus*, signaling giant turtles swimming from near Muyuw to various small and sandy islets found near barrier reefs to lay their eggs. People relish turtle eggs, so when they see the trees forming nuts they know they can access them.

Although I am inferring their group identity, this point about perceived similarity generating inclusion brings us to a last set before turning to a slightly different kind of association. Only three of the five trees are said to be a group by my informants. The two different ones are included here because they were first seen through the lens of the others. All the trees are *Ficus*; to this point in time we have not ascertained their species designations. All five are significantly sized trees prominent where they are known, which for *kwayin* and *tatoug* seems to be everywhere.

Kadimidum (Damon 246) was collected in Western Muyuw. My informant first thought it was *kwayin*. He then changed it to the name provided here. The tree is occasionally used for carving canoe pieces best described as prow boards. *Kubwag* is the key tree for gardening in north-central Muyuw, and is discussed in the previous chapter. North-central people do not confuse it with the others because they work with it so much and know that, in contrast to *kwayin* and *tatoug*, the wood rots too quickly to be used for anything. But my southeastern informants did not know it. When they saw the leaves they guessed it was *kwayin*, then *tatoug*, then they changed their minds back to *kwayin* when they saw the fruit. But then they realized they had not seen it before, that it looked different than either of these two named trees; also, its behavior as the largest tree in the *oleybik* fallow class and a tree beneficial to crops was foreign to them. *Kwayin* and the two *tatoug* seem to be known everywhere. For *kwayin*, this is partly

Table 2.7. Grouped Trees and Their Western Determinations

Indigenous name	Lae Id.	Harvard Id.
kadimidum (Damon 246)		Moraceae *Ficus*
kwayin (Damon 280)		Moraceae *Ficus*
kubwag (Damon 269)		Moraceae *Ficus*
tatoug (*auleke*) (Damon 311)		
tatoug (*wanawoud*)		

because its "new leaves," *ayshushun,* are used in place of betel nuts when unavailable. Although not considered as sweet as the real thing, they turn red and provide a jolt like the nuts. Its wood may also be used for carving prow and stern boards for the largest two classes of outrigger canoes. When dried its wood is superior for long slow fires—small burning chunks are often taken on canoe voyages so a fire will always be at hand. But while identified with the two tatoug, it is distinguished from them. Its wood is harder than *tatoug,* and therefore more difficult to fashion. Some people carve ornate forms into pieces of the wood fashioned for outriggers, so this quality counts, a value specified by saying that the "chips do not fly" (*sinay teyoyo*). By contrast they tend to "fly" with the two *tatoug.* I first learned about these latter two as if they were a single tree only to be told that there are actually two of them, one found on the beach, the other inland. An important difference between these two is that only the more inland version tends to develop buttresses, more easily appropriated for canoe carvings.

Using the *kwayin/tatoug* set, a group of men laid out for me an analysis of variation that at the time surprised me with its formalization. I provide their model here, a framework that could be achieved with many species.[23] What stands out in this analysis is the specification of difference. And these differences, some quite minute, are critical. The beach *tatoug* is one of two trees—see chapter 4 for the other—that elders direct children to use in model outrigger sailboats so that the youths experience appropriate aero—and hydro—sailing dynamics.

I suggest that all these examples show the ability to see similarity, the kind of similarity that might lead to inclusion in a specifiable class. Some analysts have used the notion of covert categories to describe this awareness. However, the names and the various categories of knowledge people find in and bring to these names emphasize discontinuous relations. It is not the point that, for example, *silamuyuw* is similar to *tabnayiyuw*; the point is that they are different.

Table 2.8. Analysis of Variation of Tree Features

tree/feature	*kwayin*	*tatoug*/on the beach	*tatoug*/inland
bark (*tapwan*)	Same		
sap (*shosun*)	same: profuse, white, sticky		
leaf structure (*igiligil* [slender]/*kapakop* [palmate])	slender	palmate	palmate
tip (*matan*) of leaf	sharp	round	sharp
wood (*kamlokan*)	hard	soft	soft
buttress (*dibanas*)	negligible	negligible	significant

A similar conclusion results, I believe, from a consideration of a bunch of trees that are not so much in the same "group" but have the same name, for they are conceived to be almost identical. One is *kudukud* (north-central)/*gudugud* (southeast) (Damon 21; 61; 174). Like *tatoug*, it is distinguished by its position inland or on the shoreline, a distinction I did not realize when I collected the first two of three specimens. But it was the reason for collecting the third, which came from the shoreline. For Damon 21 I received species identification from the Lae Herbarium, Urticaceae *Pipturus argenteus* (Forst) Wedd. The other two, both from southeastern Muyuw, were only determined to the genus level, also *Pipturus*. They are distinguished by inland versus shoreline location. What is crucial, however, is that the bark of the inland, but not the shoreline, tree is extracted, dried, and woven into string that is then fashioned into variously sized, but similarly constructed and shaped, fishing nets.[24] I do not know if there are substantive differences in the inner bark of the differentially place trees. Although, like the materials used in houses, finished nets become visible repositories of resources from across the landscape, I have no indication that representing spatial differences is the reason for preferring one location over another for net string. Nevertheless this inland/shoreline distinction is important and I return to it later.

Two other names with complements like these are *alul* and *tabdakan*. Male versus female distinguishes *alul*. The male has leaves that are *kapakop*, round or palmate. The female leaves are *igiligil* and *veyayov*, slender and long. Both are understory trees, so they tend to be found in older forests. I did not collect either of these so I have no idea if they might be varieties of the same species or different species in the same genus (or, less likely, unrelated at all). The two *tabdakan*, one of which I collected, are distinguished by leaves and environment. One has small, round leaves, and people find it in early fallow gardens/forests; the other is found in the oldest fallow, as well as in swamps (*demiavek*); its leaves are long and slender. I learned about the distinction between the two after I collected a tree by this name, so I can only presume the one I collected (Damon 176, Lauraceae *Cryptocarya*) is the latter kind.

One of the more interesting sets I collected consists of two small "trees" called *weylau*, one female and the other male. The *weylau* female (Damon 87, Sterculiaceae *Abroma augusta*), grows up to five meters tall in areas with lots of sun but at a remove from the sea. The *weylau* male (Damon 127, Malvaceae *Hibiscus*) reaches a height of only a meter or so and is located closer to the ocean. Although the male plant is not used for much, the female was or is an important source of two kinds of tying material. Before European product was available, it was the principle source for making fishing line, a vivid fact to all of my instructors even though I doubt anyone saw the practice. But the tree remains a source for belt-like cords around which women wrap the shredded coconut leaflets they use to make their skirts (*dob*). Unlike the coconut, which has a rough, barbed feel, the *weylau*-generated material is soft and very pliable. As is the fishing line, the material is obtained by pressing the plant's trunks and larger branches into the sea for a week until the outer bark begins to rot off. When the plants are removed from the sea, the inner bark is stripped off, dried, and turned into useful threads.

Two sets of facts related these plants as female and male, though as near as I can tell only one of them seems to force the complementary identity. The first concerns the barb-like nature of the female's leaves, branches, and trunks, which are prickly to the touch. By contrast, although the leaves of the male feel slightly hairy, they do not pierce the skin. How this difference generates a female/male linkage was expounded upon by one of my best informants; the paradigm was heard many more times (when it eventually stuck to me). If a man tries to hold a woman she will likely scratch or bite him. The same thing happens when Muyuw harvest the yam that they consider female, called *parawog* (Trobriand *teytu*): the tuber's barbs prick skin. Among young lovers, a woman should gouge the areas around a man's knees with her fingernails. I've seen young men take pride in appearing in public with fresh wounds around their knees. The female *weylau* is like this: if you try to take it you will be pricked or gouged.

Yet what really forces the complementary relationship between the plants is their flower structure and color. The point was immediately evident to me the first time I found the male tree, by which time I'd heard about, seen, and studied the female several times. The description I produce now comes from informants. The female tree's flower is somewhat cup shaped. When mature, it hangs down wide open. The corolla is white, but it is streaked with lavender or purple. Although not the most erotic-looking flower I saw, it was beautiful, captivating, and, because of the way it hangs down wide open, a stimulant to Muyuw eroticism. By contrast, the male *weylau* flower structure opens up, rather than down. Its corolla shows a lavender/purple color nearly identical to the female's, and its long erect stigma is understood to be a penis to the vagina-like shape of the female.[25]

Not all plants understood by simple contrasts like inland/seashore go by the same name. Two that do not are *ipwawun* (*iyapwawen* in southeastern Muyuw) and *bitok* (table 2.8). According to Muyuw, *ipwawun* are found along the shoreline and *bitok* near swampy areas. All the *ipwawun* I saw were modest-sized trees overhanging the water's edge. The wood is very light, occasionally used for small outriggers according to eastern Muyuw people, sometimes for the outrigger float according to a north-central informant. The leaves and fruit are fed to pigs to fatten them; humans eat the fruit of neither tree. All the *bitok* I have seen were massive trees, one planted and still standing at what was once the western end of Kaulay village.[26] Both trees have similar chicken-egg-sized fruit, which are a creamy and translucent yellowish-white color when mature. *Bitok* fruit differs in that it has an arresting fragrance; liquorish-like was my impression. I was first exposed to these back in the 1970s when men would occasionally come into the village after a day's work with *bitok* fruit hanging from their ears like enormous earrings, the point more a matter of smell than sight. But there is more to the story than that. Unlike in eastern Muyuw, north-central people practiced a funeral custom that emphasized their higher, *guyau*, status (see Damon 1990: 76–77). A buried person was sat upright with their head facing west. When exposed the skull would then be covered with *bitok* fruit. That is why the *bitok* tree was planted proximate to Kaulay village; fruit from the tree was raided for mortuary purposes.

Table 2.9. Grouped Trees and Their Western Determinations

Indigenous name	Lae Id.	Harvard Id.
ipwawun (Damon 24)	Hernandiaceae *Hernandia ovigera* L.	
bitok (Damon 209)		Hernandiaceae *Hernandia sp.*

Although I only have voucher specimens for two, another set consists of three similar trees (table 2.10). One is unequivocally called *ukw*. The name I first learned for the other is *silawowa* in north-central Muyuw, *asilawow* in southeastern Muyuw. When I showed the north-central specimen to a southeastern person he told me it is called *akinod*. Later I was told that these two were distinct, one called *asilawow,* the other *akinod*. These trees grow to be ascending, canopy-top trees before they spread in mature forests, and all are fairly common. A portion of land north and west of Wabunun coming into cultivation again for the first time in decades had so many *ukw* that it was referred to as *waukw* ("at *ukw* trees"). All rot relatively quickly, though all may be used as inferior wood for small outrigger canoes, outrigger floats, or sago troughs. But while recognizing the similarities, people readily distinguished these types. One axis of variation is the leaf structure. *Ukw* is the most palmate (i.e. roundish), *akinod* more oblong, while *asilawow* is the narrowest with a very sharp leaf "point." In April 1996 I hand-delivered some 150 voucher specimens to the systematists at the Papua New Guinea herbarium in Lae and was clearly confused by some of the seeds I had, *ukw* and *asilawow* included. When I returned to Muyuw I described my confusion to Dibolel. He thereupon produced a detailed description of the seeds of these three plants, plus one other. Unfortunately, I did not write down the lecture. But the description, the level of detail that was at immediate recall to him, remains a powerful image. In addition to the seed structures, and unlike the other two, *ukw* is deciduous; once a year it loses all its leaves (*ilawus*) and then grows new ones back (*iyasus*). However, these details are not just biological facts about incidental plants. All three of these plants, and in fact with one exception, every plant I collected in the Sterculiaceae family, is noted as a minor or major source of tying material. Although for most of these the substance is nothing more than strips of bark quickly pulled off a trunk for very provisional tying, such is not the case with *ukw*. Inner bark from relatively young *ukw* was used to make the most important rope for masts and sails of the two highest classes of outrigger. I photographed this being done in the 1970s; the tree still provides the paradigm for such rope now, though usually "modern" tying materials are employed.

Table 2.10. Grouped Trees and Their Western Determinations

Indigenous name	Lae Id.	Harvard Id.
asilawow/akinod (Damon 189)		Sterculiaceae *Sterculia*
silawowa (Damon 17)		Sterculiaceae *Sterculia*
ukw (Damon 51)		Sterculiaceae *Sterculia*

I return to a set of trees in the same genus, *Calophyllum*, in chapter 4, so I close this section by discussing an encounter with a closely related tree that Muyuw call *amanau*, or *yamanau* in north-central Muyuw (Damon 131, 127, Clusiaceae *Garcinia*). For several reasons I began to find the tree interesting as my knowledge expanded. This tree is most frequently found in soils Muyuw class as "wet." This means that it is infrequently found around Wabunun, but common in the falloff to the north and west of the rise that forms the southeastern sector of the island. It is also very common in the north-central portion of the island in the vicinity of Kaulay and Dikwayas villages. Although in absolute elevation that area is higher than the rise that forms the southeastern section, the soils are often "wet," and after my eastern friends pointed out the facts, I too began to notice that the make-up of the forests there consisted of more "wet" than dry trees. In Kaulay village gardens the tree is frequently used as a fencepost. Everywhere *amanau* is valued as a crossbeam to connect outrigger floats to the keel structure of a canoe. The properties that make it useful for that function, however, are canceled when it comes to house construction—*amanau* attracts bugs, so it is never used to build homes.

In color and texture *amanau* leaves are similar to the *Calophyllum* set, the trees of which were also becoming a focus of my inquiries. When I brought an *amanau* branch into Wabunun to assemble voucher specimens from it, several youths peered into my house and noted that I had another *kausilay*, one of the *Calophyllum*. I told them they were wrong, but also noted their misrecognition. Their identification followed from the fact that *amanau* leaves look like they have the kind of venation that typifies *Calophyllum*. They do not, but you have to look carefully to see that difference. In any case, about the same time I had realized that there were similar qualities in the *amanau* wood and all six of the island's *Calophyllum*. These all have wood that tends toward being a shade of red; if this quality pertains to all but the inner bark in the *Calophyllum* species, in *amanau* it is restricted to a smaller heartwood core. Additionally the tree's interlocked grains are like two of the *Calophyllum* species. My best informants were aware of the wood structure and color similarity. I go through this experience because I was so taken by the similarities among these trees that I asked my guides if the trees were in a similar group. The most experienced observers said no—they instead stressed differences, which were equally obvious to anyone paying attention. Now this account is interesting because until recently, Western systematists have had these trees, *Garcinia* and *Calophyllum*, in the same family. So while the trees share features that our grid uses to align types, fundamental discontinuity stands out in the Muyuw frame of reference.

Kweita Tayp' *and Names*

I draw this section to a close by discussing two related phenomena, one the fact of types, the other the matter of names, both complicated conceptual and perceptual problems. *Kweitan* translates to "one thing," and is composed of the classifier *kwei*, the most generalized classifier; with further discrimination a more specific classifier is used. The word often becomes an expression for "difference." Early in my research I asked an elder if there was a Muyuw term for the English word "type." He replied, *"Kweitan."* My best informants would often use the expression *kweita tayp'* when emphasizing that one object or tree was similar to but different from another. *Tayp'* is from the English "type," and the apparent duplication here is in fact consistent with Muyuw grammatical forms: duplicating all or part of a word intensifies its action or meaning.[27]

Data adduced above clearly show that Muyuw operate with models about what specific plants are. They manage with forms, ideas about forms, which match specific trees more often than not. It is equally clear that these models are generated from local experience with given plants. When they recognize empirical variation from their models, they struggle with the differences. One day while walking northwest down off the ridge that defines southeastern Muyuw—which means in wetter, redder, and more clayey soil rather than dryer, blacker, and more typical limestone soils—Ogis and I came to a canopy-top tree. Up close we could only examine the bark, fallen leaves, and the shape of the tree's top. Ogis went back and forth between thinking the tree was what he calls *amwakot* or *akewal* (Damon 126, Meliaceae *Dysoxylum sp.*, and Damon 14, *Dysoxylum arnoldianum*). Finally he settled on the *akewal* designation but pointed out that it was different because the bark was whiter than the customary reddish brown he knew.

Instances like this one confirm the point that however closely tied to tangible realities these names seem to be, they are necessarily conceptual forms that operate by means of criteria that are usually specifiable but also contingent in one way or another. And for Muyuw, a major variable in these contingencies is shifting place. Never in my experience were my informants angry or accusatory about names I received from other places that concerned trees that they either did not know or knew by means of other names. To the contrary, there is expected variation. I surveyed parts of the Sulog region with people from Wabunun; they told me right away that they were not Sulog people, and that while they knew some important trees there, they did not study the region or think they should have command of its flora. This is an important consideration for the comparative analysis of taxonomies, often phrased in terms of the number of names that systems tend to have. Drawing on speculations from Lévi-Strauss (1966), Berlin reproduces the argument that many folkbiological systems—and as a matter of their own personal competence, modern systematists too[28]—tend to have around five hundred names for what he calls generic taxa (Berlin 1992: 96–101). Everything I know about Muyuw names suggests that the range of names they use would be on this order. To fully explore this issue, however, one would have to conduct surveys of names in all of the different places on the island—or, to be more exact, islands. While there would be overlap, arguably because aspects of the environments and experiences of them are shared, data I've adduced to this point demonstrates various differences. But a project such as this would be, I think, misconceived if its purpose was to construct the set of names for the region. Our formal system of taxonomies searches for a total understanding that is, to argue against the stereotypes, askance from the related cultural systems of the Kula region, and probably most of this part of the world. In Crumley's (1994) terms, our system is a hierarchical one. We attempt to connect everything to a common value ("phylogenetic" relation, the value for the last hundred and more years). The system strives for an, or perhaps "the," original continuity.[29]

By contrast, the "naming system" here is heterarchical: there are different values, and names, for different places as those places differ. This is why informants told me they couldn't be expected to know the trees from the Sulog region; that is not their responsibility. And as one man observed after one of my tours from those distant places, "We are early fallow people" (*digadag gimgilis*) and so only know those trees. Other people fall under other categories, with the appropriate forms of knowledge. A fundamental sense of discontinuity underlies—indeed, creates—much of the culture that forms the realities these people generate. One could go from this analysis to the place of incest rules in establishing appropriate productive relations,[30] to a whole slew of rituals enacted as people depart one village or island for another place, and to certain beliefs, almost a set of predictions, about, for example, weather patterns. So, people drink seawater from one place to cap its food when they move to another place, usually taking the new place's water when they arrive. There are magical spells designed to prevent malevolent gossip from a village interfering with hunting, fishing, or other productive acts elsewhere. People believe the weather changes with the sighting of a new moon, the end of a full moon, the setting of major named stars, and, related to the stars, the deaths of important people. Discontinuity in one domain occasions it in another. Hence, ideally, garden activities take advantage of weather disturbances following moon phase changes and star setting. Such discontinuities or separations then become factors in new (productive) combinations.

This analysis does not lead to a tight explanation for plant names and other items. Yet I believe it explains why there is so little concern with establishing the kinds of hierarchized relations seen in the Western analytical system. The list of associated plants in the previous section shows that people recognize empirical affinities among sets of plants. And these affinities are the means our social system uses to configure contiguity across types. But Muyuw do not turn these facts into relations that are important enough to be named. Rather, the singularities of specific forms receive the most recognition.

Returning to the consideration of some plant names will underline these points. Names that effectively mean "tree-of-this-kind-of-place" are obvious examples of such an individuating process. *Kaydibwadeb,* noted earlier, occurs only on the piles of crushed coral limestone next to the sea, called *dibwadeb*. Another is called *sinasop kaynen,* "meadow plant" (Damon 264, Nepenthaceae *Nepenthes*). Since everyone knows there are other trees of these places, the names implicitly distinguish a unique type in a place where other types are also known. Following this kind of pattern, other names that at first seem whimsical at best may be appreciated in their distinctiveness, places in a set of tiered patches.

A problem in understanding Muyuw naming may be illustrated with one tree and how its name has intersected mortuary practices. Since these practices transform as one moves west the issue probably disappears by the Trobriands. The tree is an important one throughout this area, as well as in the international timber market. Properly named, it is something like *meik^w* (Damon 16, Leguminosae *Intsia bijuga*). In Muyuw that name is also used for the heartwood of any tree, which is thought to be harder than the wood just inside the layers of bark. The tree named *meik^w* is a specific and intense exemplar of a general condition. Coupled with the fact that it is rot resistant, the tree is the prototype for house posts and is the wood of choice for *anageg* "rudders" (*kavavis*). "*Meik^w*" is a very important tree, and will be discussed again.

Most people refer to it, however, by saying *kaymatuw,* the primary name I recorded. Literally, *kaymatuw* is "tree" plus "hard." And indeed the tree is such. Its real name is not heard because it approximates the name of an important elder, Meikdulan, who died in the 1960s. Although these names are not identical, they are close enough so that people believe they should not pronounce the tree name, instead referring to it by one of its primary attributes. Soon perhaps no one will know its "real name." I have no idea how often this process has changed names by which trees (and other things) are known, yet it is part of the reason for the island's rapidly changing language usage.[31]

A case parallel to this one concerns the name given to a tree called *akuluiy* (Damon 16 Rubiaceae *Mastixiodendron smithii* M. and P.). Found in mature forests in practically every environment but swamps, this tree is important for house construction. It is strong, not too heavy, and rot resistant as long as it is kept dry; because it rots when it gets damp it is never used in boat construction. While the name is known everywhere, people in north-central Muyuw call it *digunakay.* This is a name for a specific mushroom that grows on this tree and is particularly good eating when it does.[32] So in this example, something metonymically connected to the tree is becoming its name, on part of the island at least. Although everywhere else people know the tree is a good source for mushrooms, it is only in the north-central area where that association is becoming the tree's standard reference.

Other names for which I have accounts of their "meanings" tantalize with their suggestiveness but lead, most of the time, to no generalizing significance. In some ways the most bizarre is *kaypwadau,* a name everyone understands, but not everyone realized was in fact given to three different plants. The first one I found (Damon 95, Costaceae *Costus*) is a shrub-like plant found on the edge of meadows in the central part of the island. It grows to about three meters in height with leaves to twenty-one centimeters in length that spiral out of a semi-woody stalk. Its flower structure entails a red cone branching off the stalk that is tipped with white flowers. The second one I came across, which some people called *ulkowad,* was specifically understood to be a meadow-edge plant ("tree," i.e. *kay* for Muyuw) not well known in southeastern Muyuw (Damon 263, Zingiberaceae *Alpinia pulchra* (Warb.) K. Schum). The third (Damon 284, Orchidaceae *Spathoglottis*) is a very different plant, a more delicate item growing to some sixty-five centimeters in height. Frequently seen, it grows in highly disturbed areas typical of early fallows (*digadag*), often along the paths running through these zones. Its flowers are lovely. Informants recalled seeing them in their youth, being attracted to them, and, as they often do with such items, cutting them off to place in their hair or press into an armband. However, when they appeared in the village so decked out, elders immediately told them to get rid of the flower. The name translates to "anal intercourse." That is "forbidden." And this is why the elders told the youths to get rid of the plant. I was never able obtain an explanation of the usage. However, with respect to the second one, one informant said the name was drawn from the brilliantly red color of the seeds. While some plants and their names crystallize positive associations, others are beacons of proscription.

There may be other plants with exotic names like these three, but the only other one that I know about is both astonishing and subtle. The plant is a vine named *namonsigeg* (Damon 92, Passifloreaceae *Passiflora sp.*), one we return to in chapter 5. But for now, *namon* refers to sexual fluids—literally "its" (*na*) "fat" (*mon*). These are of course male ejaculate, but also liquids engendered in a woman at a high state of desire. *Sigeg* means to last forever. The name leads to a hope that young couples will spend the rest of their lives feeling for each other as they do in their period of youthful joy. This plant is also a dramatic landscape beacon.

Analysis of names with decipherable meanings illustrates how unique properties are recognized for specific plants and, sometimes, used to name the plant. The critical features vary from basic properties of the plant—e.g. *meik^w*—to incidentals that entail a projection of human values onto the organism, or part of it. A variant of this practice may be seen in a sort of reverse order with respect to a tree that is very significant to outrigger sailing craft. The understory tree (Damon 60, Rubiaceae *Ixora* cf. asme Guill.)[33] is similar in appearance to American dogwoods. Consistent with some of Berlin's ideas (1992), informants distinguish two varieties, one near the shoreline, the other inland. The shoreline tree has reddish bark, longer leaves, and down-hanging flowers with a strong smell. The inland tree has a whiter bark, smaller leaves, and down-hanging flowers that project a spreading effect but little smell. Both kinds have the desired qualities for outrigger canoes: hard wood and tightly interlocked grains, making the wood extremely resilient. The delicate flowers are so attractive that men sometimes cut them off as bouquets for their wives. And this gets to its name, *aukuwak*. So far as I know, this name means nothing; however, the name is appropriated for two other places. One is a small island about a kilometer east of Wabunun across from a hamlet called Sibulouboul. The island is called Simkuwaku, "Island of *Aukuwak*," because that microenvironment is good for the "beach" *aukuwak*. The second is more poetic. The tree name applies to a sector of the sea lying between Sulog point, the mountainous region of south-central Muyuw, and Nasikwabw, a piece of an original barrier reef that has been split off the main island structure, moved south, then raised enough to support a sailing community. Because of the structure of the seafloor between Sulog and Nasikwabw, waves often churn to whitecaps in the area. Muyuw think the effect is lovely, and it reminds them of *aukuwak* trees flowering beneath trees above them. In this case, the "meaningless" tree name coveys a meaningful, and aesthetic, condition that is applied to another set of relations.

After my informants, men between twenty-five and forty years of age, explained "*aukuwak*" to me they apologized for their elders' intemperate and typically nonsensical practices. Both the sea-sector name and the fact that these men felt they had to apologize for their ancestors stunned me. The "apology" was a "colonial moment," a residue of incorporated self-loathing. Among other things I came away wondering what else I could no longer learn ... because of the "silliness" of the elders.

Some tree names have unique meanings, and some meaningless tree names are turned into meaningful relations. Yet it must be noted that few trees have "meanings" that are explicitly "meaningful." So far as I know *akewal, atwabeb,* and *ayayak* mean nothing other than names for these respective trees. There are, however, some "meaningless" names that, from a comparative point of view, are interesting. One is *utawut*.[34] Although I never made a voucher specimen of the tree, with little doubt it is the shoreline "poison fish tree," *Barringtonia aseatica*. And with less doubt this word is cognate to the Proto Oceanic **putun* (Osmond 1998: 221; Kirch and Green 2001: 138). When this tree's use was described, people also said it was only half as good as "our fish poison" (*yakamey maputu*), in this case referring to a vine that, unaccountably, I also did not collect, but was presumably the rotenone-loaded legume *Derris sp.* A similar relationship seems evident in the tree group *ukw, asilawow,* and *akinod*. Although cordage of various qualities may be derived from all of these trees, far and away the most important is *ukw*. Inner bark from this tree was used to make the strongest ropes for handling masts and sails on the two most sophisticated canoe classes. Across the Melanesian and Polynesian Austronesian expansion, terms similar to this turn up for bowstrings or fishing strings (Kirch and Green 2001: 198, **uka*) and, following the Pawleys, for mast stays, **tuku*

(Pawley and Pawley 1998: 197).[35] Muyuw's "*ukw*" may come from functional relations embedded in the spread of Austronesian cultures.

Toward the Synthetic Use of Arboreal Categories

I now turn to a usage that is almost certainly part of the Austronesian repertoire most Kula Ring cultural units share, the ways by which the contrast between "tip/top" and "base" structures reality. The Muyuw words *matan* and *dabwen* may be translated "tip" or "top"; *wowun* is "base." Although *matan* connotes a point while *dabwen* invokes a rounded tree crown, the words may be used interchangeably. *Dabwen* may not be an Austronesian cognate, but *matan* (eye, point, tip) and *wowun* probably have Austronesian roots.[36] I presume the latter is cognate to, for example, *pun* as described for the Sarawak Iban by Sather (1993). There the term and contrast orient longhouse construction, the placement of houses by rivers, and the flow of water in rivers.

Primary, encompassing floral properties articulate totalizing images. The next chapter concludes by showing how particular usages and secondary tree properties individuate these totalizations.

In Muyuw, *wowun* usage is profoundly related to the support connotations of a tree's base. In Kenneth Burke's terms, *wowun* is a "stance" word, suggesting "basic," "fundamental" "the reason for" something (Burke 1969: 21–58). The standard way of asking for the reason for some act or thing is by using the expression, "*Aveiyag awowun?*" ("What is the reason?") In Muyuw usage this sense of "reason" is not identical to origin place. Origin locations for subclans are not considered their *wowun*. That origin place is *mumug*, a duplicated form of the verb "lead." Occasionally the word *pwason*, "navel," is used for a subclan origin point. Its significance seems to be that it is a point of transition, where a group "comes out" (*sunap*) of the ground. When I asked for the "basis" of a subclan, expecting to be told "blood," informants replied instead "people." And their "basis"? "Food." "Because" food, prototypically from fathers, makes people. People are substance because, like food, they are processes that do things.[37]

In some societies, plants or plant parts become major vehicles for illustrating relations. In Balinese music, associations among core and variant melodies are "often metaphorically compared with a tree, where the core melody is represented by the trunk," others "by the limbs and branches," and others still "by the flowers or leaves" (Tenzer 1991: 44). In fact, Balinese society uses botanical forms to represent key facets of the composition, decomposition, and recomposition of the body, individual and social. Usages like these are fairly common throughout the Indo-Pacific world (e.g. Traube 1986), as well other places (Fernandez 1998). Rarely, however, did I learn anything as subtle as Tenzer's Balinese case. In 2009 a Muyuw person used a "base"/"branch" reference to distinguish between someone central to a conflict—so the "base," *wowun*—and another peripheral to it, a younger sister, hence "branch," *yayag*. Usages like this, however, were rare and in themselves did not lead to productive associations. The base/tip opposition is, however, fundamental and recursive. Figure 2.1 provides basic parts and names of trees.

Base/top imagery orders islands, villages, gardens, and boats (and boat-part positioning) on Muyuw and Nasikwabw and, at least with respect to major landmass orientations, on Gawa, Kweywata, and Iwa. The "base" is the initial direction. Chapter 1 described the fundamental structures of a garden in reference to "sunrise"/"sunset." That opposition, which organizes New Year practices across the island and within each village as well, may also be phrased with respect to tree images. Sunrise, east, is the "base"; sunset, west, is the top of the tree. An outrigger canoe's base is always created from the base of the tree from which it was cut. I have no direct knowledge of these forms in the Trobriands and Kitava, but they are important (see Mosko 2009). Although Muyuw, Nasikwabw, Gawa, Kweywata, and Iwa are all defined by the same *wowun/dabwin* (*matan*) contrast, the actual direction varies. In western terms, southeast

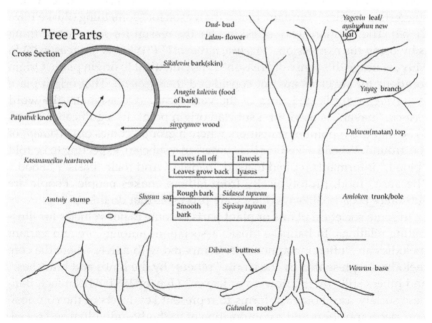

Figure 2.1. Tree Parts and Terms

(base)/northwest (top) orders Gawa; Iwa's fundamental axis is northeast (base)/southwest (top).

People in Wabunun, and generally throughout southeastern Muyuw, refer to the western and eastern ends of villages by the terms *kalamatan* and *kalatatan*. These two words derive from *matan* and *tatan* respectively. The latter term is applied to the stalk of taro plants and trunk of sago trees; both are analogous to the plant's "base." These references orient the ideal village of two parallel rows of houses, with each row running east to west. About the time I was learning this conception in 1996, I discovered two related facts. One is that the westernmost village name in southeastern Muyuw, Unmatan, is composed of the noun classifier for village, *un*, and the word *matan*. It means "village at the end of the row." The row is composed of the line of villages from the current eastern Guasopa, sometimes called "Obwilim," which is a synonym for "to the east," to Wayavat, which translates to "to the west." This implies—inquiries confirmed the deduction—that as people think of a village's east/west-oriented rows of houses as modeled on the basic shape of a tree, so do they consider rows of villages. It then occurred to me that there might have been, and people might have conceived there to be, a row of villages running parallel to the shoreline villages across the uplifted region that forms the northern limit of the southeastern region. Having found evidence of an old village directly north of Unmatan on that ridge, I asked if this was the case. It is. Then another person said the southern villages, Obwilim to Unmatan, were the *tatan*, base, for the upper, northern villages. The word they used for the upper villages, however, was *yamwik* rather than *matan*. This word refers to broad-leafed plants whose large leaves are used for wrapping things up, as taro (and sago) leaves are often so employed. When I tried to confirm this paradigm in 2014, an elder Wabunun man recalled from his youth being told to go to the *yamwik*, a village called Sinamat, to "beg" for betel nut and *gudugud* (*Pipturus argenteus* [Forst] Wedd) saplings, which would be used for making string for fishing nets. Surprised by the inclusion of *gudugud* in this story, I was told this tree appears when an old forest fallow (*ulakay*) has been cut down; if that forest is repeatedly cut at short intervals, *gwed* rather than *gudugud* grow up. As we will see from the next chapter, that area is systematically associated with old fallow classes. When in the Wabunun man's story he got to the *yamwik* village, an elder there instructed youth from his village to go down to the *tatan*, the base, to ask for betel pepper (*domwit*). Here the base/top orientation configured reciprocal complementarities, betel nut for betel pepper exchanges, between different places.

North central Muyuw informants replicated this model. The old village (names) Kaulay, Dikwayas and Lidau form the northern yamwik/

matan to the southern shoreline villages, Kalopwan, Kawuway and Kalabwadog/Kuban, constituting tatan. These fractal images (figure 2.2), formed from the tree model, dissipate by the western end of the island. There the idea of formalized rows of houses and arrayed villages no longer holds; the transformation evident there continues to the small islands heading to the Trobriands as their defining forms take on a new pattern.[38]

With the use of plant forms for specifying locational relations, I have moved from what I mean by classification to the question of categories. From here I bring this chapter to a conclusion.

One may suggest that Muyuw tree names, when fit under an overarching class, *kay/ke*, conform to the received understanding of folkbiological systems—a tendency for inclusive rankings, thus providing support for the Berlin-Atran model of folkbiological categorization. Yet it is more accurate to argue that these forms represent behaviorally distinct patches of significance, not all of which carry names, and that their place in a system of ranked relations is not very important. Names, or slots of significance, specify patches of distinctive properties rather than positions in a set of contiguous relations. Some of these crystallizations speak to higher-order concepts.

Basic floral properties provide fundamental models for the orientation of many social and behavioral forms, prime orientations on islands, relations between villages, and other primary forms. The next chapter moves this analysis to frameworks that are both more encompassing and specifying. We move from trees to forests and their specification of regional sociality.

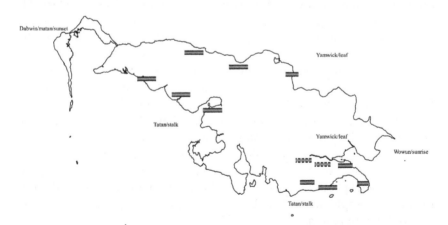

Map 2.1. Fractal Presentation of Base/Tip Contrasts

Notes

1. Although nobody claims to have seen Ogis in the vicinity of the *unumug*, only after the evening ended was I told that he saw the tree and the hearth. Seeing what he should not have seen is thought to have led to his altered state of consciousness.

2. A tree north-central people called *bovagun* (Damon 220; Rubiaceae) was called *asivay* (Damon 98, Rubiaceae *Psychortria sp.*) by a southeastern expert. Another called *katibwadan* (Damon 228) in north-central Muyuw is called *tabdakan* in the southeast; there people distinguish two kinds, one with longer leaves (in higher forests), another with rounder shorter leaves in younger forests toward the shoreline.

3. Damon 177, Lauraceae *Cryptocarya,* one of several species in this genus on the island; also Damon 282.

4. The logic of this assertion: how wood holds up to the sea's unforgiving conditions tests the materials.

5. One tree I brought to Sipum's elder brother, Dibolel, baffled him despite his acute discriminatory abilities. When I mentioned that the branches were brittle, he immediately recalled that the bark really "flew" when you chopped it, so it was called *ayyayi,* which he said was similar to *amanau.* The latter was Clusiaceae *Garcinia.* Harvard Herbarium personnel first classed it as Myrtaceae *Syzygium;* during the fall of 2004 they reevaluated and decided it was *Garcinia sp.* (Emily Wood, personal communication, November 2, 2004).

6. To my great astonishment Sipum once hacked a gnarled and extremely bent mangrove tree (Combretaceae *Lumnitzera* [Damon 157]) out of its mooring, passed it through a fire, then straightened it out and told me it could now be used as a outrigger crossbeam to connect a hull to a float.

7. Some people say "snakes" are not man but rather form their own distinctive group.

8. Young informants suggested that "food" (*kan,* lit. "food-its"; *n* is a third-person possessive) should be a basic category, one elaborated in origin mythology. However these people realized that the content of this class is in other classes.

9. Although taro and yams take different classifiers, both are "seeds" (*yopis*) and may be classed using the classifier for taro, *un-*.

10. The generic term is *yadidiu,* which when marked distinguishes among *adimwakal* and *munga,* which is "very large." Leban Gisawa, who has hosted me in and out of Papua New Guinea for the last few years, works in the Fisheries division of the government. His original specialty was tuna. According to him the commercially significant species found in Papua New Guinea waters are: yellowfin (*Thunnus albacares*), bigeye (*T. obesus*), skipjack (*Katswonus pelamis*), and on rare occasion albacore (*T. alaluga*). However, there are more species closer to shore, those used by local populations (Leban Gisawa, personal communication, September 10, 2009).

11. *Busibuluk* (Damon 123) refers to an outrigger canoe's directional motion (*buluk*). Sails are partially framed by a piece called *kunai,* often made with this

tree because of its properties. *Abunubuluk* (Damon 135, hesitatingly identified as Rubiaceae *Cephaelis*), refers to pigs (*buluk*) that put their noses (*abunu-*) into the plant's flowers.

12. It is called *bwit* on Iwa, and is considered a female. In western Muyuw it is called *kaniu*, although people there knew its name to the east and west.

13. The analogy from the clan/subclan relation would be misleading. While Muyuw people conceive their matrilineal subclans to be in clans, they do not understand clans as unique units. They are rather understood as sets—in the model four, in fact eight—that depict appropriate marriage relations. Genealogies do not link people in one clan (*kum*). Genealogical connection is not the primary framing device for subclan members.

14. In 1996 I was led to another area, west of Unmatan, and shown another tree said to be like those in this group. Called *abwibuyau*, I made no voucher specimen.

15. My Kavatan informants reported that it had white flowers. Early on I asked Wabunun people what flowers looked like on many of the *Ficus* trees. They told me they didn't know. I think they were being polite.

16. This is a new name for what used to be called *Syzygium longipes* Merr. and L. M. Perry. I thank Lyn Craven for this information.

17. This genus contains over five hundred tropical plants.

18. Trees of the Mahogany family (Meliaceae) were cut for lumber by the Neates until they left in 1978. They were again cut for timber by Milne Bay Lumber Company, which exported raw logs between 1982 and about 1995. This set is one of several commercially significant trees whose distribution results from human action.

19. My specimen notes and a further written description by David Mabberley suggest that this tree is *D. parasiticum* (e-mail message August 16, 2004).

20. A Kavatan man denied it was *ayogal*, saying instead its name was *gwas*, and that *ayogal* was the name of another tree.

21. The leaves are placed across the top of a pot. Although I have not identified these trees to the species level, I have heard Pacific botanists refer to them as the "stinky dysox."

22. A French systematist I met in the US National Herbarium told me he was glad they never made it to his interest, Tahiti, because they were too complicated. There are at least three genera from the Meliaceae family on Muyuw; Muyuw recognize their similarities.

23. My notes are full of cases like these—in some instances I have a voucher specimen determination, in others I don't. For example, there are two trees called *tobba* (Damon 96, Clusiaceae *Garcinia hollrungii*) and *baliuwew* (Damon 1986, Clusiaceae *Garcinia sp.*) that people sometimes confuse, but Dibolel framed them: *tobba* have soft wood, many roots beginning above ground, smaller fruit, and are found in freshwater swamps; *baliuwew* have hard wood, few above-ground roots, larger fruit, and are found on firmer, dryer soil.

24. This tree is one of three sources for the nets. *Gayas* (Damon 27, Genetaceae *Gnetum*) is the other main source. In north-central Muyuw, along with *kudukud*, a

Pandanus species Muyuw call *loud* is used. It is the source for the string called *im*, also used for ceremonial Kula necklaces, *veigun*.

25. Peter Stevens put these trees in different families. I then told him that the two were female and male of the same type by Muyuw reckoning. He found that very interesting. Asked why, given that he had placed them in completely unrelated categories, he said the forms had not been studied in a long time, so undoubtedly somebody would soon redo them, thus rethinking the criteria for understanding them and perhaps completely changing how we class them. The Muyuw recognition of a similarity by means of a transformation stimulated him.

26. Southeastern Muyuw people use this tree as evidence to assert that the trees in the Kaulay region are more typical of "wet" soils than their own set.

27. The word *kakin* translates to "understanding." It is a duplicated form of the verb "to see," *kin*.

28. My interaction with systematists bears this out. These experts become specialists, if not forever then over significant periods of time, in one or two genera. And while they know fantastic amounts of information about those plants, their stored data about others tends to be very abstract. With more care and using keys, Stevens could have furthered the determinations of my voucher specimens. But for many plants he just left the identifications at the family or generic level. By contrast, with the *Calophyllum* specimens I gave him, he was operating roughly on the same level of detail as most Muyuw people with those plants.

29. To the bafflement of Wabunun people, a European agent for government and mining interests was trying to connect everyone on the island by their genealogies.

30. See Damon 1983c, 1990. Separations in Muyuw kinship ideas generate the conditions necessary for the production of persons that centers the culture.

31. In the 1970s elders told me I would not have understood the usages of their elders from the turn of the century.

32. I believe these are Shiitake mushrooms (*Lentinula edodes*). Perhaps they grow on other trees, but they are only eaten off of three certain ones.

33. Peter Stevens could only identify it to the genus. Because of the importance of the tree, I asked the Harvard Herbarium to see if they could take it further. They sent copies back to Lae, instructing me to specify its identifier: K. Damas, January 8, 2001.

34. *Tawut* in central Muyuw, Kaulay village; just *ut* in Mwadau, western Muyuw.

35. The Pacific historical linguist Andrew Pawley questions this. I have followed the discussion of the verb **buku* (Ross, Pawley and Osmond 2003 Vol. I: 85). In Muyuw *patpatuk* translates the idea of knot. Considered a weak sector, knots as such are avoided for critical boat parts. I know of no linguistic or semantic association in Muyuw between "knots" in wood, and the word for "knots" for conjoining or tying something, the word for which is *sip*.

36. Correspondence with Andrew Pawley and Malcom Ross in February 2015 supports this. Ross believes *dabwen* is cognate to a common "Papua tip" Aus-

tronesian word often translated as "head" or "forehead," *deba* in Dobu, *daba/ dab'* in the Trobriands and Muyuw.

37. The idea of a person as a process has a bearing on the discussion of the anthropomorphized tree in chapter 5.

38. Damon 2005 explores these relations carrying the chaos terminology further than here: "Muyuw" and the "Trobriands" constitute two different phase states, the small islands between them illustrating confusion often found in phase changes.

CHAPTER 3

The Forest and Fire, *Tasim*, Inverted Landscapes, and Tree Meanings

"Trees" and Places

This chapter shows how people use the region's flora to organize its places. The "tree" *kaydibwadeb*—"limestone ledge tree"—illustrates how location is fixed in a basic understanding, in this case a name of a unique life form. Other examples are subtler. Traipsing around a hilly and swampy area in the south-central part of the island in March 1996, I collected an ugly-looking plant that I was first told was called *tanagau*; later the namer's father said its real name is *kaunuleyt*. When I relayed the amended name to the first person, he said he did not know the "real" name. After he had corrected the plant's name, the elder asked me where I got it. I told him, the distant hilly and swampy place called Kweybok. He then said, yes, *tanagau*, as if confirming the place made relevant the "incorrect" name.

It did. Places, positions, matter. "Kweybok" exists as both a real and imaginary location for Wabunun people. It is real; I've been taken there, in south-central Muyuw east of Sulog harbor nestled amid many sago orchards and proximate to well-known meadows. But I think it is also imaginary because it is said to be a place where people lived long ago. While there is evidence of habitation near there, Kweybok's actual location is a nearly impassable mélange of steep little hills broken by swamps and holes in the ground. Those holes sometimes contain sitting water, but more often rushing underground torrents circulating chaotically throughout this region. They merge into brooks or small rivers and empty into the south-central shoreline. At least three species of fish that live in the sea wander deep into those freshwater currents, making occasional appearances. Muyuw subclan origin myths are usually located in places that are inversions of contemporary locales—swamps, hilltops, tiny islands—and I think Kweybok occupies such a place for some Muyuw: the point is that we are not there now. Part of Muyuw peoples' presentation of their past has them living in the forest with no culture whatsoever. When people come across potshards and stone tool debris in the high forest, they take that as proof of their former state.

And this has a bearing on the name "*tanagau.*" It translates to "we commit incest." This herb-like plant—a species in the genus *Cyrtandra* (Damon 118)—is used in magic for acquiring women, kula valuables, or pigs; in all cases it acts on women because they are the ones who must decide if a woman, a kula valuable, or a pig are to be granted to a suitor. The plant is considered a female complement in a pair; the male plant is called *tanagau nakwav* (Damon 128, *Ophiorrhiza sp.*),[1] also a small herbaceous plant. *Tanagau nakwav* is found on exposed limestone ridges close to the sea. A subtle set of oppositions and correlations contrasts these two plants, putting them in a common female/male set. *Tanagau* grows far away in cool, wet places. *Tanagau nakwav* lives closer to villages or inhabited areas, and must withstand the high heat of its location—exposed limestone beaten by the sun. To be used, the two plants' leaves, "skin" (i.e. bark), and sap are mixed together with seawater and crushed coconut. The male suitor is then washed with the concoction, though a bit may also be inserted in a cigarette and given to the woman. Its magical effect wafts over and enters into her, whereupon she is overtaken with desire. This brings us back to the name *tanagau*, the significance of its place, and a common Muyuw rhetorical form. The effect of the magic is said to be so strong that if one is not careful he will sleep with a relative, prototypically a sister. That is to be avoided. Like much Muyuw magic, a major effect here is to heighten the subjectivity of the acting agent as much if not more than the person or thing supposedly acted upon. And the distance of the plant *tanagau* fixes this notion in the actor's mind: the concoction is so effective that it must be used only when far from close relatives—hence that distant location.

Another plant whose name and understanding are equally explicit and subtle is *bwaloudaskaygum* (Damon 233, Nyctaginaceae *Pisonia sp.*), a small understory "tree" with simple waxy leaves that spiral out of branch ends. The name refers to pigs (*bwaloud*), which eat the plant's (*kay*) leaves so that their (*as*) teeth turn black (*gum*). The usual word for "black," is *gunugun*, which somehow *gum* connotes. People, as well as pigs apparently, prized black teeth. Several plants were used to turn betel-stained teeth from a motley red to a brilliant black.[2] So here the plant refers to part of the behavior that becomes associated with one of its properties, the capacity to stain teeth. But there is more than that because this plant is widely known for its fixed spatial location, the *matan opweypwey*. There are a number of expressions like this: *matantalal* (roughly "end of the axe"), and the name of the village Unmatan, which translates to "village at the end." We shall have reason to visit the expression "*matantalal*," in the next chapter. The expression here, however, translates to the "end" or "point" (*matan*) of relatively dry land (*pweypwey*) right next to a swamp. The location is a liminal, ecotonal transition point.

Although the location of *bwaloudaskaygum* is implicit rather than explicit in its name, behavioral, locational characteristics are an internal component of the organism's understanding. Similar, though more paradoxical, relationships are apparent in another tree name, *niniwous* (Damon 191, Lauraceae *Cryptocarya sp.*). I was with middle-aged Wabunun men when they came across this well-known tree for the first time.[3] We were in the Sulog area just east of Sulog harbor on a peninsula called Kunsigan. Although nobody lives in this region, and there is no evidence of anyone ever having lived there, it is an important resource location. In addition to sago orchards, its natural and anthropogenic gaps are full of *legis*, a *Pandanus* tree whose leaves are used for making sails. (One named landing on the peninsula is reserved for Nasikwabw people who go there for sail-making materials.) As will be noted in chapter 4, the trees that provide the best masts also come from this region; I saw those trees when I came across *niniwous*, another peculiar resource.

While going through the forest, we came across a recently downed tree that, at first sight, my companions did not recognize. However, they soon saw seeds firmly attached to many branches and immediately realized that it was niniwous. They recognized it because the nuts, partially for their rich cinnamon-like smell, are important for medicinal purposes. If, however, they have never seen it before, how do they have knowledge of it? The answer to this question shows a way location is inscribed in tree knowledge, however unique it is to this case.

The nuts are found in very different places, on small islands that dot the barrier reefs on the south side of the main island, spits of sand found between Sulog and Nasikwabw some thirty miles to the south, and knolls of growth amid mangrove swamps that verge on miniature lagoons inside a ledge of coastal sand in the south-central part of the island. These areas are homes for a bird called *bwaboun* (probably a pied imperial pigeon), making the locations *kaylel*. *Kaylel* denotes a resource center of a special kind, in this case for the nut of the *niniwous* tree as well as nuts from one of the two *Canarium* species found on the island (*kinay*).[4] The birds roost in these locations because snakes (and lizards) cannot get to them.[5] According to Muyuw people the birds feast on the softer exocarp and mesocarp of the *niniwous* and *kinay* nuts by swallowing the nuts whole. They can neither crush nor digest the harder endocarp, so they excrete the encompassed seed back at their roosting places, the *kaylel*. Desiring the seeds to eat (*kinay*) or for use in medicinal or magical purposes (*niniwous*), people go to these areas to collect the nuts when they know they will be falling. Women of Unmatan, the first village east of a vast area leading toward the mangrove swamps fringing the Sulog Mountain area, usually take the lead in checking nut availability. Sometimes they collect them for other villages, other times they pass on word that collecting them would be worthwhile. Men readily gather the nuts from islands they land on while pursuing seafood. With sufficient experience, these facts are easily observed. Less observable is the assertion that the *bwaboun* bird usually eats the *niniwous* seed first. The nut plugs up the bird's anus, and only when it is near bursting with *kinay* does the bird evacuate its pile back at its roosting spot.

In this case, then, one ecological zone—the *kaylel*—mediates between the tree location and the zone of human understanding and use. Three places are specified in the construct: where the birds find the nuts and trees; where birds roost (to avoid snakes and lizards); and where humans are. If it is birds that connect humans with a distant zone whose tree they have never seen, it is humans who connect these discrete, partly observed patches and turn them into periodic yet enduring social relations. This is a fundamental pattern across this social system.

This chapter describes the shapes trees bring to the social structures across the Kula Ring's northeastern sector. By the end it will be clear that knowledge of patchy places—"an ecologically distinct locality in the landscape; it is problem- and organism-defined" (Winterhalder, 1994: 33)—is distilled information about distinctive human action.

Places, "*Kaynen*"

Although I was aware of spatially and ecologically distinct categories from earlier research, and quickly recorded the relevant principles again early in 1995, it was not until visiting eastern Muyuw's Kavatan in late April 1996 that facts forced a necessary reorientation. As noted previously, contemporary Kavatan arcs along the shoreline of a beautiful lagoon opening to the southeast. Leading away from the shore into interior are finely drawn bands of ecological zones. Most people class them as *aulekel* (sandy shoreline); *umon*, a sand-composed area just behind the reach of the usual wave flow and almost invariably covered in vegetation, capable of supporting taro gardens as well as its own regime of trees; *siposep*, a wet area with a completely distinct vegetation; another *umon*, in this case covered more in trees than formally domesticated plants (many of which might be trees such as breadfruit); a final, swampy *siposep* area; and then, in this specific case, the *dibwadeb*, the limestone-composed structure that now rises above the sea and leads to relatively dryer soils (*nibunab*) and, sometimes, deeper soils, *sepon*. In Kavatan's somewhat unique case, this last area begins with the oldest class of forest regrowth (*ulakay*), then the more frequently gardened area (*digadag*), which gives way to further *ulakay* to the west. Most everywhere else *digadag* come before *ulakay* before giving way to completely unworked areas beyond them.

The same regions exist along the southeastern shoreline, which I knew much better from experience. However, except for the five- to ten-meter-high *dibwadeb* rise to the dryer limestone soils, discontinuous pockets constitute the southeastern area rather than bands of distinct zones. In some places the limestone rise starts at the shoreline, hence there is no sand (*umon*). In other places *umon* is nonexistent because it has been turned into a village—as in the cases of Wabunun and Wayavat. Along the whole southeastern stretch there are clumps of *siposep*, freshwater swampy areas, some probably man-made. Although there are few trees of interest in these zones, two are important. One is used for house rafters (*anakan*), and the other, *vitivit*, is presumably a chestnut tree in the Fagaceae family whose nuts are eagerly roasted, boiled, or baked for food.[6] Until coconut plantations were built near Guasopa around 1900 and the US military put in its airstrip in 1943, the entire southeastern stretch of shoreline villages was bounded by two large swampy areas full of *vitivit* trees, both of which were conceived to be major resources. Now only minor *vitivit* patches spot the region between Unmatan and Guasopa. When in July 1996 I first learned that the row of villages between Guasopa and Unmatan was bracketed by the nut-bearing *vitivit*, my Unmatan informant told me these forests matured in the "base"/"top" order devised by the Creator. Hence the far eastern *vitivit* forest matured and was harvested earlier because the sun rises there first, whereas the one still existing and extending west beyond Unmatan matured later. Fully cognizant of the cosmology upon which this assertion drew, Wabunun sources ridiculed the idea, insisting that the flowering and fruiting had nothing to do with the Creator's order.

For each of these regions people refer to their botanical forms in terms of the expression *kaynen,* the *nen* a locative translatable to "place of."[7] So there are *aulekel* (sandy beach) *kaynen, umon kaynen, binbunab* (drier soil) *kaynen, koy* (mountain) *kaynen, siposep kaynen* (wet soil), *mwadog* (mangrove swamp) *kaynen*. One expression uses the category I translate as "weed," "*yol auyonen*," "seaweed place," a sea turtle environment through which people move on their way to other places hoping they might catch one. In one formal setting I recorded the term *wanawoud kaynen,* which could translate to "forest trees" and would presumably refer to the regions totally outside of human intervention. Although not in southeastern Muyuw, most everywhere else there is the category called *tawan* (*tawala* from Gawa to the Trobriands), referring to a region of large, high trees that grow on steep rises from the shoreline and that define portions of many of these islands.

Demiavek kaynen, the flora that typify sago swamps, are different than *siposep.* In addition to sago they have their own relatively unique flora. Much of the western side of Muyuw, west of its lakes, is a vast region hardly occupied at all. On modern maps the region is designated as if it is full of sago swamps. It is not. And I was told sago does not grow there. Wabunun people planted the tree in the swampy areas inland from the beach along the cove to their east, but they report it does not mature. In fact, sago swamps are confined to the regions on the peripheries of the mountainous center of the island. I presume some combination of geological and historical conditions created the distribution. I return to sago orchard structure later.

Beyond these forms, which primarily concern the constitution of a raised coral shelf, further distinctions are made. While reviewing land names around Kaulay village, I learned that people recognize patchy distributions of trees more or less within the same ecological setting, and exclusive of differences generated by fallowing cycles. Some of these effects are realized in names. As noted in the last chapter, there is an important tree called *aukuwak* (Damon 60, Rubiaceae *Ixora* cf. asme Guill). For a reason nobody could explain, a concentration of these grows on a small island half a kilometer east of Wabunun called Simkuwaku, "island of *aukuwak.*" People had a proportional way of talking about tree distributions around Kaulay as well as Wabunun: land zone X had more of tree 1, land zone Y had more of tree 2, and so on. Although the important tree *meik^w/ kaymatuw* (*Intsia bijuga*) is found in many different regions and fallow types, most people can designate regions where it grows in near grove-like conditions. Several places were so designated for Kaulay; when I asked Wabunun people about this tree, they specified an area, not exclusively theirs, a somewhat distinct *umon*-like region called Lawadog west of Unmatan. There is a real and imaginary path going north by northeast from Wabunun to a spring called Lulubwau. East of that path, more mature forests have many trees called *antunat* (*Palaquium sp.*); to the west of the path they are less common. The tree, in any case, tends to be found more toward the shoreline, because, according to one knowledgeable source, it seeks seawater. The Euphorb *akobwow* (*Macaranga tanarius* [L.] Muell.-Arg) tends to be found in the early succession areas, *digadag,* of a region near Wabunun called Bweybwayet but not in the same fallow stage only a kilometer west of there in the region called Wasimoum. I have mentioned the high concentration of the *ukw* tree leading to a region being named for it. A patchy distribution of the older successional *tabnayiyuw* is also recognized. A zone defined by an exceptionally high concentration of *akuluiy*, runs for several kilometers from the Sinkwalay River southwest toward sago orchards. Those who know about this concentration call it a *venay* or a *lapuiy* to reference it. *Lapuiy* are lines from east to west (usually) in Muyuw gardens used to define the garden's smallest unit of food, a *venay*. Although this tree can be found in greater or lesser numbers in many ecological zones, this concentration was easy to experience—if virtually impossible to explain—one day as we walked in and out of it.

While these differences are articulated landscape features, people do not worry about explaining them, even though explanations might be hypothesized for some of them. *Antunat*, for example, is the tree of choice for the strakes of the middle-sized class of outrigger built especially in southeastern Muyuw. In my time they were built in any village between Kavatan and Unmatan. However, until its collapse around 1900, the Kweyakwoya region, which once existed across the northern ridge of southeastern Muyuw (see Damon 1983b), was the craft's primary producer. And that might be why the *antunat* tree was concentrated there. This hypothesis, however, would not really explain concentrations of other trees even though all are very useful: the bark and inner bark from the roots of *akobwow*, for example, are the prime source of caulking material Muyuw people use for all their boats. Another important tree for outriggers grows to enormous sizes only along Muyuw's western and northern shorelines, as well as on some of the small islands like Yanaba, Gawa and Kweywata, and Iwa. Later I return to this tree, *kaboum* (*Manilkara fasciculata* [Warb.] H.J. Lam).[8]

Relations like these are not, however, just observed. They are conceived and so turned into ways for talking about social relations. A prime mechanism here is an encompassing opposition detailing how trees are distributed over the landscape. Trees are either *kalow* or *man ankayau*. *Kalow* means "fall down." This refers to those trees that reproduce primarily by their seeds falling and then sprouting where they are. There is often a source, a "mother" (*ina*) tree. And, at least in the prototypes, the trees tend to be found in groves. In the next chapter I discuss several *Calophyllum* that fit this description. In all likelihood, the swamp-defined groves of *vitivit* that bound the line of villages from Guasopa to Unmatan, and appear in much smaller wet areas between these places, were distributed by humans. But that is not how they are conceived now; rather, they are understood as *kalow* trees that come up where they are from their own seeds. By contrast, *ankayau* refers to seeds that are spread far and indiscriminately by *man*, a category that, in this context, usually suggests what we mean by birds but also includes small mammals. "Birds" eat tree fruit then excrete their seeds at unpredictable distances from the mother tree. It is said that nobody knows where the tree came from, other than from "bird" activity.

The prototype for this relation is a tree called *aunutau*.[9] The tree turns up in a wide range of areas, across all fallow types and the island's geological variation.

So fixed an idea is the *kalow/ankayau* contrast that subclans are distinguished by whether they are like "*kalow*" trees or "*ankayau*" trees, the former having easily traceable histories while the latter do not (probably because of warfare, slavery, or defaulting on ritual payments). One significant elder from a latter group named his truck "*ankayau*."

The Fallows

If the patchy distribution of some trees might be first ecologically determined, or so removed from human agency as not to be considered anthropogenic, many other divisions are human products. The issues here bring us to the relations between forests and gardens, ideal fallow patterns across these islands, and the all-important concept, *tasim*.

The three fallow classes, *digadag*, *oleybikw*, and *ulakay*, were introduced in chapter 1. The categories are known and used all across the northern side of the Kula Ring. It is understood that these categories refer to differences in soil quality, numbers and sizes of trees, the kinds of trees that typify each area (including the kinds of trees that grow up when it is cut down), and, consequent of all of these factors, the quality of the food grown in them. Muyuw people can determine what fallow class yams and taro are grown in from their taste and texture.

Trees from the genus *Dysoxylum* (e.g. *akewal*, *tabnayiyuw*, and *silamuyuw*), and *Sterculia* (e.g. *ukw* and *asilawow*) are not found in *digadag*, only *ulakay*; *Rhus taitensis* (*gwed*) is never found in an *ulakay*, and only once did I find a lone and enormous *Alstonia brassii* (*atulab*) in what was classed as an *ulakay*.[10] More trees are found in *digadag* than *ulakay*, but those in the latter area are much taller, in the vicinity of forty meters high, and thicker, depending on age, ranging up to fifty centimeters in diameter at breast height. *Digadag* canopy tops are on the order of twenty meters high, with few trees more than fifteen centimeters in diameter at breast height before they are cut again.[11] There is also a question of repeated usage. Repeated cutting inside the five- to fifteen- or twenty-year period is what makes a *digidag*, and in the last analysis it is this repetitive practice that changes an area's classification, creating a relatively unique biome. Once in Kaulay, I noted that a certain patch of forest was an *oleybikw*, because of the way it looked and abutted another area that happened to be one. However, my guide then immediately corrected me because the region had so few *mamina* trees in it. That meant it was an *ulakay*, because most *oleybikw* there are filled with the important *mamina*.

If these categories seem to hold the same meaning moving west from Muyuw, in fact the relations between villages and forests in these places differ. Consequently I review the situation on Muyuw first before turning to the other islands.

Distinctions among different fallow regimes become clear in those cases where the types are confused. One day I traveled west of Unmatan to study the floral changes evident from there as one headed off the southeastern uplift into a mixture of salt and freshwater swamps, meadows, sago orchards, and high forests leading to the rolling hills broaching the Sulog area. My guide led me to a place called Pwanlikilok. *Likilok* is the Muyuw name for trochus shell, *pwan* translates to "buttocks," and referring here the broad base of the conically shaped shellfish. This geological formation looks like a "trochus shell" rising up out of a swamp. It is a circle ten to fifteen meters across and several meters above a different ecological niche. Because the place is frequently used as a camp for nearby sago orchards, it is often cleared. It has a suite of trees, although small ones, that reflects its contexts, surrounded by *oleybik-* or *ulakay*-aged forests[12]; its top also evidences trees regularly found in *digadag,* which my guide then, Deneig, pointed out to me before I realized it. I recorded twenty-five plants on or near the top. Six were from *digadag,* one a grass sometimes found in digadag, almost always in meadows.

Although almost every place uses all the fallow classes, virtually every village or set of villages understands itself to follow, ideally, one fallow regime. There are some exceptions to these generalizations that I turn to shortly. Also, and ultimately, the ordering principles entailed here extend beyond the issue of garden fallows. However, the basic facts are simple. Coastal southeastern Muyuw concentrates gardening in the shortest fallow regime, *digadag.* The north-central Muyuw standard is *oleybikw,* the middle-length fallow. And western Muyuw people say they key to the oldest fallow, *ulakay.* Running along these divisions are those places that are not really thought of as gardening regions at all. This includes Budibud, to the east of Muyuw; the Sulog region in south-central once devoted primarily to stone-tool production and sago; and the sailing villages on Nasikwabw, Yemga, and Boagis. Nasikwabw and Yemga are distinct islands, Boagis is located on the spit of land extending south on the area's western appendage, Mwadau Island. It follows from this that in place of gardens, coconut trees are intensively planted in the Budibud area, along with other fruit- or nut-bearing trees like breadfruit. Sweet potatoes and tapioca have allowed Nasikwabw and Yemga (and Iwa) to garden during the twentieth century far more than in the past, while only since the 1980s have Boagis people taken gardening seriously; their model is that provided by Mwadau village to their north. Boagis's immediate surroundings are swamps and compressed coral limestone outcrops; to garden people paddle north into the region just south of Mwadau's customary area.

Over the course of my time on Muyuw, someone has always tried to employ traditional Muyuw gardening techniques in the Sulog area, always with problematic results. In fact, people tend to live off nearby sago orchards, or the labor of friends and relatives elsewhere along the southern shoreline. Wabunun people once joked that they knew when it was time to harvest their yam gardens because, invariably, Sulog people would arrive to "beg" for food. This is not an isolated joke about Sulog gardening endeavors.[13]

In chapter 1 I noted two apparent anomalies with regard to the use of the *gwed* tree. North-central people knew the ideas but ignored them; and my western informants denied the claims with respect to the tree altogether. These gaps follow precisely from the organized fallow ideals: *gwed* trees are all but nonexistent in *oleybikw* fallows, especially in north-central Muyuw. They are completely gone from *ulakay*. But there are at least two more sets of relevant facts. First, as also noted previously, the tree that occupies roughly the place of *gwed* for both these areas is found only in their respective ideal fallow classes: the *Ficus kubwag* in the case of the Kaulay region and the *Dysoxylum silamuyuw* in western Muyuw. *Kubwag* were the largest trees in the Kaulay regions where I found them, and until I received the biochemical results reviewed in chapter 1, I operated on the idea that the trees were favored by virtue of their size. Based on other "magical" practices, a like-for-like reasoning principle seemed in play.

Yet there is another factor here for the Kaulay-to-Dikwayas region. The favored tree for floats for the largest class of outrigger canoe is one called *vayoun*,[14] and it derives from *oleybikw* fallows. Although a detailed discussion is reserved for chapter 6, note that a contributing factor to the unique north-central Muyuw fallow ideal is that the practice generates a *regionally* significant product.

The same explanation leads to the western Muyuw preference for the oldest fallows. As noted earlier, that pattern was a surprise because, it seemed, *digadag* gardens dominate all of the regions west of there and I was expecting the area to be more like them. But the apparent anomaly disappears with further information. According to Muyuw people, temperature and moisture content of their fields varies with the fallow class. The garden soils derived from old fallows are said to be cooler and moister than either *oleybikw* or *digadag* (partly because of what has happened to it by being in that state, partly because a garden cut out of an *ulakay* remains contiguous to the same category of forest). And this becomes an important factor when these regions are hit with major droughts, as they are with every ENSO event. The following observations follow from this set of relations.

Mwadau Island is much smaller than the main Muyuw landmass, so it has a much smaller underground freshwater lens. People must be more careful about protecting their crops, as both food for the present and seeds for the future, than the people to the east. They also do not have sago orchards as an emergency food source. Based on the dynamics of their own area, it is clear that working *ulakay* has significant advantages. But more important than western Muyuw's specific condition is its relation to the islands to the west.

The small islands from Gawa to Kitava rise straight up out of the sea, nearly two hundred meters high in Gawa's case, declining to less than a hundred for much of Kitava. Although there are tiers to the rises, with each tier being gardened, most gardening occurs in the oval concavity centering each island. As is the case with the tiers, the gardens are all early fallow. And when I asked people about *oleybikw* and *ulakay*, I was told they no longer had any. This surprised me because the steep rises for all these islands are dominated by tree kinds and sizes that I knew of as *ulakay*-like in Muyuw. It was then that I learned the category *tawala* (*tawan* in north-central Muyuw). So, as viewed from the water, all these islands look like they are densely forested, and of course these forests produce valued trees that can only be found in older successional stages. However, the gardening on these islands is all done in areas where the growth is early fallow. Since the flat areas on these islands are too high to be moistened by the islands' underground water lens, droughts severely stress these communities. This explains why Mwadau Island peoples focus on *ulakay* gardening. Because the small islands necessarily produce vegetables in the most precarious circumstances, the Mwadau region moves toward the opposite pole. And this, I suggest, explains why Iwa people told me to go to Muyuw if I wanted to know more names about types of yams and taro. Those names have to do with variations among types, thus knowledge requisite to reproducing seed stocks. That is Muyuw's obligation, for all the islands from at least Gawa to Iwa. During the 1997–98 El Niño event, the small islands lost all of their vegetable crops. I returned to Wabunun during July 1998 after the drought was over. Although they had lost much of their taro and were harvesting diminished yams, people told me exactly how many "baskets"—or, sometimes, the *venay* unit of their gardens—they had sent west to Gawa and Iwa. The situation in Mwadau village was different. People told me they were having a good harvest. In spite of the severity of the ENSO drought, it did not seriously affect their yam gardens (probably a result unique to Milne Bay Province). Yet they had sent so many yams off to Gawa, Kweywata, and Iwa that they were not counting them. Knowing that Muyuw people count everything that passes back and forth in interisland or intervillage exchanges, I protested; the response was that the issue was starvation and death, and that the return would be future Kula activity. Although there is more to this story yet, in short the informant skipped from the immediacies of a lower ranking exchange sphere to the one that totalizes everything—the *kula*. Mwadau area garden practices reproduce *the conditions necessary to maintain the small islands to the west*. The 1997–98 El Niño exposed the limits of the conditions for reproduction of the region and made plain one of its features of organization: it is organized for those limits. Just as north-central Muyuw's fallow regime is partly designed to function with respect to regional realities, reproducing a critical canoe part, so western Muyuw's practices receive their sense from a larger set of regional relations. Landscapes are regional relations. However, this is an idea that did not become fully fleshed out until my focus shifted to outrigger canoe structures, to which we return in chapter 6.

Tasim

Early in 1995 I learned another new, important fact about how gardens were related to fallowing stages and processes: they are cut with respect to plots of trees that should be left uncut. The uncut areas are formally named with distinctive ecological functions. These areas are called *tasim*, the word understood and employed all across Muyuw. West of Muyuw, its conditions are approximated, though I do not know if the idea exists for it came into my focus after I surveyed those places. Passing by Koyagaugau in the southeastern sector of the Kula Ring in 1998, a Muyuw friend showed me how *tasim* worked there. However, in 2002 I figured out what he was using to map that landscape, and I became unconvinced that the local practice conformed to his model.

The "*tasim*" form presupposes a clear-cut distinction between villages, gardens, and forests. Such relations are visible in the Trobriands, and I shall comment on their patterns shortly. On the small islands of Gawa, Kweywata, and Iwa there is no clear-cut distinction among villages, gardens, and "forests"—these relations are all mixed up. The steep inclines marked by high forest on those islands, however, replicate relations that *tasim* effect on Muyuw. One of these concerns the harvesting of alternative forms of protein. Among others is the bird *Merops ornatus* (*kelkil* in Muyuw, *vakiya* in Iwa), discussed in chapter 1. It flies into the high trees where it rests until it dives into the more bug-infested *digadag* areas. In Muyuw the bird plays off the same relations but is not eaten. These relations imply the bird's migratory patterns take advantage of—are caused by?—anthropogenic relations.[15]

The *tasim* idea is well formed. According to the indigenous explanations, the word is composed from the verb meaning "to cut" or "chop," -*tay*, and the word for "island," *sim*—thus an island-like structure made by chopping around an area. The uncut trees remain high like an island stands above the sea. Many *tasim* in southeastern Muyuw, and at least a few in central Muyuw, are further defined by the fact that they are *sasek*. Although coral limestone is found everywhere on Muyuw—and other islands—some paces are so filled with the material that there isn't enough dirt to garden effectively. These are *sasek*. As shall be detailed in the next chapter, one of the crucial *Calophyllum* species is related to these units. Once while exploring an *ulakay* area, my guide and I came to a virtual grove of that species. He then stopped, looked around, and said the area was a *sasek*. In north-central and western Muyuw, *tasim* were not always fixed to *sasek*.[16]

Recall from chapter 1 the Creator-inspired rectangular order of a Muyuw garden, an order fixed with the deliberately marked constructions at the intersections of the east-west and north- south paths, the *pwason*, navel. Recall too that these garden spaces are likened to boats. People consider these "boats" moving around the landscape, diverting from "islands," *tasim*, like real boats must dodge protruding reefs and islands. While this is not the only model people have of gardening activities over time, the image should be kept in mind for later chapters that outline technical knowledge about boats. Profound questions of order lurk here; it is no accident that the Creator who brought this order arrived in the largest class of outrigger canoe, the *anageg* form.

These structures resemble what are often called "sacred groves" in Southeast and South Asia (Uchiyamada 1998 and Freeman 1999), and they are part of a productive order said to derive from the Creator Geliu. However, the primary way Muyuw talk about them gives them an ecological rather than overt "religious" sensibility.[17] Before listing their positive associations, I must note that now, at least, they are an ambiguous part of the landscape. They are havens for birds, especially cockatoos that prey on crops like sweet potatoes and tapioca. These crops have probably only been significant for a few hundred years, but the way birds prey on them seems to have intensified over the course of my research. By 2002 some people were refusing to plant them; others, especially, youth, were beginning to think they should ignore elders' commands to preserve *tasim* to dispense with the bird problem. For other reasons, Muyuw people who have lived away from the island for years encourage a more open than mottled landscape than was once, and for many elders remains, the ideal.

The negatives aside, southeastern people spoke positively about these areas for the following reasons. First, *tasim* keep the land cooler and moister, a good outcome. Second, they furnish a seed store for *oleybikw* and *ulakay* trees. These are thought to be stronger than *digadag* trees. Although plenty of useful plants come from *digadag*, more come from *oleybikw* and *ulakay*. For the latter, of special note are support timbers (e.g. from the tree *aku-luiy*) for houses and more significant trees for boats. Vines for tying fence stakes and yam and village houses are found in *ulakay*, so having these in closer proximity is advantageous. Third, their presence serves to block out the sun. This has two effects, both conceived positive. It is said that both planted crops and new trees strive to get more sunlight, so they both grow faster and taller by virtue of the existence of *tasim*. By growing taller, crops are presumed to produce larger or more food. And by the trees having to grow faster and taller, they are thought—whether the issue is becoming a mature *digadag* or transiting to older forms—to more quickly reproduce the soil's strength as well as more useful resources (fence stakes, yam poles, various trees used in boat construction).

The last item explicitly associated with *tasim* is the island's famous endemic cuscus, *kwadoy* (*Phalanger lullulae*). According to Muyuw, these animals' preferred foods are the young growth of *digadag* in general and in particular nectar from the vine called *weled* (Damon 72, *Flagellaria sp.*) and the *gwed* tree. After both plants flower, it is said, the animals become fatter and more desirable. A substitute for fish when winds make fishing problematic, the animals are a source of meat and grease. Cuscus are nocturnal animals and sleep in high trees during the day; at night they descend into the fallowing areas. People hunt them during the day by climbing up "*tasim kaynen,*" the trees of *tasim*, which are the animals' favored locations for sleeping. A list of the trees people look in first includes those regularly found in *tasim*: the *Calophyllum kausilay,* two strangler figs noted in chapter 2, and a few other canopy-top trees. People carefully examine the places where they are likely to be sleeping—among various long-leaved epiphytes in the canopy—and then use long poles to knock them down to a waiting group of men or boys.[18]

Tasim found in southeastern Muyuw tend to be smaller clumps of trees, several hectares or more apiece. As the pictures in the website accompanying this book show, they sustain a mottled, patchy appearance to the southeastern garden region. In north-central *tasim* near Kaulay I sought out the same units with the idea of conducting plant surveys in them. They certainly exist, and plants were surveyed in a ten-meter square in two of them. However, by the time of my 1996 return, the second and third new visits, I realized the landscape had a different shape. From talking to people about their past and projected future garden plans, it appeared that people often worked down lines of (usually) *oleybikw* forest, leaving equally long avenues of uncut forest swaths between the implicit avenues. Although twelve months apart, two episodes with the *Syzygium* tree *mamina* solidified this model. In 1995 my older informant, Vekway, pointed to a long line of *oleybikw*-sized trees (the canopy top approached thirty meters), saying that when they all flowered they were a beautiful sight. In the survey of tree stumps conducted in that garden, there were many *mamina*. But the movement of the presumed new gardens was not to the south toward that apparent grove, but rather west continuing along a line of *oleybikw* already traversed for five or more years. The next year, several kilometers north of the location visited in 1995, I walked through a recently used garden area in the second or third year of fallowing. People were still regularly visiting it to hack firewood out of the drying boles, often *mamina*. It would be fifteen or more years before that tree would even begin growing back into that area, but it was evident that the garden was still being harvested—for energy. That night I diagramed a model of strips of gardens paralleled by strips of *oleybikw*-forested *tasim*. People looked at each other and said, "He knows" ("*Bwikakin*"). The next question was obvious: because many active gardens moved along these apparent lines either to the west or to the east, was it the tradition to follow these directions, an orientation consistent with the culture's paramount sunrise/sunset axis? To my surprise, informants denied that this is what they were supposed to do. While the empirical information rehearsed was correct—some people were moving to the west, others to the east—they said nobody did that in accordance with the Creator-inspired orientations governing much other behavior. One man was and had been cutting gardens out of a *digadag* class of fallow, just on the edge of an *oleybikw* class that he would now and again infringe upon to expand the region he could call *digadag*. He was cutting his gardens by following a roughly circular path. He told me his father (then recently deceased) had told him to stay there and garden in that fashion. (By 2002 he had abandoned that area for an older *oleybikw* area bordering on *ulakay*.)

While Kaulay people mixed gardens and *tasim* as in southeastern Muyuw, perhaps because of the different fallow ideal the whole looks different, often long rows of uncut trees. A similar minor change turned up in western Muyuw's Mwadau village. As already noted, the ideal is to cut gardens out of the oldest fallow class. Although several examined gardens were *digadag*, others were classed as *ulakay*. Having less time there to survey than elsewhere, however, more of my information comes from direct questions than casual conversations and extensive walking. Yet from the latter I quickly noticed many gardens tending north or south over consecutive years. In one place sufficiently large for people to be going in haphazard directions yet where the north-south movement seemed evident, I outlined what was going on, trying to make my diagram completely random. I was immediately corrected. Not only were the people moving north—after several years of cutting to the south, they were doing so by following what is supposed to be the plan there. My informants understood this way of cutting gardens to be following the Creator-devised plan for the area. As noted earlier, in western Muyuw the Creator takes a north-south detour up and down Mwadau Island's axis, so the garden path running east to west back east runs north to south there. So in western Muyuw there is a tendency to generate rows of gardened areas alternating with rows of fallows turning into the *ulakay* class, both orientations fixed to an ideal north-south axis.

Before we turn next to the small islands to Muyuw's west, a note on the Trobriand situation is appropriate. Although my visit in 1995 occurred before I was aware of *tasim*, from Trobriand aerial photos and excursions made with Linus Digim'Rina's central Trobriand village Okaibom, the situation seems clear: tall, economically significant trees surround often-circular villages. But beyond that band of trees there tends to be vast garden areas mostly in *digadaga* fallow stages. By Muyuw standards, these are exceptionally well defined in terms of fields and garden units. Radiating lines of coral limestone collected into piles or lined up into tracks through the years outline the forms—the *takulumwala* noted in chapter 1. These are rarely broken by *tasim*-like clumps of higher forest, though an occasional isolated tree stands out. Only toward the shorelines are high *ulaka* forests evident. Along the higher eastern ridge of the main island, these areas are called *tawala*.

As in the Trobriands, in Muyuw there are clear-cut distinctions between villages, forested areas, and gardens. The only difference between the two places is the mottled appearance of garden regions, broken as they are by *tasim* in Muyuw. Kitava closely resembles the Trobriand pattern because quite distinct villages lie just inside the stockade-like line of limestone that encircles the top of the island. But moving from the villages toward the oval center, one quickly passes into open fields dominated by kunai grass (probably *Imperata cylindrica* [L.] P. Beauv) rather than *digadaga* regrowth evident on most other islands. However, there is one pattern evident in Kitava hardly at all witnessed on either Muyuw or the Trobriands. It consists of small, invariably fenced gardens inside of or contiguous to well-defined villages. And this pattern is the essence of Gawa and Iwa (and presumably Kweywata).

As noted in chapter 1, before visiting Iwa in 1995, I was told by Wabunun friends the place was so bereft of trees that people were forced to save their yam poles from one year to the next. My visit was in August, amid replanting of new yam gardens. So I quickly found and photographed stacks of *avatam*, poles left from the previous year's yam gardens for the next season, thus confirming Muyuw projections. But the impression the Iwa landscape created was opposite of what I expected: it seemed jammed with trees. Many trees that would seem to be "wild" on other islands are planted on Iwa. Both of the *Calophyllum* I collected on Iwa were planted. Inside Iwa "villages" were little sets of stone among which many plants were set. One of these, in fact a set of megaliths called Duwag toward Iwa's northern end, had twenty-two different plants, some medicinal herbs, others trees, still other plants providing different qualities of tying materials. As gardens were full of tall trees, so villages took on the appearance of forests. Moreover, as in Kitava, though on Iwa more pronounced, there were small fenced-in gardens in what appeared to be the two organized villages. Later I learned that the organization into villages was from government intervention, and that the traditional pattern was for individual houses or several houses to be dispersed throughout the oval-shaped island, a pattern confirmed in 1996. Then scattered about the island I found hard and rot-resistant *bwit* posts, *dibedeb*. They had anchored ceremonial centers for the houses and hamlets that at one time were more widely distributed throughout the island.

Iwa is an extreme case of intensively manicured landscapes, a place of arboriculture rather than horticulture and a familiar pattern among small Melanesian islands (see Kirch 1989). This is the other side of the water issue that becomes critical with El Niño droughts. With far deeper roots than vegetable crops, Iwa trees are more drought resistant than yams and taro, and they provide a product that is exchanged with the other islands. Visitors (from the Kula Ring) to Iwa expect to depart with liberal supplies of tree products, most especially the soft-shelled *saido* tree nuts (*T. Cat-appa*), a variety the island's proud inhabitants claim is unique to Milne Bay Province. I received the same assertions about its uniqueness from Gawa and Muyuw people. *Saido* trees everywhere else (with one exception!), including those on Gawa and all across Muyuw, are hard shelled. Once cooked, they must be beaten with a stone or axe to be opened.

This brings us back to my 1998 visit to Mwadau. My informants told me that they had given so much vegetable food to Iwa they neither counted it nor expected an explicit return. A major factor in not worrying about a return was the severity of the crisis. But the other side of that exchange was that every time somebody passes back to Muyuw from Iwa, they carry baskets of *saido* nuts, only apparently unreciprocated—until the next drought.

Gawa—and I was told Kweywata—partially resembles Iwa in that it features relatively large numbers of fruit- and nut-bearing trees planted inside its considerably larger oval and concave interior. However, Gawa's more important development of tree resources relates to its primary function for the whole eastern side of the Kula Ring: producing the largest class of outrigger canoe, *anageg,* or, as it is said in the southeastern corner of the region, *kemuyuw,* "wooden thing from Muyuw." I shall return to these relations in the next chapter.

A conclusion needs to be drawn from these facts. Although there are neither *ulakay* nor *oleybikw* gardens on the small islands of Iwa, Gawa, and Kweywata (or Kitava), the islanders' primary productive processes do not derive from *digadag* fallows. Since their tree crops tend toward the permanent, it may be suggested that their reality is the next logical step beyond western Mwadau's *ulakay*: permanent tree coverage. This accounts for one of the details Iwa people told me when they explained why they planted *gwed* next to their crops. While some people provided the model I received most everywhere else—the tree puts sweet things into the soil—several people said the practice was to grow a *digadag* fallow along with the garden, the logical limit to the interval between horticultural practice and continuous tree coverage.

All of these places perform variants on slash-and-burn agriculture. In contrast to the Asian systems that mold land with water, here spaces are made by means of fire among growing trees for horticultural, and other, activities. These fires are availed by chopping down a set of trees from an area the size of the intended garden. But a condition for that fire is a sufficiently long gap of dry weather to facilitate a burn. We experienced difficulties with this during the—presumably—La Nina phase of the early 1990s ENSO. Both 1995 and 1996 were so wet—conditions that have continued on and off up through 2014—that people complained that they could not burn, and therefore plant, in customary ways. But there is more to this hoped-for discontinuity than this. As explained elsewhere (Damon 1990, chap. 2), Muyuw believe periods of wet and dry weather follow the setting of significant stars or constellations. About twelve important named stellar units spread throughout the annual cycle. With each heliacal setting, when the "star" is no longer visible at dusk on the western horizon, a period of stormy wet weather follows because the star releases a power. An attentive gardener should capture this power, realized in rain. So an area is cut to take advantage of an expected dry period before the next heliacal star setting and power-induced rains. Planting likewise should take advantage of this periodicity so that a new crop is in just before a star sets. It then gets drenched by the next rainy period. Although more than once a stormy period was explained by some star having just set, I met nobody who took this star pattern literally. Yet it is a figurative model of actual and necessary discontinuities intrinsic to the production process. As the landscapes are mixes of growth, *tasim,* and plots of land under cultivation or in fallowing stages, so the weather patterns are understood to be patches of difference. Productive action is conceived and coordinated with respect to these discontinuous points, the imagery of which enshrines the most complex artifact the culture creates—its boats.

Sinasops and *Demiavek/Dum*:
The Inverted Landscapes of the Sago Orchard System

Idealized fallow regimes and *tasim* were the first unexpected forms I learned about as my researches started to become synthetic. I now turn to another practice that gradually crept up on me—and is creeping still. I first experienced this practice as meadows in high forests; then I was pulled into the relationships of these landforms to sago orchards. The meadows are called *sinasop,* and most of them are named. I walked through an early mining-era and mining-caused meadow in 1974 and then in 1982 but discounted it as a significant (Muyuw) social fact. I experienced the meadows first in 1995. In early 1999, Cornell soil scientist Shaw Reid flagged the two soil samples I had taken from them as extremely acidic and peculiarly loaded with aluminum. Only in 1998 and 2002 did I began to appreciate them in relation to forests and, often contiguous with them, sago orchards, *demiavek* in eastern Muyuw, *dum* in central Muyuw. My 2009 return facilitated a more careful examination of both the meadows and adjacent sago orchards and provided an opportunity to collect soil samples from them.

In 1998 I realized that many meadows are anthropogenic.[19] In 2002, when I insisted my companions and I walk through one, Bungalau, I discovered a sago tree growing in the middle of it. No surprise to them, it is called *Geliu an yabiy*, "the Creator's sago tree." It must not be harvested. It grows, matures, and dies, then another grows back; it was dead in 2009. With the meadows directly or indirectly contiguous to sago orchards, the appearance of this tree correlates but also opposes these forms: sago trees outside the meadow but inside named and defined orchards may be and are harvested, while this one, outside an orchard but inside a named meadow, must be left by itself. Its isolation is like the garden forms erected when a new garden is designed; also from the Creator, they must not be touched during a garden's life cycle. This relation to gardens is not my imposition. In 1974 and 2002 my companions and I camped on the edge of one of these meadows, and each time they found "wild" yams growing on meadow borders that they harvested, cooked, and we ate. But the interrelation is not just that; sago orchards as food sources temporally and sociologically complement gardens. Meadows, sago orchards, and gardens are a dense network of social facts realized in biological form, the description of which is the purpose of this section.

Lying beyond my grasp is one more issue: the relationship to water. Sago is a swamp tree, and all the Muyuw orchards are in swampy land or ravines coming off the island's central mountains, keeping the trees in, usually, saturated soil. The southeastern orchards through which I've traversed are in the Kweybok area noted earlier, a mélange of rolling land meshed by underground streams that sooner or later turn into small rivers—*lituk*—that evacuate at the south-central shoreline. These rivers form the usual transport mode into many of the orchards. Tidal flows affect much of this area backing up and draining a vast portion of the region. Although the hydrological dynamics in central Muyuw are different—most of its orchards are too high to be influenced by tidal dynamics—people note that strong rains flush water from the meadows immediately into the ravines supporting the orchards. When I asked a Kaulay instructor if there was a burn-and-fallow stage with the sago orchards analogous to that with gardens, he said no because the continuous water flow into and through the sago beds keeps the zones in continuous production. After I passed through the southeastern orchard and meadow area in 2012, I was told the area near Kweybok and proximate to the meadow called Bungalau has many holes open to rushing water that eventually merge into a stream called Alanay. That stream flows through much of the orchard area, one orchard named Alanay after it. One of the holes is called Nabulkwakwit. *Kwit* means "octopus," most people agree. When I heard this it just seemed like noise; I do not think it is, but is instead part of a set of conceptions that embed this area in meanings, and a consciousness about hydrological conditions. A strong hypothesis is that the flows of water and nutrients through this region are similar to those that define Asian irrigated rice agricultural systems, thus one complementary contrast to and for the cultural systems in the northwest and north.

An emerging literature on sago (especially Ellen 2004 and 2006; McClatchey et al. 2006) guides the organization of my description; in another work I address this literature in more detail. In this section my purpose is to describe sago orchards and meadows, their functional interdependencies, and their relationships to the *tasim*/garden relationship. This relationship completes, by complementing, the configuration of social relationships chapter 1 outlines for gardens. I suspect, however, that I am describing a particular instance of a much more common pattern so that our appreciation of traditional sago production must be encased in wider contexts than is currently the practice.

I had to learn that *tasim* entailed relationships to fallows, fallow lengths, burning cycles, and gardens, and of course the production and consumption sequences and transformations all these implied, and modeled. Boat imagery integrates these unique forms. Analogous, if inverted, relationships pertain to the meadows and sago orchards, but I had to learn much about each before I realized they were in relationship to one another.

Perhaps an anecdote that echoes through my mind will serve as a representative: My 2009 host, Dennis Dibolel, the son of my primary host over the last twenty years and grandson of a major informant and teacher, Aisi, was a successful student on the island in the mid-1990s; by the early years of this century, he was away in a school in Alotau, Milne Bay Province's provincial capital, training to become a mechanic. Occasionally his mother would tell me that if he failed out of school and its opportunities in the *dimdim* (white man) world, he would come back to "make sago." Raised as I was in the post–*Coral Gardens* world, and having made my living describing coral gardens in another Kula Ring sector, I thought gardening and fishing was the central practice, and sago-making and its associated arboriculture peripheral to it. Yet my prejudices might reflect an East Asian Austronesian intrusion on an otherwise original, and Melanesian, tree-defined world. Dennis's mother may have had the historically and logically precedent point correct: sago-making here is central. It is what men do.

First some basics about Muyuw sago orchards and their harvesting process. The orchards I have seen in central Muyuw, and at least several in the southeast, follow meandering streams. The orchard called Alanay arcs around the edges of a low hill called Tadimuk; it was easy to see how frequent downpours would course down the hill into the orchard. Although not readily visible, orchards are distinguished by a *wowun/kunun (matan)* distinction that is conceived to be identical to the axis that defines a garden. Gardens clearly follow the sun, going from east to west, but this seems rarely to be the case for orchards. However, they are divided in ways similar to the garden divisions called *venay*, but take a different naming device, *kali-*: so *kalitan* (1), *kali* (2), *kalitoun* (3), and so on. Other trees mark these clumps of sago trees, which are often varieties going by the same name. These are pragmatic understandings, for orchards or units of orchards move back and forth among clan (and subclan) units following the movement of pigs (often in mortuary rituals) and people.

Muyuw sago trees probably grow to no more than thirty meters in height, a panoply of fronds growing up from the end of a trunk, botanically speaking the plant's stipe and the source of its carbohydrates. In Muyuw orchards tall forest trees—some functioning to section the orchards—overhang the sago palms; aerial photos do not readily reveal the orchards because the taller trees enshroud them. This is an intended effect designed to protect sago trees from squall-generated gusts of wind that slam the central part of the island. It is "forbidden" to cut these tall trees. Trees frequently observed in the orchards are *akuluiy*, *akisau* (*Elaeocarpus*?), and the extremely tall *Syzygium ameleyu*. Some of the *Calophyllum* trees central to the next chapter are found in this region and assist in this protection—"*kayamat*," "wait on, protect," is how one person described the purpose of the *Calophyllum* called *dan*. When I asked my Kaulay, central Muyuw people to map the orchards lying about Mount Kabat—to its west, south, and east—I inquired about the location of that *Calophyllum* tree; they readily located concentrations proximate to and inside orchards.

Sago is a *kay*, the life form category I translate as "tree." Although I often thought Muyuw used several words interchangeably for the sago tree, the word form *yabiy* is its specific term, and the word form *kabol* is the very starchy flour-like food obtained from it. The organism is a palm-like tree in the Arecaceae family, genus *Metroxylon*. Scholars debate how many species are in the genus, but the tree is clearly a product of Southeast Asia and Melanesia.[20] It is this region's contribution to the world's stock of human carbohydrate sources. The literature (e.g. Ehara et al. 2000) suggests that this tree grows for some seven to twenty years (one of my informants guessed it took about thirty years for one he planted to mature), then flowers, seeds, and dies. This variable maturation process is not fixed to solar standards. As with other plants, crops, and trees, Muyuw do not count years or months as a way of talking about growth; rather, they continually monitor the maturation process and have their own terms for its sequences and relevant time demarcation. Every individual or group returning from sago swamps discusses the appearance and condition of the trees, focusing on what might be harvestable in some immediate or distant time frame. For while there is a bell-shaped curve to annual sago making—the peak occurring after yam gardens have been planted by December—almost always somebody is harvesting a tree somewhere. In 2009 I asked two different people from Kaulay village how many they might cut during the heaviest months of use, and they reported "about forty"; Ogis reported that Wabunun alone, a complex village of some five hundred people, cut "one hundred" from December to March or so of 2010–11. Muyuw are casual with numbers and these figures strike me as high for serious gardening villages. Yet the unequivocal report here is "many" trees are downed every year. I'm sure that smaller and differently situated villages like Kavatan in far eastern Muyuw and Unmatan, proximate to orchards in southeastern Muyuw, proportionately cut more (because both places are less serious about taro and yams). In his classic description, Barrau states that about twenty-five trees might be harvestable per acre per year (Barrau 1959: 156). Although I did not keep precise count, in three orchards I walked through in July 2009 nothing close to this number were coming to maturity, though in two of the three trees had been harvested from them during the previous austral summer. That said, people know they do not always harvest trees at the optimal time—said to be right after the seeds have formed—and there are terms for harvesting them before (*aktugoad*), when (*kaowu*), and after they are mature, when the available starch begins to dissipate (*bwikayyo*, "flown away"). White cockatoos pecking at a tree's fronds are the "sign" for optimal harvesting times. I was told, for example, that if a tree's flower structure, which shoots up out of the top of the highest fronds on the bole, develops in December, there will likely be seeds and optimal times for harvesting the tree in September or October. Again, however, external time markers are not significant for these calculations; instead, perceivable qualities, and birds drawing from those qualities, define the peak.

Although seeds are sometimes fertile, the tree is usually propagated by means of asexual reproduction, new trees budding off more mature ones. Hence a downed and harvested tree will have a small clump of maturing trees surrounding its stump. By means of their budding suckers, Muyuw have tried moving the plant to more convenient locales—usually without success. In the orchards, however, the tree is rarely transplanted—the several new suckers are allowed to mature next to the harvested "mother" tree. But people recall moving trees from one island to another. Muyuw sago trees are the spiny type—spines grow out of the fronds and are understood to be part of the Creator's "curse" on the island. But some Misima Island sago plants are spineless, and one friend claims credit for moving that variety to, especially, the southeastern orchard called Upwason. It is one of the two orchards in the southeast that is understood to have water for processing through all but the most severe droughts brought on by El Niño droughts.

The problem in sago production is filtering the starch out of the tree's pith. Muyuw do this like most everyone east of the Moluccas, by pouring water over pulverized pith while kneading it with their hands. But first the material has to be pounded out of the tree trunk. To do this the selected tree is chopped down, its top severed from the bole, and the bark split along a straight line and forced back away from the inner wood. That inner wood is then beaten into coarse sawdust by a lone individual standing in the bole's center, gradually moving toward each end. He flakes off the material with an arcing sago-pounder (*labus*), on the end of which is tied a well-formed hunk of igneous stone (*kiligil*). This piece overlaps the wooden handle so that its sharp edges shave slices off the bole. It is regularly chipped so that its edges remain sharp. The stone comes from the Sulog region that was once a source for many stone tools in the greater Kula region. Although the people living in the Sulog region may produce more of this material than others elsewhere,[21] most of my informants produce their own and have to be familiar with the hafting dynamics to keep the hitting edge sharp. The block of stone is always tied onto the handle with *yit*, the usual tying material for things not exposed to the ocean. The wooden handles are finely carved[22] and often pass between generations of men, usually but not always within subclans. The reason for the inheritance is always an active, enduring relationship between the parties, not the subclan identity.

In addition to the sago pounders, some people carve and take to the sago swamps troughs (*kas*)[23] for collecting the sago starch. Often, however, with the exception of the pounders, all the tools used in the subtle production process are created at its site. These include troughs frequently constructed from bark pulled off a small *ukw* tree (*Sterculia sp.*); bucket-like structures for scooping water out of a stream or a hole to the water table (if a stream isn't convenient); a slide made from the long petiole of a sago frond set so that its bottom, its widest portion, slopes down from about chest height; and a series of screen-like structures and a convoluted set of small crisscrossed sticks made from the inner sago bole. These forms create a filter to slow down the water falling into the trough. This enables the final settling of the starch out of the water. As the trough gradually fills up with dense edible sago, water spills over the ends and sides. When the trough is full people usually shove the sago into cloth bags obtained from purchased flour or rice. Formerly the sago had to be fired three times[24] before it would be wrapped up in containers made from sago leaves. Coming in two sizes, the traditional containers are still frequently seen because they distinguish sago to be given away in mortuary rituals (usually the larger) or held for individual consumption or for exchanges elsewhere (the smaller), usually Budibud (see later in this chapter; also Damon 1990: appendix 2).

To the meadows: Throughout the whole of eastern Muyuw, small meadows filled with dense grasses and ferns dramatically break the high forest canopy that seems to dominate the whole island. The famous meadow, Klibulouboul,[25] lies in juxtaposition to some of the sago orchards on the north side of the Sinkwalay River in southeastern Muyuw. East of Mount Kabat, orchards used by people from Kaulay and the far eastern villages of Ungonam and Kavatan also have meadows in their vicinity. One of these centrally located with respect to the Kaulay orchards is called Dilupwaup; I have not seen it. Bungalau and Salayai, the southeastern meadows examined repeatedly, center the sago orchards regularly visited by all southeastern Muyuw villages. Others, some of which are mining related if not caused by the practice, flank the sides of the Sulog Mountain rises and dot the eastern region leading up to those mountains. By 2007 I had been told of similar structures—both in form and function—in north-central Muyuw. In 2009 I studied two of them, Digumwamwan and Ukulel. Because of intravillage disputes, one of my Kaulay informants lived in one for a few years extending from 1999. Logging and mining activities have cut through both and have led to recognizable changes to their plant communities, especially the existence of numerous *Pandanus* trees of type called *yagal*, common in very early fallowed areas and the permanently wasted soil patches, *kadidulel*.

Except for lower sago and betel orchards that also infiltrate these areas, the all but undisturbed canopy of the surrounding high forests approaches forty to forty-five meters in height. The pictures accompanying this section show how higher trees encompass meadows. In one, the first row of higher trees includes sago trees growing in a swamp just beyond Bungalau's eastern border along with a *dan* illustrating the typical shape of that *Calophyllum* species towering over the other trees. A unique member of the island's set of *Calophyllum*, this tree is a part of relations configured by the meadow/sago orchard construct.

Meadows, with growth approaching only a meter or two high, can be seen when passing the Sulog Mountain range by boat and from aerial photos taken of the island (those in my possession are from the 1960s); they can be spotted from Google Earth. As noted earlier, I walked through several while taking an old mining path north from Sulog Bay toward the European center Kulumadau in the 1970s and in 1982; people explore and exploit one of those, Bunyavat, searching for trees suitable for *anageg* outrigger floats like those found around Kaulay. In contemporary memory, the Bunyavat meadows derive from the mining epoch that ran from 1895 to the eve of World War II. I had presumed many of the other meadows, especially those that flank the mountains, resulted from early twentieth-century European mining practices. Approaching the Bungalau and Salayai patches in 1995 and 1996, I was struck by how peculiar, out of place, the spaces were. I presumed those far removed from any known European involvement were the consequence of soil patches produced by the island's geology. Partly because of that and partly because my initial soil samples were designed to be taken from as many different environments as possible, I collected a few samples from them. Without knowing their sources, Cornell University's Shaw Reid flagged the meadow soils as so full of aluminum (164, 485 parts per million) that they were practically toxic; their pH values are also the most acidic (5.84 and 4.45 respectively). Samples I took in 2009 are consistent with the 1996 data. I acquired twenty soil samples, two from a forest area fifty meters removed from a meadow, seven from three different sago orchard/swamps, and eleven from four different meadows. Using slightly different counting methods, the nine forest and orchard samples averaged 586 parts per million (range, 210–940) while the eleven meadow samples average 2,719 ppm (range

2,169–3,307). These results sustained the presumption that the areas were intrinsically peculiar. So either the soils are unique or the conditions that make them drive down—rather than up, as in the case of the gardening/firing regimes—the pH, and other nutrient numbers, found in other forested areas. The acidic nature and high aluminum content of these soils is a sign of extreme leaching processes, undoubtedly facilitated by absence of forest cover.

Although there are meadows on the island that are the result of mining activities, I eventually discovered that these are intentional, anthropogenic forms and that they are linked both spatially and conceptually to nearby sago orchards.

The meadows seem to be on slight rises; those at Sulog are on slopes; the ones in central Muyuw first visited in 2009, Diguwamwan and Ukulel, are just off the immediate flanks of Mount Kabat, standing above the myriad of streams that come off its sides. The ones east of Sulog, however, are situated differently. The region of their location rolls into the Sulog terrain (proximate to the "Kweybok" referred to earlier) where there are many little rises. However, only a few of these hills are meadows, and, unlike other knolls, they are named. They are quite stable. The 1960s aerial photos I have of Bungalau and Salayai demonstrate shapes and locations identical to my surveys from 1996, 1998, 2002, 2009, 2012, and 2014. That is untrue of any garden area surrounding any village I have frequented since 1973. New gardens, successional regrowth surrounding villages and paths leading to the gardens, change so quickly that within a year or two the landscape is nearly unrecognizable to the frequent visitor.

As for vegetation, ferns dominate on meadow rises, while in the very wet areas, a sharp grass called *yalayyala*, or *leleiy*, *le* in Iwa, probably kunai, mainly grows. The predominant ferns in Bungalau in 2002 were, in order, *kokoyit*, *signavan*, and *koliu*. The same plants dominate Diguwamwan and Ukulel in central Muyuw, plus the recent intrusions. The fern *signavan* (Damon 186) grows in a particularly dense way, forming an incendiary mat; long before I could conceive that these were anthropogenic meadows, Sipum had pointed out to me how flammable the ferns made the meadows whenever there was a "big sun," the term for an extra-long dry period, typically generated by ENSO events. *Kokoyit* (Damon 292, *Gleichenia sp.*) and *koliu* (Damon 293, only determined to be a fern), although classed as ferns, are both vine-like and so classed by Muyuw. They are important resources. *Kokoyit* have two thin but differently sized black strands running throughout the otherwise green stem. These are extracted and used to make bands with simple if sometimes intricately woven designs. The black bands are *sasi* (the name given to the plant in central Muyuw), armbands; *kaykwas*, calfbands; and *palit*, waistbands. Many people still wear armbands, so every time I went near one of these meadows, one of my companions would extract a *kokoyit* bundle. For themes discussed in chapter 5, note that in the past Muyuw bodies were more visibly bound up than they are now; the meadows were a prime source for binding material.

Koliu[26] is a "swamp plant," not just a meadow plant. It has two root structures, people say, one to hold on to whatever surface it climbs, and the other going into the ground. The second is the preferred—and still used—material for tying strakes to keels and strakes to one another in the two highest classes of outrigger canoes. The root is dried in the sun or over a cooking fire. Only by drying is it said to achieve its full strength. Before it is used, it must be soaked in seawater for a few hours, after which it can be pulled tight. It is considered neither as strong nor as long-lasting as new artificial materials (mostly nylons), but it is said to have an advantage over them in that once pulled into place, it will not stretch.

In two of the eastern meadows, Bungalau and Salayyay, only a little kunai grass (*lileiy*) grows, but more is evident in the two central Muyuw meadows. This material could be used as an alternative thatching resource, replacing sago leaflets, or, especially in the past, used to create an inverted broom-like tuft tied to the two ends of the top of a roof, especially in new houses constructed for rituals. It is the roof's *antenna*, the term for a flower inserted under an armband used for self-decoration.

There are also isolated *digadag* trees and others that need open sunlight to transform from seedling to sapling and beyond.[27] The existence of these trees is the sign that were it not for periodic burning, these meadows would revert to the surrounding forests.

Until 2002 I only went a few meters into Bungalau and Salayyay. I wanted to go into them every time we approached one, but my guides kept diverting me until I finally insisted. Then I "understood the point" (*kakin matan*): the dense matted growth is almost impossible to walk through and can slash feet and legs; most of Bungalau hasn't been burned since before 1998, so it was impassable in 2014. Hence, although their edges are occasionally used as campgrounds when men are harvesting sago, people generally circle the meadows rather than enter them. However, wild pigs readily go into them, knowing that they provide refuge from hunters. Muyuw report that female pigs brood in them. So common is this understanding that when men go hunting for pigs, they cruise around the meadows. Pigs of course forage in the forest for roots and nuts—most especially the nuts from *Canarium* and the *Calophyllum* called *dan*. If a pig is roused, there is a race to see if the dogs can get the pig before it gets to a meadow. Dogs refuse to go into the meadows as well.

The well-conceived relation of pigs to these meadow patches makes them inverted instances of the high *tasim* amid the much lower garden and fallow growth discussed in the previous section. As those islands of high growth provide havens for periodically hunted cuscus, so the meadows shelter wild pigs. Moreover, I was told that Bungalau and Salayyay also provide refuges for pigs when much of the region is flooded following extreme rain. One informant told me that in such times, scores of pigs congregate on meadow crests. And, I was somewhat mysteriously told in 2002, in times of drought a small, slow-moving stream flowing through Bungalau never goes dry, and so waters animals during extreme droughts. The stream area even remained wet during 1997–98's severe ENSO event. This flow of water meanders above and below ground generally in a northern direction before it appears first in the hole called Nabulkwakwit. A few meters from there it emerges as an open stream, flowing into what is called the Alanay River and its associated sago orchards. Wabunun men used to push outrigger canoes very close to Bungalau by means of these streams.

I closed the previous section by referring to the patchy, cyclical oscillation of wet and dry conditions. Understood with respect to star times, people use the cycle to model garden activities. That pattern facilitates intra-annual productive activity. It is encompassed by a pattern that repeats inter-annually. The NOAA index here shows alternations between El Niño Southern Oscillation (lighter above the line) and La Niña (darker below the line) phases. Although there is significant regional variation, El Niños are usually a time of drought, the La Niña a wetter period. From discussion with Muyuw people in 1998 and again in 2002 it became clear that long drought phases, most obviously created by ENSO circumstances, become the contexts for generating the meadows found throughout the high-forested areas of the island. Whatever the origin of these forms, they have to be reproduced by regular burning. When I returned to Muyuw after the

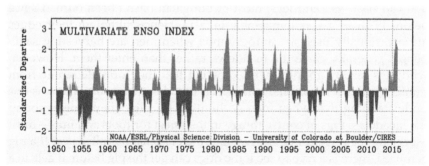

Figure 3.1. Multivariate ENSO Index. Image provided by the NOAA-ESRL Physical Sciences Division, Boulder Colorado from their Web site at http://www.esrl .noaa.gov/psd/.

1998 ENSO event, Sipum told me that during that "big sun" he burned every meadow he came across while hunting pigs, except for Bungalau; he looked at it and decided that it did not need burning.

The meadows constitute part of the region's overall fire-defined landscape. Their peculiar soils might result from their nutrients continually running off elsewhere—perhaps especially to the neighboring sago orchards.

But there is more to these forms than this. When I returned to Muyuw in 2002, one of my plans was to obtain soil cores from various places on the island. Because by then I knew that these meadows were anthropogenic, at least one was to come from them, and it was to be Bungalau. We headed there with the idea that we would stay overnight, and, since it had been dry—there was very minor El Niño in process—we burned off a patch of the meadow's western end (*matan*). Bungalau is constructed around one set of small rises going more or less east to west and another, perpendicular to it and to the west, sort of going south to north. The stream that never goes dry divides the two areas. And it was there I found a single sago tree of the variety Muyuw call *musilimuiy*. As noted earlier it is called *Geliu an yabiy*, "the Creator's sago." Only one grows there and it must never be cut and harvested. It grows, flowers, fruits, and dies as do other sago trees, but then it sprouts up again. Carcass-like remains of the two previous trees were easily visible. This brings us to the orchard/meadow relation relative to garden/*tasim* forms.

Muyuw consider sago trees feminine, on the basis of their swelling prior to fruiting; their reproduction (new trees hive off older ones); and the spines that are found on most varieties (these are dangerous so men move with care in the orchards). Male/female imagery is refracted through a series of activities related to the plant from production to consumption. Without question, the male task of making sago involves a male/female combination similar to the one involved in large traditional Muyuw fishing nets. Nets are considered female because they are very productive, heavy, and difficult to move. Just as the feminine nets become productive by the application of male capacities—labor and speed—so does a sago tree result in produce by male activity on the female plants. Going to the sago orchards is a male ritual; young boys are initiated into it long before they can do any of its productive labor. By means of their ambling amid the trees, raised bedding in the swamp camps (*sitow*), and streams, they become familiar with these places, not unlike the situation back in their villages where they are also expected to spend hours ambling in gardens and along shorelines and their reefs. People often speak of outings to the sago swamps as desirable activities, and Sipum excitedly talked about how charged they would get a day or so into what might be a week or two away from the village. It seemed clear to me that he was describing a sugar high, for in the sago swamps men subsist on nothing but (often just fire-burned) sago and whatever wild pigs or fish might be caught nearby. After one of these affected descriptions, Sipum's wife added, changing the tone of the experience, "Yes, and you all come back sick." Sago is not a nutritious food, so consuming only it leaves the body depleted. Nevertheless, producing it is considered a condition for a male being a man.[28]

To make sago, a man leans over an upraised frond so that with the whole force of his body he pushes himself into the sago mash to squeeze the liquid out of it. The frond and the system of filters used to slow the falling water and to help separate it from the carbohydrate are considered female (see also Damon 1990: 130). The *kabol,* the carbohydrates separated out from the liquid, are considered children that result from this process. When sago leaflets for thatching roofs and house sides are collected to carry back to the village, they are carefully placed so that the bundle is well balanced and individual leaflets can be readily pulled from it. Each leaflet has an inside and a backside, paired and opposed together. When the inside is up it is in the female position; when it is down it is in the male position, the positions redounding to how the leaflets will be used.

Such productive processes and conceptions fashion the practice's subjectivity; so too do the uses of sago tree products. First, briefly, sago leaflets are collected and bundled so that they may be carried back to villages where they are used as the prime thatch material for all significant dwellings (in central and eastern Muyuw). Often a day or more is set aside to unpack the leaves and sew them onto two-meter long stripes of *Pandanus* bark. Although everyone in a village might do this work, in theory it is women's work. And back in the sago orchard, men who bundle the leaves do so knowing that elder women in the village will pay attention to and comment on the bundle's constitution.

Second, sago flower is the essential exchange item for Budibud products. Budibud is the set of small low-lying islands to Muyuw's southeast, and is entirely planted in coconuts. In descending order of value, it produces pigs, coconut-leaf skirts, sleeping mats, and coconuts. Exact accounting defines these exchanges (see Damon 1990: appendix 2, 231–34). Exchanges between Muyuw and Budibud were—and to a lesser extent remain—coordinated to the shift of winds over the annual cycle. Budibud boats arrive in Muyuw toward the end of the southeast wind, sometime in November or so, and return with the more variable northwest winds of, approximately, February. Virtually all Budibud items are used to produce social relationships. The simplest of these transactions concern elders awarding juniors sleeping mats in exchange for sago camp labor. But more importantly these products went—and still go—into mortuary rituals and kula exchange. Mortuary rituals distribute resources across generations, thus serving to reproduce or alter specific constellations of resources and their attendant production relations. Names generated by the kula valuable exchanges are the ultimate measure of wealth in the interisland system. So, for example, a Budibud sleeping mat exchanged for sago may next be exchanged for a clay pot from the south. The sleeping mat/clay pot conveyance converts into the exchange of kula necklaces for kula armshells (*veigun* for *mwal*). In short, sago orchards are part of a capitalization process, underpinning relations that extend far beyond an orchard's activities. These are not so-called subsistence practices; rather, these relations are about the way Budibud as an environment, by means of Muyuw's sago orchards and sago-making technologies, fits into all of Muyuw's fashioned relations to the wider region. The orchard/meadow relation is a position in a set of related, ranked positions.

Third, Muyuw gardening villages used prestations of sago for their complementary relationships with Nasikwabw, Boagis, and Yemga sailing villages. The sago went along with garden vegetables, both produced in exchange for the boat labor that the sailing villages employed for their communicative responsibilities connecting the northeast and southeast corners of the Kula Ring. Sago also figured into payments for the *anageg* outriggers produced principally by Gawa and Kweywata islands to Muyuw's west, though sometimes the sago went to these places in the form of clay pots Muyuw people acquired from the south by means of Budibud sleeping mats.

Fourth, in Muyuw itself sago is a prized food for all significant rituals, both as a delicacy whose consumption marks stages of most rituals and as an item bequeathed in them. Every ritual should begin with men stirring large clay pots full of *mon* made from dumpling-sized clumps of sago, often mixed with shredded coconut.[29] These delicacy-like servings form the "eye" of every significant ritual—they open them to view—the debt-structuring processes defining subclan relationships and their resources.

Sago is the prime source of carbohydrates during lean times of the agricultural cycle—usually from December to April—and emergency food in droughts that derive from ENSO events.[30] Wabunun designates at least two named orchards—Upwason and Lipsiyes—as places that can be used through all but the most serious droughts; that is, they have streams or underground water tables that enable washing the sago mash. Although having to eat it is often considered a sign of sloth or ill luck, people also like sago, and sometimes a man or a small group of men will head to the sago swamps just because so many people have "a taste" for the food, whether it's boiled in *mon*, baked in an earth oven, or griddled into a pancake-like form.

These facts place the associated meadows like Bungalau in the midst of a major resource zone. A few sago trees in fact grow from and are visible

in the swamps just to Bungalau's east and north. In 2002 we left Bungalau heading west to meet with a group of Wabunun men working in an
orchard an hour or more away to discover the last set of facts concerning
sago's use as a food. The last meter or so of each end of each harvested
sago bole must be left untouched. These remains are to provide food for
the "wild" pigs that flourish in the area. Whatever other relationships tie
the meadows and orchards together, their complementary connections to
wild pigs—refuge and food source—make them a set, and one effectively
identical to the *tasim*/garden form.[31] The low meadows attract and sustain
a modest, periodic meat and grease source from pigs like the tall *tasim* in
garden areas maintain cuscus. But whereas the garden's significant carbohydrates come from the cut and burned forests of the garden areas, it is
the largely uncut forests and their sago orchards that provide the starch
next to the meadows. And each inverted form partakes of the other. As
noted, wild yams (*kuv* of a type called *gayieg*) grow to small degrees in the
meadows. Both regions are organized by fire. Not only does fire maintain
a meadow's existence, but when a sago tree is perhaps halfway to maturity, fallen and dried fronds are collected around its base and burned with
the idea that the fire will make it stronger. People liken this process[32] to
the burning preparation of a garden area and the smoking of infants. Outrigger canoe imagery is used in both places. Imagery for the garden was
briefly described in chapter 1, and with respect to *tasim* I have discussed
it earlier in this one. The bole of a downed sago trunk is likened to a *wag*,
canoe/keel; once it is split, the sides pulled away from the pith are likened
to *budakay*, strakes tied to keels forming the sides of a boat. One carefully
beats that pith to loosen the heartwood with the sago-pounder so as not to
"puncture the bole (keel)," *katibol wag*. When all but the two ends are finished, the expression *kalibog wag* is used for the next process, which consists of folding the separated skin and bark back over the hollowed-out
portion of the trunk. This is so wild pigs have to furrow into the enclosure
to eat, making them easy targets for a nighttime return. And if one returns
to a sago orchard at night to pursue wild pigs and succeeds in obtaining
one, the expression is, "*O nuwan wag!*" "Oh, inside the boat!"

Finally, orchards partake of and complement garden-defining structures. They are organized with "bases" and "heads" similar to a garden's
"base" and "top" and, like gardens divided into brother-sister associated
venay, they are internally divided by readily distinguishable clumps of
taller and thus older and shorter and younger trees, the latter budding off
the former. These sago clumps may be but are not necessarily differently
named sago types. But they are specified as a unit and distinguished by
the word *dal*, exactly the same word Muyuw use to specify the category
translated as "subclan" throughout this region. One woman said that as

a younger tree rises after its mother has been cut down into food, so her daughter rises to take her place when she finishes giving life. As gardens form a manifold exhibiting all of the encompassing relations to Muyuw culture—*clans* related by marriage, gender and age categories configured in complementary times—sago orchards model the content of those relations, the continuous replacement of old life by new life. And of course this is no idle ideology. Kaulay men showed me how, by the way their children's marriages, sago orchards moved back and forth among intermarried lines of men related by their mothers. And about the first thing I ever learned concerning the mortuary practices in southeast Muyuw was that sago orchards went back and forth between units based on affinal pig debts generated in the rituals. Social time is embedded in sago orchards not only by means of the ways sago trees and products flourish within and between annual cycles but also by means of how the tree provides a concrete image of temporal replacement within the subclan unit. The social and biological form of this tree is an experience of time (see Munn 1992).

The meadow–sago orchard region is a created network of interrelated life forms. While interesting questions about concepts, homologies among concepts, and forms of understanding lurk about these collected *and* collective facts, from the point of view of the main themes in this chapter one idea stands paramount: these relations are patches of life forms, in fact embedded layers and tiers of arranged patches. As one may argue that individual plant names express discontinuous life arrangements, so we see larger collected life forms similarly organized by and through their distinctive productive conditions. I now engage the next argument about these forms by focusing on the formal social and symbolic use of trees.

Trees and the Construction of Social Discontinuity

The productive infrastructure availed by this area's flora organizes a set of regional relations. This is the institutional base for the activity that originally brought this area to the attention of the world, Kula. While ranking its participants, the kula establishes relations of contiguity among participating islands. Everyone I have interviewed about the institution understood it as composed of two sets of shell wealth continuously circulating among the islands. And by the actions of specific persons, this ensemble creates specific relations of cooperation and competition between people whose actions establish their very existence. When I was on Kitava in 1996 examining *mwal*, it was easy to ask from whom they came. All came through Muyuw, often from people I knew. The signs of recognition easily led to further social relations. Most people in Kitava have never been to

Muyuw and know very little about it; yet specific kula valuables and their histories turn the unknown into the known. Two of my Wabunun instructors, Dibolel and his younger brother Ogis, have never been to Kitava and would not be recognized if they went there. Yet the moment I mentioned their names a whole series of dramas constructed through the medium of shell valuables brought them to my Kitava acquaintances.

Elsewhere I have detailed exactly how this system works from Muyuw's point of view (e.g. Damon 2002). Yet kula, both as conceived and as practiced, is not the only device by which people imagine and create contiguous relations. I kept learning new ways people visualize connections across these islands. Ties between the eastern Muyuw Kavatan and the Trobriand Kavataria were noted in chapter 1. While taking a list of field names in Kaulay I noticed that one west of Kaulay looked like a Trobriand village, Liluta. Informants said the names were identical and pointed to two names east of Kaulay, Samarai and Guliguleu, that referred to the well-known places toward the southeastern corner of the Kula region. When I was told that one of the original names and locations for Wabunun was *Obulak*, after a tree that grew there, Wabunun people said that there was a place/village in the Trobriands (*Oburaku*) by the same name and was named for the same reason. Iwa is full of names (and practices) that recalled relations to other places. Overlooking a path at the northwestern tip of the island is a rock called *Kweyum*. It looks toward Gawa and Muyuw to the east, to which the whole northern portion of the island tilts. A story describes how the rock used to be one of two sisters. Long ago the two women were at this point, but they decided to go to Kitava leaving from the opposite side of the island. As they went to their boat, one said to the other, "You return" (*Kweyum*), which the woman did. She then turned to stone (a lump of compressed coral). So a strategic point oriented in one direction is named so it recalls places to the opposite. Although most of Nasikwabw's connections today are with the islands of Bwanabwana, meaning from Gaboyin or Koyagaugau to Tubetube, people there speak a language closer to that on Misima and Panaeati. Their origin stories claim that they came from "Misim", arriving with sago and other cultural accoutrements.[33] But when these were being set into place on the northwestern end of the island somebody saw the Sulog Mountains off in the distance. So all the good stuff goes there, including the ability to produce stone tools. Nasikwabw people today draw on the sago (and other) resources that abound at Sulog, so this story orients them both to a conceived past and to today's realities. That past is viscerally present. In addition to significant implanted rocks whose storied names recall these events, potshards and fractured igneous stone litter the location as if Sulog's resources were once there.[34]

In Gregory's terms (1982, 1997), relations of reciprocal dependence, of the conceived past and the present, organize these places. Every place is defined by bits of others. This is the sociological condition the following data bundles. If initially these materials inscribe distinctions, they become the means and sometimes reason for tying together discriminating differences.

Fence Posts (Gaduiy *or* Tut)

I surveyed fence posts in Iwa, Mwadau (western Muyuw), Kaulay (central Muyuw), and Wabunun (eastern Muyuw). The term is *gaduiy* in southeastern Muyuw, *tut* in Kaulay. Since a posted tree trunk less than five centimeters in diameter, sometimes partly rotten, is not much for making identifications, I learned subtle discriminations that people regularly make.[35] But with one possible exception, few surprises loomed from these investigations.

In the two Iwa gardens I examined, fence stakes were drawn from all environments Iwa people distinguish—sandy shoreline, steep rises, oval flats—and include trees representative of the three main fallow types (*ulakay* trees grow on Iwa's steep rises). Of 821 fence posts in one Iwa garden, 67 different trees were employed (nearly twice the number in other surveys; in one Wabunun *digadag* garden there were 47 trees represented). The tree with the largest number, *bwaka* (not identified), had 130 (16 percent). The third highest number 58 (7 percent), were *meikw* (*Intsia bijuga*), a very hard tree often selected for house posts and, from Gawa and beyond to the west, yam houses. Thirty-five stakes were *gwed*, and all nut and fruit trees were represented. These included Iwa's prized *saido* trees. None of my other lists contain this tree, uncommon elsewhere. In Muyuw the vast majority of fence stakes came from the surrounding fallow type. So gardens cut out of *digadag* fallows had fence stakes typical of *digadag* trees (sprinkled with trees of nearby *tasim*). Iwa is an extremely small place, so drawing trees from everywhere is easy, and reusing stakes from one year to the next is a necessity.

North-central Muyuw's Kaulay, however, presented an exceptional situation. Unlike anywhere else there was almost a prototype for fence posts, the *Calophyllum*-like tree discussed earlier, *amanau*. This is the tree most frequently found in soils Muyuw class as "wet;" thus they are infrequent around Wabunun. Yet they are very common in the north-central portion of the island, and in Kaulay are classed as a *tasim* tree. I surveyed one side of a garden's fence in this area to find that of thirty pairs of fence stakes, forty-four out of sixty (73 percent) were from this single tree. In addition to fence posts, the tree is frequently used as an outrigger crossbeam, *kiyad,* for connecting the main part of a canoe to its outrigger float. I return to its properties later, but given them, whenever Wabunun people go where the tree is more common, they frequently cut appropriately sized ones and carry them back to the village for important uses. The tree is well known. How this functional specification for boats is connected to its fence stake use in Kaulay is uncertain, and convenience may be the only reason.

In Wabunun I was told that there are no selective criteria for fence posts: people barge into forests surrounding the garden and cut "everything that blocks your eyes." This was an exaggeration because I was often criticized for cutting trees that will "rot tomorrow." Nevertheless, fence stakes replicate the fallow class from which the garden is cut, and in Wabunun this should be, though it is not always, the early fallow *digadag* class.

Everywhere fence stakes hold together logs that are gathered from the trees cut and burned to make the garden. They are, in a sense, not chosen, but follow automatically from the fallow class in which the garden was cut.

Houses

As noted elsewhere (Damon 1990: chap. 5), Muyuw houses are not the important consolidation of values that is the case in some other societies, most especially in Southeast Asia (Cunningham 1964; Waterson 1990). Gardens in fact embody such systematic and encompassing information. Nothing that I learned during the course of this research changes that conclusion, although I shall modify this point in chapter 6's discussion of boats. So I shall reserve for another publication a detailed accounting of Muyuw house construction and the trees used for them. It must be said, however, that much knowledge is embedded in houses—this was one of Cunningham's original points—known by both males and females. Trees are selected for specific properties such as durability, resistance to insects above or below the ground (in the case of posts), and things like tensile strength. This information is specific to houses. So, for example, there are trees used in house construction because they are strong, light, and easily shaped that cannot be used in boats because the particular tree kind (e.g. *akuluiy*) would rot quickly with a boat's exposure to the elements. Trees or botanical materials from most ecological zones are represented in each and every house. Realizing this by 2006, I asked if houses were composites of forest or ecological types. Although my informant understood what I meant, the facts led to no response that suggested something significant imparted by the principled selection of suitable trees for different house parts. Yet exactly what trees are used where varies according to the dominant areas near a building or village. Many house parts in the Mwadau employ the mangrove swamp tree *wil* (Damon 237, *Kandelia*) because such swamps are close. Forms in this social system are usually transparent, so while the content of house structures convey much information, and are thus hardly meaningless, significant representations of social relations are elsewhere.

I did not research changes that may exist in the uses of materials resulting from colonial rules. Some, however, are obvious. Because sawmills and timber activities have been in the Kulumadau region for decades, many houses in Kaulay and Dikwayas have boards, obtained from the mills or fashioned by these people with an axe, in place of more traditional materials like black palm (*manakut*) for flooring. Virtually all houses are now on stilts, whereas the prime living area used to be at ground level. There are certainly matters here that are important, yet they do not bear on the major issues of this work. Other than size differences—Muyuw houses are much larger than those to the west and continue to become more so—I know of no differences in house or house materials among these regions that speak to other meaningful differences; greater information on Trobriand houses may change this conclusion.

Finally, as is often the case in Austronesian societies (e.g. Southon 1995), Muyuw can view their houses as boat-like, and there are parts in both forms that seemingly share the same structures, materials, or functions. I return to this topic in chapter 6 but note a few examples here.

Some of the structures used to make house flooring are recognizably the same as those used to make the platform between the keel and the outrigger float of the *anageg* class of boat. Related to this, the tough and relatively thick outer bark of black palm (*manakut*) is one of the preferred flooring materials in eastern Muyuw. Almost everywhere it is used to connect crossbeams (*kiyad*) to outrigger floats in mid- and smaller-sized boats; the part is called *watot*. But my informants drew nothing of significance from this apparent similarity, and some pointed to a critical difference: more mature, tougher bark would be used for flooring, while younger, lighter, and more supple bark is used for the boats.

A thin, old-fallow, canopy-top tree called *akidus* (family Rubiaceae) is used for a critical part in roofs and one of several springs fundamental to the operation of *anageg* sailing craft. In roofs it is used to create a structure called *poiyi* and functions to counter the effects of wind on the roof. In each house two *poiyi* run from one corner of the roof to the other, forming an X where they meet at the top center. Often two trees are used to make each of the two crosspieces, their narrow ends tied in the middle. The pieces function as springs, so if the wind pushes the roof structure up at one end, the *poiyi* is tied and balanced as well as positioned at the other end to keep it down. These pieces are arced, and only wood with the *akidus* tree's enduring tensile strength can be used for the part. The same properties, thinness and preservation of tensile strength when permanently arced, make the tree the only one suitable for the *anageg* boat part called *nedin*, described in chapter 6. Most Muyuw people know this coincident usage between boats and houses.

The woman of a house is likened to the *kavavis*—rudder or leeboard—of a boat. Shaped like a large paddle or oar, a *kavavis* is a rudder-like instrument used to define a boat's direction. It is said that without a good woman or a *kavavis*, neither a house nor a boat will maintain its purposeful direction. For *anageg* the form must be made from the tree called *kaymatuw* or *meikw* (*Intsia bijuga*).[36] This tree is chosen because it is hard, can be finely shaped (it takes on the subtle characteristics of an airplane wing), and may sit in the sun's heat without splitting. The same tree is preferred most everywhere for house posts (*kakol*). These parts share features with *kavavis* and the analogy with women because they provide the house stability (they anchor it), but the significant feature for their use is the wood's weight and resistance to insects that quickly rot most other trees planted in the ground. Highly durable, these parts are perhaps the only ones regularly transferred from one worn-out house to a new one.

Yam Houses

Although yam houses, *seygous* in Muyuw, usually something like *bweyma* from Gawa and Iwa on to the Trobriands, might seem to be less significant than the structures people live in, they stand out because of their shifting position across these islands, minor shifts in form and terminology. The structure transforms from being a relatively flimsy building of temporary, and mostly private, significance in eastern Muyuw to an increasingly central, centralizing, and enduring structure in the Trobriands. The changes appear by central Muyuw. By the westernmost Muyuw village, Mwadau, the public nature of the form is apparent since they are located in the village. By Iwa, yam houses are complex ritual artifacts. Both their appearance and the primary materials used to make them formally display important, public, values. I convey these transformations by versions of the diagrams I constructed in 1995 and 1996 available in the photographic essay that accompanies this text.

I begin with a yam house made near Wabunun. The structure was oriented more or less to the sunrise/sunset axis. The pictures and diagrams were taken in March 1996, well after the house was all but emptied of the yams from the harvest of the previous year (roughly July and August 1995). But it was still used as a place to rest, cook some meals, and orient activities for the coming harvest. It would not, however, be used again to store the yams harvested in the coming months, and some of its timbers were rotting. It was in a garden cut out of the *ulakay* fallow class, so virtually all its wood was typical of that older class. There are no "rules" for which kind of tree goes where in these structures, so for the most part the pieces in this one were there by convenience. People will say they "chop without reason" when selecting for these parts and I was unable to obtain tree names for some of the parts. However, there are limits to that. Nobody would make posts for yam houses from trees that would rot quickly when staked in the ground. The parts called *kakol* and *ipobal* should be "real trees," that is, relatively substantial. That said, one of the posts in another yam house I studied was taken from the tree called *akuluiy*. Although it's a very fine construction wood for other house parts, nobody with enduring interests would use that tree for making a post. Since eastern Muyuw people have no enduring interests in their yam houses, there was nothing wrong with that usage. Moreover, unlike village houses, and unlike the case frequently to the west, most *kakols* for Wabunun yam houses did not have bark and inner wood stripped or sliced off. They were not made to be attractive.

Many of the names and functions for parts of this structure are identical to those found in human houses. The piece called *poiyi*, noted earlier, functions like a spring, so that if the wind hits one side of the house its force pushes the other side down. Two different trees were employed for this piece, *losiwas* and *yed*. The former was barely suited to the task while the latter worked pretty well, though it would not be used in a house. *Yed*'s ability to bend appropriately with wind makes it a good alternative mast

Table 3.1. Wabunun Yam House

Part No.	English Name	Wabunun Name	Trees used (when known by my informant, Gumiya)
1	posts	*kakol*	*alubwalieb, ayelev, atwaleu, atwaleu, atwaleu, atwaleu*
2	crossbeams (north-south)	*ipobal*	*ayelev, alaviluw, alaviluw, ayelev, alaweluw, akuluiy*
3	flooring	*atavat*[37]	
4	east-west dividers	*lapuiy*	
5	bracket beams	*sol*	*ababuyav, guam*
6	rafters?	*algivasiw*	*akigel*
7	wind brace	*poiyi*	*losiwas, yed*
8	rafter supports	*kaval*	
9	roof beam bottom	*mamwan*	
10	roof beam top	*avatoun / livkeyway*	

for some sailing craft. In any case, the whole yam house was tied with the inner bark from a young *ukw*, though that material had not been woven into a fine rope. As is the case in virtually all of eastern Muyuw, this house was plastered shut with woven coconut fronds. These barely last the six months they are supposed to cover stored yams from sight. In the same region, and in central Muyuw as well, human houses are covered with sago leaves, and these tend to last for half a decade and more.

Situated in the garden, incorporating some of its terminology (e.g. *lapuiy*) and following its main lines of orientation, this structure can also be considered "like a boat" (*mawan wag*). A few of the names used in this structure—*ipobal*, for example—find similar functions in some sailing craft. Analogies between boats and yam houses are explicit. Note the sector called *kaynikw* formed by the east-west (*lapuiy*) and the north-south (*ipobal*) pieces. The same name is applied to the divisions of Muyuw sailing craft formed by the craft's ribs and crossbeams. In a boat, this section holds things like coconuts, sago, and vegetable food. The unit might best be translated as "hold," as in a boat's storage area. In a yam house it holds yams, ideally one kind of yam per unit. People often have more kinds of

Table 3.2. Kaulay, Central Muyuw, Yam House

No.	English Name	Kaulay Name	Recorded Trees
1	posts	*kakol*	*gwed, meikw, bwit, mamima*
2	bracket beams	*sol*	*gwed, kamayeya, atulab*
3	Crossbeams (north-south)	*talapowa*	*atulab, kamayeya, alasisova, geguiy, atulab, yawlawal*
4	underflooring	*digagweleywa*	*bovaguna, tuv, asibwad, gayas, manakut, akuluiy*
5	flooring	*katavata*	*asibwad, malawa, gwed, kaliviluw, gwed*[38]
6	rafter supports	*altukwav*	*kakigel, akuluiy, yawlowuk, lidad*
7	rafters (inner)	*anakan*	*nilga* (an exceptionally light wood), *yamulula*
8	rafter supports	*kaval*	*kakigel, ayovay* (an exceptionally light wood)
9	rafters (outer)	*algivasiva*	*kakigel*
10	lower ridge board	*ugwa*	*keibw, aymatuw*
11	upper ridge board	*vatau*	*akiduse*
12	wind brace	*poiy*	*akiduse, amanau*
13	roofing/siding	*lokwat*	*Kubal, pandanus leaves.*
14	sago leaf holders	*kiva*	*manakut*

yams than units, so the ideal separation tends to dissolve. And by the end of the storage period and beginning of the new planting season, sprouting yams tend to get all entangled with one another. The intermingling shoots from the still unplanted yams model stages of the human life cycle, providing an experience of time complementary to that of sago trees. I have written about this symbolism elsewhere (Damon 1990: chap. 5).

Like eastern Muyuw yam houses, those in north-central Muyuw are placed in gardens oriented east (the *pwan*, back or butt) to west (*wadon*, mouth). Table 3.2 lists the part names and trees I recorded for about five houses. The names are similar to those in eastern Muyuw, but a few differ, as do the trees used to make some of them. And while the roof is similar, the foundational structure differs significantly. The new feature here is that there are only four rather than six posts (*kakol*), and that the piece called *sol* goes east to west across each pair with the base and top of the piece extending prominently in front and back of the structure. This usage differs from eastern Muyuw. I was also surprised that many *sol* were *gwed*. One dedicated yam gardener said he wanted his to be *gwed*, and it was apparent from the context of his speech that he was following a principled reason rather some sense of tradition. *Sol* are prominent parts of the structure, and I wondered why inferior wood was used for them. *Gwed* is light and will not last many years. But lightness defines the parameter here. Many central Muyuw yam house posts (*kakol*) were made from small *meikw*, the post of choice across much of this area for human houses. Yet while many of these people told me they would cut new posts every year, they also said that they regularly lifted the container portion of the structure off the posts used for one year to carry them to the next. And this explained the reason for choosing *gwed*: its lightness. Four men can easily shoulder an empty structure to transport it the few score meters connecting an old garden to a new one. I presume this reasoning also explains the choices of many other pieces of wood in this list (*asibwad, nilga, ayovay, alasisova, gayas*), all of which come from *digadag* fallows. Since most of these yam houses were also situated in *oleybikw* gardens, these materials were transported from one place to another, unlike the situation in eastern Muyuw where people use what is at hand.

The Mwadau village in western Muyuw (Nayem) comprises the westernmost set of people on the Muyuw landmass. And unlike any of the others, their yam houses are erected in their residential areas. The structures are behind houses, unlike the pattern from Gawa to the west. Other than that, the yam houses are very much the same those of central Muyuw, as is evident in the posted line drawings. As in central Muyuw, the "mouth" of the structure should face west, and as elsewhere it is divided into units called *kaynik*, harkening to outrigger structures. But people seemed to be a bit more emphatic about the separation of yams in Mwadau. Recall from chapter 1 that Mwadau gardens are divided more than those to the east, with east-west lines splitting *venay* to produce units called *gub*. Each *gub* should have a different kind of yam, and these differences should go into the yam house's units. Further, I was told that it is acceptable, if necessary, to mix different kinds of *parawog* (Trobriand *teytu*) or different kinds of *kuv* (Trobriand *kuvi*), but *parawog* and *kuv* should never be mixed. More fuss seemed to be made over these relations than evident elsewhere, a major reason for which is that this is becoming an emphatically public structure. Not only is it inside a village, but in Mwadau men grow *gub* sections in their own gardens for their sisters' yam houses, and the sisters, reportedly, carry the yams from the brothers' gardens to their houses.

Table 3.3 provides the terms and trees I collected in Mwadau. In more detail than the other two tables, this one shows how the immediate surroundings of the region define the wood content of the forms. *Ganag*,[39] *kaboum*, and *igsigis* (Combretaceae *Lumnitzera sp.*) are shoreline trees, where *meikw* might also be found, and *wil* (Rhizophoraceae *Kandelia sp.*) is one of the region's two primary mangrove species. The only other ones in this list are two *digadag* trees, the well-known *atulab* and *wageo*, the eastern Muyuw *kwakwis* (Euphorbiaceae *Macaranga sp.*). From this selection it might be suggested that not only do Mwadau people bring the produce of their gardens into their villages, they also bring their shoreline nature into their villages.

With one exception, most of the names in this list, and the correlative positions, are the same as those in the more detailed diagram of the Kaulay yam house. The exception is *kayyetau* for what was, structurally, *sol* in Kaulay. Although it had been some months between the times I gathered

Table 3.3. Mwadau, Western Muyuw, Yam House

No.	English Name	Muyuw Name	Trees
1	posts	*kakol*	*ganag, kabom, meikw*
2	bracket beams	*kayyetau*	*igsigis*
3	crossbeams, north-south	*tatapow*	*atulab, wageo*
4	flooring	*katavat*	*igsigis*
5	inside bracket beams	*kayyetau bunutum*	*wil*
6	rafter support	*kaydudeo*	*igsigis, wageo*
7	rafter supports	*aval*	*wageo*
8	rafters	*kanakan*	*wil, wageo*
9	lower ridge board	*kakunumwan*	*wil*
10	upper ridge board	*kukunumwan yevatau*	*wil*

these terms and the Iwa terms I review next, when I heard this one I asked my source if it was like those to the west. He said yes, knowing the significance of that term. For at least some of the islands to the west it connotes a full moon. As was the case with the Mwadau garden, it is as if a structure becomes part of a time system.

Although *bweyma* are hardly the structures found in the Trobriands, especially the chiefly houses, Iwa people pour their attention into them. Covered with woven coconut fronds or *Pandanus* leaves that rot relatively quickly—the state most were in when I visited in late January 1996—they are reworked every year for their ritual filling. My diagram shows a hoop-like structure in the *tabod*, the front of the house. Composed of a thick vine, that form is taken down when the yam house is filled to direct a conically shaped pile of yams, received and meant to be seen as gifts, during the course of harvest ceremonies.[40] Before, during, and after that event, this form is a visible analogue of elaborated social forms. One of the elaborations concerns the age of some parts. Almost spontaneously people told me of deceased people from whom they received pieces like *kakol, kaytaula,*

Table 3.4. Iwa Yam House, *Bweyma*

Part No.	English Name	Iwa	Trees used
1	posts	*kakol*	*bwit, meikw, kayleylavakoya, kovasila*
2	bracket beams	*kaytaula*	*saido*
3	crossbeams	*pou*	*ibilog ven,*[41] *yovayi, tawaku*
4	joist-like structure	*tavelulu*	*meikw, nuy*
5	flooring	*bwabukela*	different kinds of wood
6	posts for kaytaula extension	*kakol*	
7	posts for kaytaula extension	*kakol*	*lidaduma, meikw*
8	rafter support	*talkukwava*	*bilag, gwed*
9	rafter support	*kalodeyu*	*kaybas, kaybas*
10	rafter support	*kalvalu*	*abuluk, kaybas*
11	upper ridge board	*kakulumwala*	*natu*
12	lower ridge board	*kakulumwala*	*kaybas*
13	rafter	*kavilag*	*yawula—bilag*
14		*alaguvas*	
15	siding and roofing	*yoiyu*	*nuy yegavin—molay*
16	front end	*tabod*	*lagim*
17		*duledule*	

and *tavelulu*. Although in one case a *kaytaula* was from a *saido* tree, Iwa's famous soft-shelled *Terminalia catappa*, most were *meikw*, should have been *meikw*, and were said to represent endurance through time.

A foray into day names related to the phases of the moon underlines a reorientation one sees in the Iwa material. I recorded twelve such names one evening with a group of men. They exhibited hesitation or reticence in providing them to me, and I was not sure if this was because the names are not used anymore or if people were not comfortable explaining their meanings to me. In any case, here is a representative sample of the twelve. They are sequential but not on a day-by-day sequence:

1. *Sakayleyula*: The day of the new moon, "Kayleyula" is said to refer to an islet west of Kiriwina, the connotation being that the new moon is so far over the horizon that, while it is invisible to Iwa people, it is visible to those far to the west.

2. *Sakwaluiy*: Kwaluiy is said to be a landing (*kwadeu*) on Kitava. Since the moon is closer to Iwa, Kitava people can now see it.

3. *Sigmwagim*: This is a landing on Iwa where the new moon is now visible.

4. *Matasie*: The first quarter moon. This word combines that for "eye," *matan*, with that for a small snail, *sie*. A snail's eye is the stone-like plug on the end of the live part of the animal, which the first quarter moon resembles in its D-shape.

5. *Ulakaydawaga*: *Kaydawaga*, according to my informants, refers to women's work in Kiriwina with banana leaves. Women put the leaf out on a board then rub it until it is straight. This is evening work, and the idea is that the moon is high enough to allow for visibility. Iwa people do not make banana-leaf bundles but know enough about *kaydawaga* to turn it into their own time unit.

6. *Ulakaywa*: *Kaywa*, I was told, means to bend one's head to look up. The moon is now high.

7. *Ulikimwakimwa*: I was told that *kim* means to get bigger, which the moon is.

8. *Ulakaytawla*: This is the full moon, and also the word for the yam house beams that cross the posts lifting the house off the ground. The connotation is that because the full moon rises on the horizon as the sun sets, it is horizontal to the ground. Its light bounces off these beams making them visible.

9. *Vawasa*: This "day" refers to the time when there is a little darkness after the sun sets before the rising but waning moon lights up the night. The concept is also used for the rest period between two work tasks divided by a time interval.

10. *Simsidugugwad*: On this kind of day, young children (*gwad*) will play in the village center then go to sleep before the moon rises.
11. *Sikoiyola ulaalata*. I was told that *ulata* refers to teenaged children—*ulat* in Muyuw—and *koiylola* can connote their playing together, including courting. Teenagers do this during this time.
12. *Vamolila*: On this last named day, it is so dark that the aforementioned teenagers go directly to their sexual encounters with no fear of being seen. In this construction I was led to believe that the *va-* is a causative particle, *moli* "fat," the euphemism for sexual fluids.

This list suggests that temporal references are becoming the dominant organizing vehicles for culturally inscribed activities.

Unlike anywhere in Muyuw, neither Iwa yam houses nor people's houses follow an external orienting principle. While yam houses tend to be in front of people's houses, there was no consistent compass setting to those I measured, and people denied that there should be. It is interesting, however, that their backs and fronts are defined by a *tatan/matan* contrast. When I was on Iwa I was not aware that *tatan*, meaning the stalk of a taro leaf or bole of a sago tree, was considered analogous to the terms translated as "base," or "trunk." Hence I did not explore that term while in Iwa.

The yam house discussion returns to one of chapter 1's points, a sequence of transformations across these spaces. These differences are partially observed by the actors in this system; Wabunun people told me how uncomfortable they felt talking with their Gawa kula partners with the latter's yam houses, and yams, staring them in the face. The yam house structures, and the irreverent to more determined selection of wood used to make them, roughly correlates with the positional significance of these places. However, the actual language of differences people understand is written in a different key. The medium is still wood, but the vehicles are exchanges of ritual firewood conveyed in one or more context-defining practices.

"Ritual Firewood"

Muyuw fire or smoke practically everything. Garden areas are burned in periodicities mimicked by an encompassing phasing found in forest meadows. The bases of growing sago trees are lightly burned. Outrigger floats are torched before they are attached to the outrigger canoe. For their first birth, women used to be enclosed and smoked over a fire so long that they turned white; now they are still smoked a little to warm them from the loss of blood at childbirth. Given this extensive use of fire, it is not surprising that people across these islands employ wide-ranging knowledge about wood and fire for defining relations to one another. Only a few trees, for example, are appropriate for smoking an infant and its mother. Bad trees whose smoke may convey bad properties are avoided. People often asked me for newspaper to roll their cigarettes. However, they were wary about receiving paper I had used for drying voucher specimens because they knew I rarely understood the properties in the trees I was collecting. They feared they might absorb what leached into the newsprint.

Although people will burn anything when nothing is a stake, they specify more selected usage by the term *kaykwan*. The word refers to wood gathered for specifiable purposes. And trees so used vary by a contrast that runs from *punushan* (hot/lasts), e.g. *kiyay/yay* (*Casurina littorale*), *atuwaman* (Damon 26, *Chionanthus ramiflorus* Roxb.); to *gwey* (cold/stops), a word that literally means "weak," for example *nilga* or *gwed*.

Although *kaykwan* and the contrasts into which it fits usually imply a wood whose smoke insinuates its beneficial properties into a body, there are many other uses for the term. A species of ebony that grows on Nasikwabw is slightly different from the one that grows on Muyuw. Although it has been used for export carvings for several decades, it was only used traditionally for carving betel nut mortals and pestles, sometimes spears, and as *kaykwan* for sailing from Nasikwabw to Misima. When more or less dry, a single hunk should burn for the whole trip, a day or more depending on the wind. In many places the shoreline *Syzygium kakup* is frequently but not exclusively called on for smoking a woman and her just-born infant.

Different activity zones have different woods for their ideal firewood. When men succeed in chasing down a wild pig, especially but not exclusively near the meadows and sago orchards, they look for a tree called *asimatan* (Damon 8, Euphorbiaceae *Aporusa nigro-punctata* Pax and Hoffm) as the wood of choice for singeing the pig in preparation for carrying it back to their villages, or for carving it up to cook and eat. This tree is the best of three that are classed as *demiavek kaykwan*, "sago swamp firewood." The other two are *ayyeyey* (Clusiaceae? Damon 286) and *amgwalau* (Damon 279, Bignoniaceae). All three become *matamat*. Conveyed by the duplicated form of the word meaning "die," this term implies that the wood is extremely dried out so it holds a hot flame. These trees readily transform into this condition and are identified as such when dead. This is not just a matter of definition, but also discrimination, since other trees might stay wet or rot to nothing and ineffectiveness. *Matamat* is one end of a pole referring to the water content of trees. Although it is the class for quick-starting fires for wild pigs, that preferred for normal cooking is called *atayyag*, meaning half dry, half wet. If too wet, most wood does not burn; if too dry, it flames up and out immediately, not generating heat long enough to cook. But water content is not the only issue. Other trees are known to be very good for starting fires by rapidly rubbing a stick in a preset groove in their wood (*auseli, Premna*, Damon 56); still others, like the aforementioned Nasikwabw ebony, hold a simmering fire while on a boat—these include the *Ficus kwayin* and the distinctive *anag* (*ganag*), which is freshly cut when used for this purpose. A common expression for trees that are good firewood is that they hold a flame when carried across a village, from one house to another.

For a cultural regime so closely defined by fire, the knowledge just reviewed arises out of daily life. Yet I gradually learned of another practice that, while built on this foundation, turns selective trees into a model for half the Kula Ring. This practice is not known west of Iwa and its extent to the south is not clear.[42] I call this a system of ritual firewood. I closed chapter 1 by pointing to the fact that the beliefs about *gwed* and *atulab* trees were models for kinds of behavior expressing a truth on a different level than the actual knowledge people had of the trees. Those trees are to gardening practices what the system of ritual firewood is to a larger organization of places. As one Muyuw man noted, with the possible exception of the Sulog region, almost every village has access to the all trees used in this system. Yet the trees used are specific. They articulate discernible differences defining social realities, the distinctive social patches across the region.

I refer to this practice as "ritual firewood" because it is most visible during rites of passage. These rituals are the defining moments for affinal obligations, tied to the word *sinvalam* whose central forms in Muyuw collective life I have described elsewhere (Damon 1983c; 1990: chap. 4). Although men also participate in these practices, gathering firewood is women's work, so these activities I now describe fall to them. Often it is presumed a capable man has downed the tree from which a woman will later extract firewood. Once he learned I understood this custom, my best elder Wabunun teacher told me with beaming pride that just after he married his Sulog wife he cut down Sulog's unique firewood tree.

A woman's initial experience with the practice reveals the basic form. Over a few weeks or months after a couple formally marries, the new wife should systematically cut, bundle, and carry to her husband's female relatives the wood appropriate to her place. The wood given is one side of a transaction called a *takon*; for the transaction's other side, the wood's recipient gives back a modest return—now a few *toea*, the Papua New Guinea term for divisions of its prime currency unit, the kina. Although their contents vary, *takon* occur throughout the Muyuw life cycle. The initial wood gift demonstrates a woman's knowledge of trees and relationship behavior. If a woman does not deliver this wood, she risks being rejected as a spouse, and that means rejection for the mortuary obligations that center adult life. The exchange begins the realization of the marriage tie while formalizing a whole set of expected reciprocal obligations of increasingly greater value. Demonstrating competence in these matters is important. In 2002 I asked a young Wabunun woman to help the elder in whose home I was eating. The next day she brought a bundle of *nilga* (Damon 129, *Mimosa/Acacia*), a light wood that burns very fast with little heat—"weak" firewood. By then I knew what I was doing, so I told her that it was horrible wood. She was embarrassed, and the next day she returned with *atuwaman*, Wabunun's ritual firewood. In itself relatively insignificant, the practice marks basal knowledge, illustrating responsibility and competence. Once the wood exchange establishes real relations, more significant affinal exchanges follow. The most important of these are pigs and vegetable food in mortuary rituals. Since such rituals always entail serious cooking, during the weeks leading to them, villages may be loaded with firewood. Although men sometimes assist in this gathering, most of the wood women collect is the village's ritual firewood.

In everyday practices, however, people do not exclusively burn just one or two kinds of trees. Previously I noted wood that is often said to be good firewood, virtually none of it in the categories now reviewed. When I asked Kaulay women what they like to burn, they provided this list: *amkwanev, yamulul, asibwad, tuv, katibwadan, atuwaman, gigaway.* None of these are Kaulay's special wood, the *Syzygium mamina*, which is not thereby excluded from everyday use. Its bark and heartwood are the focus of attention. The former is coveted soon after the tree is down, the latter two or three years later.

When I first learned of this practice in Wabunun I dismissed its significance because I knew the gifts were transcended by obligations of greater value. As variation from one place to another came into focus, however, I realized that this apparently insignificant practice demarcated regional differentiation. Through this use of trees, the holistic understandings of this area, partly conveyed through "base"/"top" tree imagery, transform into interrelated parts. Although few people know all its details, the principle understanding is widely recognized—that places have special firewood that defines and distinguishes a critical aspect of their locale relative to other places. This is, so to speak, a conscious model, the details of which some people take pride in totalizing. Along with the sweet tree/bitter tree contrast through this paradigm, I began to understand the acute nature of Muyuw modeling devices.

Table 3.5 lists the places for which I have information, going from east to west then south to Koyagaugau. The trees are identified by indigenous name, known systematic identifications, and the tree's locational significance. I then provide a small commentary on each tree. I conclude this section featuring a few of the trees and places.

Table 3.5. Ritual Firewood Trees by Place

Place	Tree	Comments
Budibud	*kakup* *Syzygium sp.* shoreline tree	Growing only on the shoreline, this tree reflects Budibud's position and function in the larger region—its actual location and functions are largely determined by coconuts and their products.
Kavatan	*kakup* *panapon* *Guettarda sp.* Vit, (Sapotaceae?) shoreline tree	These trees are shoreline trees reflecting Kavatan's situation.
Ungonam	*amwakot,* *Dysoxylum sp.* *kaga* *Pomentia pinnata* high forest tree	Although now perched on the shoreline south of Kavatan, Ungonam as a social entity remains where it once was inland, north of the Sinkwalay River. These are high forest trees marking that inland position.
Guasopa	*asibwad,* *Timonius sp* *digadag* tree *kakup, Syzygium sp.* shoreline tree	Both a fishing village with immediate access to the eastern end of the lagoon on the south side of the island and a gardening village, Guasopa uses two trees, one from a *digadag* fallow system, its ideal gardening regime, the other fixed to the shoreline.
Sinamat	*amwakot* Meliaceae *Dysoxylum sp.* *kaga* *Pomentia pinnata* high forest trees	These are high forest trees *ulakay,* or older, reflecting the village's old forest position. It was part of the Kweyakoya carving district. My source said Sinamat are "in-the-woods people," *wanawoud gimgilis.*
Wayavat	*atuwaman* *Chionanthus ramiflorus* Roxb *digadag* tree	Wayavat, like Wabunun, is a major gardening region, ideally cutting its gardens out of *digadag* fallows where this tree grows.
Wabunun	*atuwaman* *Chionanthus ramiflorus* Roxb *digadag* tree	"*Digadag* people."

(continued)

Place	Tree	Comments
Unmatan	*kaga* *Pomentia pinnata* high forest tree *hakup* *Syzygium sp.*	At the western end of the southeastern row of villages, Unmatan is right next to high forests where *kaga* are abundant. The village is also perched on the *dibwadeb*, the rocky (limestone) ledge at the edge of the ocean—hence *kapuk*.
Sulog	*akoyu* *Garcinia sp.* high forest tree confined to Sulog region	This hard tree, difficult to cut down, is common in Sulog's untouched forest region but nowhere else.
Kaulay	*mamina* *Syzygium sp.* *oleybikw* tree	This tree appears in large numbers in the *oleybikw* fallows of this region, and its usage reflects that fact.
Lidau	*mamina* *Syzygium sp.* *tuv* *Ternstroemia sp.* mature forest understory	Although recently resettled, this old village area is west of Dikwayas. The reason for the appearance of *mamina* is the same as for Kaulay; *tuv*, however, is an understory tree in mature forests; I do not know its rationale.
Kowuway	No information	
Kalabwadog/ Kuban	*wil* (?) *kandelia*	These villages—sequential places for the same group of people—are located amid the mangrove-lined southwestern shoreline. The region is low swamps.
Mwadau	*aydidawiy* *Cryptocarya sp.* understory tree in high forest; never found in areas repeatedly gardened	This tree is chosen because it derives from the garden fallows, *ulakay*, that typify Mwadau gardening practices.
Boagis	*kaboum* shoreline tree along west and north coast of Muyuw	Wood from this tree is the most appropriate for the most important part of the *anageg* outrigger canoe class. Boagis is a sailing village, its prime work maintaining and sailing these craft.

Place	Tree	Comments
Gawa	*obulak* in Muyuw; *kul* in Gawa *Syzygium sp.* shoreline tree	Gawa people live on the concavity some two hundred meters above the shoreline. One of the prime *anageg* builders, the tree, which as firewood has to be carried up to the living area, expresses their life, living on top but oriented to the horizon.
Iwa	*asibwad* *digadag* tree *tawaku* *Terminalia megalocarpa* planted nut-bearing tree	These trees express the twofold nature if Iwa life, managing *digadag* gardens and fashioning a landscape with many trees producing nuts and fruit, the best of which to burn, *tawaku,* serves as a ritual firewood.
Nasikwabw	*vit* shoreline tree Sapotaceae? *bwit* Loganiaceae *Fagraea* *atuwaman* *Chionanthus ramiflorus* Roxb *digadag* tree	Nasikwabw is a sailing community with meager gardens from *digadag*-class soils
Koyagaugau	*meikw* *Intsia bijuga,* found in many environments, including (sandy) shorelines and *tasim.*	One person told me this tree was Koyagaugau's ritual firewood because its hardness tested the women. Others pointed to the fact that the tree is used to make the large paddles used to steer *anageg* (*kemurua* in Koyagaugau), of which traditionally Koyagaugau had many.

Once I started systematically asking people questions about this wood, their answers to me had an air of obviousness. Since I pursued the issue only from late 1996, most people realized I already knew respective villages, their activities, and environmental contexts. They expected me to be able to go from a tree's ecological setting to the role the village played in the larger scheme of things. These are especially evident where there is a relationship between the tree and the ideal fallow regime (Guasopa, Wayavat and Wabunun; Kaulay-Lidau, Mwadau).

The significance of the shoreline trees for Kavatan should also be relatively clear. As noted earlier, Kavatan houses are now perched on the sand right next to the water, whereas they were once several hundred meters inland. Yet Kavatan's existence was and remains related to the bays to its north and northwest and the expansive lagoon to its east. In both of these areas the Kavatan people work the water for seafood. Although they consume much of that seafood themselves, they also take their product from both areas to gardening villages like Kaulay and Dikwayas. Recall that Kavatan's New Year's ceremony entails moving from the village location down to the shoreline. Although Kavatan people go deep into the island's interior to acquire sago, one major reason for that is to pass it on to Budibud people, connected to them by water to the east. There are more than a few marriages between Kavatan and Budibud, and Kavatan was the only place Muyuw people told me they could also speak the Budibud language. Kavatan's trees index their activities, and are consistent with their mediocre gardens.

For Boagis the choice of *kaboum* cannot seem straightforward to an outsider; however, everybody in the eastern half of the Kula Ring knows that the most significant parts for a canoe are made from that tree. Everyone knows Boagis people are first sailors, and anybody who has ever seen Boagis village has been struck by the three enormous trees overhanging it, two of which are *kaboum*, the other *meikw*. For Boagis people the three clumped trees define the place where they bury their deceased. This harkens to practices in keeping with their southern Kula Ring identities, for, like the case of Nasikwabw, the first language there is closer to Misima's or Panaeati's.

Gawa's tree is also striking. When I was told about it I already knew the tree's shoreline environment. I immediately wondered why anyone would carry heavy loads of wood on one's head up Gawa's steep two-hundred-meter-high cliffs. People responded that this was the point—marking locational competence (see Munn's [1976] discussion of verticality). Iwa's two trees make the same sense. The island is a completely anthropogenic landscape. One of its two trees is from a *digadag* fallow, the other a planted nut tree (*tawaku* on Iwa; *bwanabon* in north-central Muyuw; *utuwol* in eastern Muyuw; *Terminalia megalocarpa*).

Several other cases are more subtle. I was first surprised by the fact that people laughed about Sulog's tree. As noted, one of my best teachers emphasized his ability to cut it. The issue is this: Sulog people hardly ever garden, and therefore hardly ever cut down this hard-wooded tree—cutting it is a struggle. The tree plays a joke on the place's regional significance. Recall the aforementioned Wabunun quip that they knew when it was harvest time because Sulog people would come to "beg" for food.

This is the negative side of the original rationale for that part of the island, where its inhabitants manufacture stone tools for trade with other regions for vegetable produce and live by sago when not exchanging them. Sulog thus forms the logical complement of those practices in Iwa concerning *gwed*: there people plant *gwed* seedlings next to their crops so that the interval between gardening and fallowing falls to zero. Sulog's moves the fallow length to infinity. Trees thus act as a calculus of relations crossing the boundaries of their specific locale.

Wil, a mangrove swamp tree, for the Kalabwadog/Kuban area also contains more than the first obvious message—these people live in a low swampy area where this tree is very common. But the more complex point was conveyed when people contrasted this area with Mwadau's ideal *ulakay* garden regime and its consequent relations to Gawa, Kweywata, and Iwa. And this returns the discussion not just to Iwa's firewood trees but also to the islanders' absence of knowledge of yam and taro names. Mwadau is supposed to replace Gawa and Iwa's yam stocks following severe droughts. Hence it ideally grows its yams in those areas that stay in the best condition during the worst droughts. Recall that *ulakay* soils are thought to be cooler and wetter. Complementing this, Kalabwadog/Kuban is supposed to be able to replace their taro stocks; it does this for villages to the east as well, as it did for Wabunun following the drought of the 1973 ENSO event. Because these people are situated in this low area, their soils remain moist during severe droughts and they will not, therefore, lose their taro plants. The *wil* tree represents that place's ability to reproduce another region's social existence.[43]

Sinamat and Unmatan's use of the old (or no) forest fallow tree for their firewood—*kaga* (*Pomentia pinnata*)—brings up a final point. Sinamat is the remaining representative of the line of villages that crossed the northern rise in southeastern Muyuw, a region called Kweyakoya; Unmatan is the westernmost village along the southeast coast. These two lines of villages remain tied together by the taro and sago tree terminology for "top" and "base," *yamwik/tatan*. But their firewood trees differentiate them. When I raised the top/base distinction with a friend, he immediately invoked the point that people from his area, the "base," used to be sent to the "top" to acquire the *gudugud* tree for making fish net string. That tree only appears when old fallow trees, like *kaga*, are cut down. Unmatan's situation marks a different point. It is understood to be an old village site like all the other places along that southeastern shoreline. Potshards and stone-tool trash line this area no less than the others. Moreover, approaching Unmatan the ground becomes increasingly littered by obsidian as if this area once specialized in using that cutting material (for making sails?). Yet Unmatan people to this day cut their gardens out of the oldest fallow class and reportedly, and by means of my regular inspection, have no *kadidulel* zones, those patches of ground cut so frequently that their soils are exhausted, zones commonly found to the east. Intentionality, not the length of use, has generated this system's spaces. Firewood trees mark these intentions.

That Teili *Tree* (Terminalia catappa)

Individual tree types, uniquely classed items that correspond closely to the Western category "species," mark relations *between* villages and their different activities. In doing so, however, it is clear that these are not individual trees. They are, rather, summations of differentiated information about relationships among plants and people. They are elements of a calculus expressing the conditions of existence in this region. And as such they organize as they manage its production and reproduction. More generally these facts should provide caution about creating models for Austronesian societies predicated on the replication of similarity.[44] For it is not just at the level of origin that these societies specify differentiation as a model of reality (Fox and Sather 1996).

Located amid understandings differentiating a vast socioecological landscape, I now return to the *teili* tree that initiated part 2 of this book. The casual visitor to Wabunun may wonder why I begin this account with a tree that should not be seen, however clearly it exists in a village's culturally determined mythology. This is partly because another tree, *sinisin* (Damon 258, Boraginaceae *Argusia sp.*), was for decades front and center on Wabunun's beach. That tree served as a shady respite there—it was always surrounded by sailing craft, outriggers for my early months on the island in the 1970s and 1980s and dinghies with their twenty-five- to fifty-five-horsepower motors by 2002. In addition to providing welcome shade, the tree's leaves supply properties for fishing magic, thus imaging another of Wabunun's chief occupations, plying the reefs inside and outside the lagoon that runs some twenty kilometers from Guasopa to Sulog. But that visible sign takes us to the less apparent one, and what lies just over the reach of the trees that cover the rise right behind the village: Wabunun's magnificent gardens. But what does a nut tree have to do with those gardens? Moreover, that nut tree is uniquely from another island, Iwa, left there by people saying, "Do not forget us."

The reason for this should now be clear. This system is organized to deal with the troughs of its existence. Its social and intellectual relations reproduce its conditions without problem when everything is fine. Yet when things turn for the worse, such as during El Niño droughts, "big suns," the system is tested. Continuous communication flows among the small islands to the west, Iwa, Kweyata, and Gawa, and western Muyuw villages. And as just argued, the ritual firewood marks the necessary conditions of reproduction. In 1998 after the earlier *El Niño* my Mwadau acquaintances said they had shipped to the small islands to their west "too much to count" (*nag tavin*)—because the issue enjoined the very existence of the higher-level flow of kula valuables. In eastern Muyuw, however, the small islands are at least once or twice removed. They are out of sight, and by virtue of their distances transformed into semiautonomous regions considerably removed from those near Wabunun. Yet their existence is necessary for Wabunun's, and vice versa. His yams configured by *venay* divisions in his garden, Dibolel told me exactly how many of them he had sent to the west. And so that tree, which might generate a destabilizing transformation, keeps fixed in the memory, not the past, but an unpredictably destabilizing future. An instance of captured knowledge, the tree is a particular case of distilled information necessary for varying times and distant places.[45]

Notes

1. This name means "*tanagau* is its wife," *nakwav* meaning "its wife."
2. Southeastern Muyuw people blackened (*kalpwasisa*) teeth by first washing them with liquid generated by soaking part of the *vitivit* plant, the Pacific chestnut. Then they chewed one of two plants, *bwadloudaskaygum* or its shorter complement, *gamagaskaygum*. (*Gamag* translates to "people.")
3. Astonished they had never seen a well-known tree, I quizzed people from Unmatan, spatially and socially closer to its biome, to see if they knew what the tree looked like. They too professed ignorance, though like Wabunun people they knew it well.
4. Muyuw's *Canarium* (Damon 149, *babeu*, and Damon 296, *kinay*) are understood respectively as male and female, the former a small hard nut, the latter larger and softer. People eat both. Both species are sources of highly sought pitch (*seyak*). In its more liquid form, the pitch was used as a skin-toughener for men who make string by rubbing the inner bark on their thighs. When congealed (*g'eu*), it was used for lighting fires, likened to kerosene, and was especially important when men went to the sago swamps that flank the Sulog region. The pitch continues to be used in self-decoration. Called *sabayal*, it is combined with potblack to produce a gelatin-like paint. Gathered pitch was/ is exchanged to acquire *kitoum*, an important category of article in the Kula system. According to Yen (1990), *Canarium* were moved along various Melanesian islands by humans, either Austronesians or their precursors. Although it is not known if humans moved these trees into the Sulog region, knowledge of them was moved. Close to where we found the *niniwous* tree is a range of mountains called Lukidus; from these emerged an important subclan, Dillukidus, or Udunay. Wabunun Dillukidus informants told me their subclan mates

live in Vakuta, and one of their landings is called Okinay, in remembrance of kinay trees near their origin spot.

5. Muyuw are the sources for these ecological observations. After being told about them, I began to see that birds roosted in trees with smooth bark and long boles.

6. My voucher specimen could only be determined to family level. Undoubtedly the tree is *Inocarpus fagifer*. According to French (1986: 180) the tree grows in swamps or near rivers. Muyuw say it is fixed to these swamps with one inverting exception—a small grove lives on Mount Kabat's summit, more the moral than geographical center of the island.

7. After using the expression with everyone I encountered from 1995 on, late in 2002 two southeastern people told me that "*kaynen*" was the Wamwan, i.e. north-central Muyuw pronunciation, and that if I wanted to speak real Muyuw, I should say "*enan.*" I stick with *kaynen* in deference to my north-central friends, who often accuse me of speaking Muyuw rather than their language.

8. This determination was arranged through the Harvard Herbarium by the Lae Herbarium, by K. Damas, January 8, 2001.

9. The voucher specimen (Damon 256) for this tree was one of my best, well dried with flowers. Stevens only got as far as its family, Vervenaceae. I suggest the genus is *Gmelina*.

10. The experience of these fallow regimes changes over time. During 1973–75 many Wabunun gardens were being cut in fallows called *oleybikw*, last cut in the 1930s and 1940s ("when the Americans were here"). By 1996 there were few *oleybikw* to be cut around Wabunun, so virtually everyone was cutting either *digadag* or *ulakay*.

11. Muyuw wisdom states that *ulakay* closer to the shoreline are taller than those inland because they absorb the sea's power. I believe the generalization that trees closer to the shoreline are taller holds except for those in the mountainous area stretching from Sulog north to Mount Kabat. I was told that before the colonial order moved many villages closer to the shoreline, most of the land, especially along the south coast, was lined by *Casuarina* trees, which tower over other trees, rising to fifty meters or more.

12. The place is also at the junction of dry, wet, and swampy areas. The surrounding forests were not simply *ulakay*, as that term would prescribe tree sets as few as tens of meters east.

13. As a traditionally inhabited region, Sulog collapsed about 1870. No one living there is considered from there, although "real" Sulog people live elsewhere. But the active stereotype about the past remains that people worked stone, and traded it for vegetable food in an exchange called *takon*; if and when this transaction was not possible, they resorted to nearby sago swamps. By today's standards, "Sulog" is not enviable.

14. Damon 270, the tree is one of hundreds of legumes found in Papua New Guinea; I never found one with a flower, so the determination remains at the family level, Fabaceae.

15. I thank Jared Diamond for a discussion of this point.

16. I do not know these parts of the island as well as southeastern Muyuw; however, in 2009, areas around Kaulay village that I knew to be *tasim* but did not think were *sasek* were in fact so defined.

17. Freeman's critique neglects the significance of water for South Asia, a reason to question his criticism of ecological interpretations. Because groundwater is at least part of the understanding of Muyuw *tasim*, these constructions have family resemblances with those of Southeast and South Asia.

18. The history of the endemic Muyuw *Phalanger* (*P. lullulae*) remains unknown. See Heinsohn (1998) for a discussion of the creation of favorable ecologies from eastern Indonesia to the Solomons. Muyuw is a special case of Heinsohn's "captive ecology" (see also Bellwood et al. 1998).

19. Some meadows result from mining and deforestation activities of Muyuw's first colonial encounter, circa 1890–1940. The current mining camp called Bomagai is located on one of them.

20. Ellen (2004: 78), drawing on Yen (1990), notes the relationship between sago orchards and *Canarium* trees. Muyuw's two *Canarium* species grow in the same region as sago, only on higher, dryer ground. Yen raises the possibility that these forms were developed and spread together.

21. In 1996 Simon Bickler asked a Sulog resident to create one of these forms. See Bickler 1998, chapter 5, "Suloag Peninsula and Stone Tools of Woodlark."

22. For these handles most people prefer two shoreline and short trees, *kwanal* (Rubiaceae *Scyphiphora*; Damon 158) or *giyagi* (Lythraceae *Pemphis acidula*).

23. North-central Muyuw people told me they use a tree they call *malau* (Sterculiaceae or Tiliaceae) for the trough; in eastern Muyuw, *abanay* (*Podocarpus sp.*) is the tree of choice.

24. The southeastern Muyuw sequence: the first time, the burned crust was tossed aside, eventually becoming wild-pig food; the second time it would be fed to dogs; the third and final time, men and boys would eat it. Then it would be packed up. Men relished this material, and would often break into smiles when telling me about it. Central Muyuw people noted the sequence differently: first, *melol* (for dogs, *awuk*); second, *sivyuwen* (for people, *gamag*); third, *sivotonun*—*simwal yopway*, wrap up.

25. The meadow's name is well known because until 1975, the name of what most people thought was the highest ranked *mwal* in the Kula Ring derived from it; the valuable remains important, but no longer number one.

26. *Koliu* leaves have venation similar to *Calophyllum*, so Stevens's fern identification was a surprise.

27. One *digadag* tree is *auseli* (Damon 56, *Premna*). Invaluable for starting fires, its leaves and inner bark give off a smell like Vicks VapoRub, and it is used as an inhalant like that Western medicine. The other tree is *auduvid* (Damon 55 and 116, probably *Euodia* and *Melicope sp*). Only used for modest construction purposes, its sap might be chewed as gum when it hardens. *Akilim* (*Endospermum medullosum*), and *apul* (*C. apul*) grow in various meadows and both need open light when young; they may become canopy-top trees in mature forest. The former is a pioneer species, so it springs up in mass whenever *ulakay* are cut.

Ants, which live inside branches of young ones, clear surroundings to avail light when the trees are not growing in an opening.

28. Central Muyuw women are charged with transporting sago back to the village, so most have seen the swamps. By contrast, very few women in southeastern Muyuw have ever been to or seen the swamps. Some informants also knew that sago-making was not gender-marked on other islands as it is on Muyuw.

29. White flour is now as common as sago in these affairs.

30. Although sago remains a significant nutritional source, because of timber and gold-mining wealth people now consume rice, flour, and other "Western" food. During my first research period, bodies waxed and waned over the course of the annual cycle; they no longer do so and not a few Muyuw people are quite hefty.

31. Central Muyuw, unlike people to the east, also eat frogs (*kumeu*) associated with meadows. They say frogs spend much of their time amid or in trees of the high forest, but when it gets too wet they mass in the meadows; people go after them then. Although I learned about frog hunting in wet conditions from my Kaulay informants, Wabunun sources were the first to tell me about these central Muyuw culinary practices. Southeastern people do not eat frogs and find the custom among Kaulay people not unlike the way, proverbially, the English regard such customs of the French.

32. This Muyuw practice is at odds with McClatchey et al. 2006.

33. Sporadic sailing between Misima and Nasikwabw continues. In 2005 or 2006 a Nasikwabw outrigger canoe sailed to Misima and returned with twenty bags of betel nut; within the next two years a Nasikwabw man sailed Wabunun's *anageg* to Misima for betel nut and pigs.

34. In 2002 there were a few new houses near here, contemporary Nasikwabw's gardening area. But most people lived along the shoreline of a minor cove opening toward Muyuw. The area is plastered right between the ocean and cliffs rising to the island's flat top. Perhaps people could not have lived on the cove until the so-called Little Ice Age sufficiently lowered the ocean to expose it. Based on Nunn's work this would have been about 1300 AD (Nunn 2000; Nunn and Britton 2001). By late 2006 people were being forced back up.

35. Peter Stevens's discrimination made with *Calophyllum* was comparable. He distinguished two otherwise similar trees on the basis of their stem colors after they were dried.

36. *Kavavis* of mid-sized outriggers are shaped like large paddles. Several *Dysoxylum*, especially *tabnayiyuw* (*Dysoxylum papuanum*) and *silmuyuw*, are the ideal because of their exceptionally cross-grained wood.

37. Considered analogous to the term *dukuduk*, the black palm pieces that make flooring boards in eastern Muyuw houses, this term connotes a bottom divider with the function of protecting whatever rests on it. The same word, *atavat*, is used for the leaves from a tree called *alsivasova* that are put on the bottom of a clay pot to help keep the food from burning or sticking to the pot. In a survey of fence stake names, a female informant used this pot-use to name the tree.

38. This is the flooring upon which the yams will be directly set. In eastern Muyuw this part was called *atavat*, and as noted earlier its sense was given as a kind of protection. I was told a similar tale about *gwed* uses for this part in central Muyuw: it helps prevent yams from rotting.

39. Boraginaceae *Cordia sp.* has the astonishing ability to hold a fire in a charcoal-like state immediately after being cut. Sailors often use it to maintain a fire while at sea.

40. This resembles Trobriand practices, but I've not seen any of this.

41. Driftwood.

42. Neither Shirley Campbell (Vakuta) nor Linus Digim'Rina (Okaibom), both of whom have read (Damon 1998) or heard my accounts of this material, know this practice from their Trobriand experience.

43. People who have spent their whole lives working in Muyuw's western orbit will say this village's work is *mwagould*, bêche-de-mar, sea cucumbers, for the China trade.

44. One usage of similarity as an organizing principal is Lansing (2006, 2007), whose computer modeling of Balinese rice practices presumes that people copy those next to them. Whether that is so in Bali is an empirical question. Melanesia manages itself through organized differences, a nichification process not unique to this region.

45. Social structures designed to mitigate, among other experiences, ENSO events, were part of the social organization of the Indo-Pacific. Davis (2002: 280 ff.) discusses eighteenth-century Qing Dynasty responses in China. Thomas Trautmann (personal communication 2014) believes everywhere from Egypt to the Orient was organized for these events.

CHAPTER 4

A Story of *Calophyllum*
From Ecological to Social Facts

One way indeed in which signs can be opposed to concepts is that whereas concepts aim to be wholly transparent with respect to reality, signs allow and even require the interposing and incorporation of a certain amount of human culture into reality. Signs, in Peirce's vigorous phrase, "address somebody." (Lévi-Strauss 1966: 20)

Part of the problem faced by ethnobiologists was that they were narrowly comparing Linnaean taxonomy with particular folk ... domains and not looking at language more broadly.... These are all cognitive processes (means, agents, instruments) which help us first comprehend the world and then negotiate our way through it. They do so by acting on and through existing sets of beliefs or representations (the medium) and influencing the generation of new ones; indeed, they are the co-ordinates which determine how much of what comprises belief is expressed and represented. (Ellen 2006: 176, 185)

"Beach *Calophyllum*"—a European term for *Calophyllum inophyllum*—are astonishing trees to contemplate. They often grow up out of the sand or across craggy coral outcrops at the shoreline. They seem to flower all the time, the blooms turning into what botanists call "drupes" or "stones," near golf-ball-sized seeds that may fall into the sea and float considerable distances—this may be why the species is found across the entire Indo-Pacific. The thick leaves are long ovals with uniquely rounded ends and deep, semigloss green coloration, the appearance and striking feel characteristic of the genus. According to systematist Peter Stevens, the "decussate, entire leaves with dense parallel venation" make the genus "instantly recognizable, even in the absence of fruits and flowers" (1974: 349; see also 1980: 117). From the Greek meaning "beautiful leaf" (*kalos* + *phullon*), Linnaeus's original designation fits. On Kitava, Iwa, and Muyuw the names for these significant trees are *leyava*, *kakam*, and *kwakwam*.

When these trees grow next to the sea, they send massive limb-like trunks out over the water, their secondary and tertiary branches arching up to the sun. Although I saw enormous ones on Muyuw—one on the north coast west of Kaulay's lagoon that was close to three meters in diameter at breast height—and Gawa, many looked, and in fact were, pruned. So it seemed that the truly largest were on Iwa and Kitava. One at Kitava's northwestern tip had a limb at water's edge nearly two meters in diameter. One could only marvel at how its root structure must extend inland proportionate to the tree's reach over the water. Muyuw think *kakam* roots maintain their beaches; elders decry youth who wantonly destroy them.

I was quickly drawn to this tree and, by the end of my research in late August 1995, trees related to it that Muyuw understand as a "group." Four of the forty-seven voucher specimens I collected during those early months were *Calophyllum*. I sent all of them to the Papua New Guinea National Herbarium at Lae with the proviso that they would identify them for me and send two or three of the four copies on to Peter Stevens at Harvard. Although I received the identifications from Max Kuduk, Senior Research Officer at Lae, Stevens never received the copies meant for him. Yet I had learned enough about their local understandings to realize that I needed extra input on their forms, variation, and known ecology. So I wrote Kuduk asking for the authority on the genus. He wrote back immediately. I did not realize that that person was none other than Stevens, who by then had published a massive amount of material on the genus (*op.cit.*). When he learned that the trees were becoming central to my return, Stevens said I should recollect them. But I should not worry too much about their quality for he felt he could identify species from individual leaves, their shapes, and angles of venation. Although it was difficult to find flowers on any but *kakam*, I nevertheless took the best care possible, and by the end of my 1996 research period I had gathered new samples of all of the types Muyuw distinguish, though the number now had increased to six. Surprisingly, Stevens had difficulty with three, and he doubted descriptions of two species' ecologies that contradicted his evidence. He questioned my description of another species' mangrove-like roots, yet he was genuinely interested in the local knowledge and use of the set. This interest was sustained when I started comparing notes with Clifford Sather's knowledge of the genus from research among the Iban of Borneo. All species were again recollected in 1998 and 2002, and two species with flowers in 2006–7. From the 1998 collection Stevens thinks that he has settled one of the errant species and decided that another is a new one, but he continues to vacillate on a third. Amid this uncertainty, I became a mediator between two sets of experts: the people of the Kula Ring and Stevens representing the position as their Western authority.[1]

Much of Muyuw's culture, its place in the Kula Ring, its position in the cultural and perhaps biological history of the entire Indo-Pacific region, and its interaction if not collision with the West can be seen through the differing uses of *Calophyllum*. Many of these orders come to a climax with respect to one of the six species that, across the set of islands, is endowed with human attributes—it is personified. This chapter documents these relations.

Although *Calophyllum* began to capture my attention during the initial 1995 research foray, it was not until after *El Niño* 1998 return that the present understanding came into focus. The timber rush that characterized activity in Papua New Guinea from about 1975 to 1995, and on Muyuw from about 1980 to the early 1990s, contributed to this new focus because members of this genus are important export timbers. The trees thus formed a principle point of mediation between Muyuw culture and the encroaching Western world. This "West" is now driven by the rise of East and South Asia that is peripheralizing the region once again. Specific qualities of the genus—principally "usefulness" and prevalence—led the world market to Woodlark (Muyuw), among many other places in Papua New Guinea. Yet I believe the genus became important as much from its human as its natural history. Demonstrating this possibility is one of the purposes of this chapter. A final factor for this focus was that the genus turned out to be a principal focus of Peter Stevens's work as a systematist who, before I realized his *Calophyllum* expertise, had agreed to identify my voucher specimens. Identification chores that seemed to be an annoyance became a matter of curiosity when I handed over *Calophyllum* descriptions and voucher specimens. By late 1996 we had an identification problem on our hands, and a dialogue developed between Stevens and my informants. His previous work challenged what my guides taught me, and they accepted the challenge. Among others, this chapter tells the story of my mediating role between two vast bodies of knowledge and experience. Although contemporary "economic" issues concerning the genus are simple, both the Western botanical knowledge and the trees' indigenous cultural significance have proven to be more complex. Indeed, when I first realized the genus's significance, many botany experts were not aware that *Calophyllum*'s family name had been changed from Guttiferae to Clusiaceae; and in 2009 the classification was changed again so that Calophyllaceae is now a family designation[2] (see Stevens, Peter F (2001 onwards)). My experience with the set has been enjoyably humbling. Presuming there will be an end to the Western version of the story of *Calophyllum*, that time may be distant. As for the cultural significance of these trees, this chapter shows how they are central to the outrigger canoes of the region, creations of great technical and cultural sophistication behind which stands the histories of Pacific peoples.

The Genus *Calophyllum* and
the Recent History of Woodlark Island

Trees in the genus *Calophyllum* stand out in the recent history of Woodlark Island because they were often cut for lumber by the Neates until their mill closed in 1978. The tree then became a major export by Milne Bay Logging Company (MBLC) until that enterprise ceased operation in the mid-1990s. MBLC cut and shipped many other trees, mostly to East Asian markets. These included the previously noted species of *Dysoxylum*, *Syzygium* (of which the Muyuw *ameleyu* was the largest), *Sterculia* (*silawowa, ukw*), and in significant numbers *Mastixiodendron smithii* M. & P. (*akuluiy*), *Endospermum medullosum* (*akilim*), and *Intsia bijuga* (*meikw* and *kaymatuw*). *Pomentia pinnata* (*kaga*) is found in high numbers and grows to considerable size, but like other large trees—such as *Eucalyptopsis* in the Sulog region[3]—it rapidly becomes diseased. At least one and perhaps two species of ebony appear on the island, but they tend to be undersized and grow in exceptionally isolated places—places where people have never gone. Valued extremely highly by the international market, they are often cut and sold legally or illegally. Among others, German, Japanese, and Middle Eastern people flew to the island to look at its ebony while I was there between 1991 and 2002; in early 2008 a small, mysteriously named logging firm called Opus Diwai more or less snuck onto the island to extract some sixty ebony boles. Yet cutting ebony is a risky proposition: one never knows if a bulldozer might be destroyed while retrieving a single log. So the *Calophyllum* stand out because at one time they were not only valuable wood, based on international timber market considerations, but many were also regularly large enough to cut, near or larger than fifty centimeters in diameter at breast height, and relatively healthy. This is true throughout Papua New Guinea and Island Melanesia, where industrial-scale timbering has taken hold since the 1970s. From the point of view of the Western market, the genus plays a role similar to *Shorea* species found to the west in Indonesia and in nineteenth-century British India—a large, high-quality wood, found in large numbers.

Both within the regional setting of the Kula Ring and between it and the encroaching modern world, *Calophyllum* play a role that invites a comparison with sugar in the West (Mintz 1985). Sugar is a plant totally altered by human intervention, and consciously reproduced by human action. It became the very model for capitalist, large-scale production because of the way it is produced, its ability to cheapen the cost of labor power, and its capacity to act as a class palliative. The consumption of *Calophyllum* has not transformed modern class relations, but it will become clear that these trees figure prominently in the organization of transparent relations between people in this regional setting, the analogue of class relations in the West. And they had to be made.

Sugar comes from plantation agriculture. By contrast it seems that Western extraction of *Calophyllum* follows from its mere finding. However, this may be incorrect. *Calophyllum* seem to be a landscape phenomenon, a product of human invention. This possibility raises intellectually and socially significant questions about this fascinating but problematic genus. And this possibility brings us to critical details about one member of the genus. Mintz, and others, have argued that sugar cane and sugar production became a model for and testing ground of capitalist organization, meaning an intellectual and practical fulcrum for the production of wealth in the modern world system. In the different regional system that is the focus here, people locate wealth in their production of people and relations among them. This difference raises the question of how we should consider the anthropomorphized tree that centers many of the indigenous concerns with the whole set. The previous chapter closed by suggesting that specialized tree knowledge and practice serves as a calculus of relations across social boundaries. In describing a more limited set of trees here I continue this argument by showing how these trees are positioned as a model of and for production relations in a network that says something about something.

The Category Calophyllum *and Its Significance*

In his 1966 book *Timbers of Sabah*, P. F. Burgess includes data from *Calophyllum* trees obtained from New Guinea "because Calophyllum is likely to become an important export timber from New Guinea and the Solomon Islands." In his masterful "Revision of the Old World Species of Calophyllum (Guttiferae)" (1980), Stevens sustains Burgess's prediction while telling how he got involved with the tree:

> When I arrived in Lae, Papua New Guinea, in May, 1970, as a junior and rather raw Forest botanist, the question arose as to which economically important group of plants I should revise. M. J. E. Coode suggested the Guttiferae, although he later admitted that it was hardly the group to give to somebody with little taxonomic experience, or even to anyone with whom one may later wish to communicate in a civilized fashion. However, a revision of the Papuasian species of Calophyllum L., the genus of the Guttiferae with the most species attaining loggable size, duly appeared (Stevens 1974a) (Stevens 1980: 117).

Subsequent history has confirmed these statements. According to archaeologist Jean Kennedy, so much *Calophyllum* (*euriphyllum*) was taken out of

Manus in the 1980s, under peculiar circumstances, that she was asked to do contract archaeology over the cutting areas.[4] As noted, much was cut on Muyuw. By 1993 the editors of *Plant Resources of South-East Asia* could write that the future for the tree was good, for it may take over the places of formerly important tropical timbers, notably red meranti, (*Shorea* spp.). Several Papua New Guinea species (*C. Pappuanum* and *C. pauciflorum*) were "considered a decorative substitute for dark-coloured mahogany if suitably stained, and for all kinds of mahogany if transparently coated" (I Soerianegara, I. and Lemmens, RHMJ (eds.). 1993: 114). They write that "bintangor"—a trade name for the wood derived from a Bornean language—"often produces rather decorative figures on flat-sawn boards, and the distinctive colours of the timber are attractive for decorative purposes, such as furniture, parquet flooring, solid door construction, and for veneer and plywood" (118). In 1995 *Calophyllum* were the second highest exported timber from Papua New Guinea.[5]

Calophyllum is not significant only for timber. As a synonym for its old family name, Guttiferae, dictionaries gave "St. John's Wort family"—in other words, a group of plants of long-recognized medicinal usage. This sense also fits its few New World species as well. Bryan Finegan of Tropical Agricultural Research and Higher Education Center (CATIE) writes that

> the species we know best here is C. brasiliense which is occasional to common in Central American rainforest at least up to the north of Honduras. Besides being valuable for timber, it is recorded as being of high medicinal value by R. Chazdon and F. Coe and its bark is mixed with that of Coutarea hexandra (Rubiac.) to make an infusion used as a vermifuge [to kill and expel parasites—fhd)] and as an antidiabetic in French Guiana ...[6]

In a related apocryphal story about an adventure with a member of the genus, a biologist returned from Borneo to Harvard University in the late 1980s with extractions from trees that were to be analyzed for potential medicinal use with regard to cancer and HIV. Materials from one *Calophyllum* tree had positive results. The collector returned to Borneo to obtain more voucher specimens of *C. lanigerum austrocoriaceum* only to find that the forest was gone and no others of the same kind existed. In the ensuing panic, Stevens, then a curator of the Harvard Herbarium, was called in, produced a tentative identification from the existing voucher specimen, and conducted a search to see if any live ones could be identified. One was found in Singapore's botanical garden, biochemical research continued, and dozens of voucher specimens from this genus were collected and surveyed: Stevens was asked to identify about two hundred specimens in 1997 and 1998. For people like E. O. Wilson trying to make a claim for preserving biological diversity for future human use, the *Calophyllum* case is a prototype (Wilson 2002: 123–24).[7]

Taxonomy, Evolution, Distribution, and Asia-Pacific Names

"*Calophyllum*" is Linnaeus's creation. While the old family designation Clusiaceae has a worldwide distribution, *Calophyllum* does not. Those growing in the new world show little variation. According to Stevens they "may be derived from a single ancestor originally from the old world" (1980: 153),[8] and he suggests that the tree "may have originated in the Indo-Malesian tropics" (53). Fossil and pollen records from Java, Sumatra, and Assam suggest the genus has been around since the Miocene—so perhaps for 23 million years.

When I query Stevens about species totals, he refuses to give firm numbers. Although he has named or renamed many species and thinks I may have collected one or two never subjected to European appraisal, he does not believe that botanists have a firm idea of what species are. The "number of species" is not a meaningful question to him, but that is another story. Several species show a very wide distribution, most notably *C. inophyllum* and *C. soulattri*. Species seems to be concentrated so that the area including Sulawesi to Samoa forms one region, while the Philippines, Borneo, and India constitute another. There are perhaps two hundred species listed in the genus, fewer toward the fringes of its area, more toward its center. Madagascar has approximately twenty species, Central America has four, and South America has eight. There may be sixty or more in Borneo, perhaps the center of distribution if not the origin point; Papua New Guinea has about thirty-five species.

From the small number of species in Madagascar and on the nearby Africa shoreline, and, according to Stevens, their possible derivation from the single species *C. Inophyllum*, I surmised that they were taken to Madagascar by Austronesian peoples around 300 AD. Stevens doubts my conclusion because there are too many species for that short a time period. He may be correct, yet the form's systematics are not analyzed with respect to its many human uses. Given the six species described in this chapter, I wonder if the genus might be likened to a domesticated crop, its speciation partly human induced.

That its variation might derive partly from humans invites a quick comparison of a sister plant called *Mesua ferrea*, the "naga tree." Often called something like *penaga* in Sanskrit-based or -influenced languages, the tree held an important role in Buddhist temples and rituals. Part of this role had to do with how its flowers—magnified versions of *Calophyllum* flowers—were used for body adornment. However, I suspect the tree also figured in irrigation systems tied to Buddhist monasteries and transplanted irrigated rice agriculture because the tree moved with those social forms (see Randhawa 1969: 106; Burkhill 1935, 1944; Gamble 1902), a situation that parallels some of the usages of Kula Ring *Calophyllum* species.

This point brings us to the case of names. Although no anthropologist seems to have investigated how this genus is understood by a non-Western culture, there is information from Borneo and the Philippines, and some facts suggest that a "name" refers to something more than just a tree.

In Sarawak the trees seem to be known by binomials (Anderson 1980). It is as if *"bintangor"* references the genus grouping, the species level ascribed individuating names. For example, *bintangor madu* is *C. soulattri*, whereas *bintangor paya* is *Calophyllum hosei*, one of several Sarawak peat swamp *Calophyllum*. At one level, genus recognition is not surprising given the "instantly recognizable" nature of the leaf structure. So similarities among some names suggest cognation, along with a widespread understanding and use of trees in this group predating European interest. Given the attractive properties of this genus and the regional characteristics of this part of the world, this is hardly surprising. In this regard I find it noteworthy that the *Calophyllum* known to grow in the peat swamps of Borneo also take local names, meaning the trees were sufficiently well-known and used that their properties demanded categorical differentiation. Very speculative support for this is suggested by one Madagascar name, *vintanina* (*Calophyllum spp.*). If this name derives from any of the Austronesian Malagasy languages, and so sharing much with the primary language family of Southeast Asia and the Pacific, one would have to suggest that this name connotes something having to do with femaleness (n.b. *vin* and *ina*).

I have reason to believe that the Kula Ring situation is not exceptional. Struck by distant similarities between my own material and some of Sather's descriptions of Iban houses (Sather 1993), I questioned him about knowledge and use of *Calophyllum* in his area. His answers are complicated, but a few general points will suffice to bring my point home. Sather reports first that "Iban males ... are extraordinarily acute at tree identification and constantly attend to trees even when they leave their home river." Second, "in many areas—from mid to down river—the [*Calophyllum*] bark is used for making rice bins, hence *much symbolic significance*" (emphasis mine). People can stand in these rice bins, he told me. In our various correspondences Sather has spelled the name for this *Calophyllum* as *senaga, chenaga,* or *chinaga*. Perhaps this name is cognate to the term the forest literature often uses, *penaga*, and is therefore possibly related to the name for the tree known in India as the naga tree, *Mesua ferrea*, thus becoming associated with transplanted irrigated rice agriculture, Buddhist monasteries, and the complexities of Indic cosmology.[9] Although these associations are speculative, they generate the hypothesis that however much the botanical history of this genus dates from the Miocene, for thousand of years it should be considered part of the Indo-Pacific human record.

The Uses of Calophyllum

Wherever found *Calophyllum* have been used for medicinal and construction purposes. Most of the species have quite extraordinary sap, often profuse and sticky, sometimes clear to milky white or faintly yellowish green. Both steroids and alkaloids have been isolated from several species (see Mehrotra et al. 1986). For *Calophyllum calaba* L. var. bracteatum (Wright), Stevens writes: "In Cambodia ... the latex is used for shampoo (*Martin 1505*). ... In Malaysia ... the latex is used as a fish poison in fresh water. On Bangka the latex is apparently used to cure (?) ulcers" (Stevens 1980: 267). *C. soulattri*, found in most of the genus's range, has several medicinal or medicinal-like uses:

> In Bangka "getah malang-malang" is used to poison dogs. The bast ... is given to horses in Djakarta once a month to keep them in good condition. An infusion of the root is rubbed on to alleviate rheumatic pain. Oil from the seeds is used like that of the Calophyllum *Inophyllum*; the sourish fruits can be eaten.... In the Caroline Islands fresh bark from the shoot is used as medicine for women who have just given birth (ibid.: 287).

Calophyllum inophyllum's medicinal uses are perhaps the best known (Perry with Metzger 1980: 173–74), covering a whole range of maladies. I was amazed to see an almost instantaneous relief from eye inflammations, some kind of conjunctivitis, "pink eye." People crush fresh leaves, express the resulting profuse sap in fresh water, then bath their eyes with the water, a practice widely followed in Polynesia (Whistler 1992: 129–30).

Whatever *Calophyllum*'s medicinal uses, construction practices captured my fancy. The genus was an important timber tree before the recent emergence of industrial-scale logging in Melanesia. British India early identified several species as a valuable source for "sleepers," i.e. railroad ties. In southwest India, Kerala boat builders, who once built craft for the Indian Ocean trade, used at least two different species for their craft. One, predictably *C. inophyllum*, was used for keels and pulley blocks (see Person, R. S. and H. P. Brown 1931), while another, listed as *C. tomentosum*,[10] grows inland and was "much prized" for things like masts. This distinction, *C. inophyllum* used for one set of complex things, a more inland tree for others, turns up in the Kula Ring as well. Person and Brown also tell us that this "straight-grained" tree was tested at Patna for opium chests and used for ceiling boards. This note harkens back to the time when British

India savaged India's complex agroforestry systems partly to sell opium to China. *Calophyllum* are to South Asia, and Southeast Asia including the Philippines, what "oaks" are and have been to Western Europe and North America—major sources of wood for boats, rail transportation, and furniture (Reyes 1938: 258–61).

As a genus, *Calophyllum* is noted for species whose grains are interlocked. Experienced loggers in Papua New Guinea told me they sometimes had difficulty selling it because the grains destroy saws. Muyuw's logger had contradictory encounters. Some Philippine log purchasers refused his *Calophyllum* because of the interlocked grains while others, also from the Philippines, sought out the tree because those grains were precisely what they wanted to work with. Intimate understanding of its properties conditions its use. These properties, just beginning to appear in the Western literature, are fundamental to how the trees are appreciated and used in the Kula Ring.[11]

> Now, the characteristic feature of mythical thought, as of "bricolage" on the practical plane, is that it builds up structured sets, not directly with other structured sets, but by using the remains and debris of events ... (Lévi-Strauss, 1966: 21–22)

The "Group"

We now turn to the six types of *Calophyllum* that many people distinguish across the eastern half of the Kula Ring. Exactly how many know all of this information is uncertain. I gradually learned what is reported here beginning within days of my 1995 arrival. I never stopped checking what I knew, never believed I learned all there was to know, and never had the chance to recheck everything with everybody. On Kitava and Iwa in January 1996, I asked about the four species I learned about in 1995, but never returned there to discuss the whole set. Three species of *Calophyllum* grow on Iwa and Gawa, but I do not know the extent of those people's knowledge of the additional three that grow on Muyuw; Gawa people realize, however, that once their boats arrive on or near Muyuw, different trees are found for masts.

Some of what I learned was stimulated by the determination puzzles presented by my interaction with Muyuw people, Max Kuduk of the Lae Herbarium in Papua New Guinea, and Peter Stevens. Stevens is the umpire here, and he changed most of the identifications I received from Lae even though they worked from the key he constructed.[12] Two species he has named are *C. obscurum* and *C. vexans*, the epithets defining the ambiguity of the classing process.

Table 4.1. Muyuw/Western Identifications

Indigenous name[13]	Tentative determination	Grains *kasilu/sidumwal*	Uses	Location	Size in high forest
kakam leyava, C. Trobs, Kitava and Iwa (flowers = *kakamwa*)	C. *inophyllum* L.	*most interlocked*	*occasional keel; curved parts; leaves as sail models; maintains beach*	*Shoreline*	Not found: small to large on shoreline
kausilay in E. Muyuw, *kosilay* in. C. Muyuw, *kowo'silay* in Gawa and Iwa[14]	C. *leleanii* P. F. Stevens.	very interlocked	primary keel and *anageg* siding	sasek/garden edges to high well drained forests	large only
apul siptupwat in Iwa and Gawa	C. *apul* P. F. Stevens[15] C. *peekelii* L.	not very interlocked	negligible in Muyuw; *tadob* keel	high "wet" forest	large only
Dan	C. *vexans* P. F. Stevens	not very interlocked	wild pig food	freshwater swamps	small to large
aynikoy	C. *soulattri*	no interlocking, heavier	masts and poles, thinner	higher/dryer "mountainous" areas	small to large
ayniyan	C. *goniocarpum* P. F. Stevens	no interlocking, lighter	masts and poles, thicker	lower/wetter mountain areas	small to large

Table 4.1 presents the set arranged from top to bottom according to the most important way Muyuw class the trees, the degree to which the wood is said to be "interlocked," *kasiliu*, or "straight," *sidumwal*. Other forms of valuation would alter the ranking. Were the issue hard/soft, *dan* would be at the bottom in place of *ayniyan* and *aynikoy*; if the criteria were heavy/light, *apul* would be heaviest.

After nearly a week in the Trobriands, I arrived in Muyuw on July 15 1995. On July 20 I had my first conversation about the tree *kakam*. I had noticed them on Budibud, and on Kavatan's beach near another *kakam* I asked if there were any inland. "Yes," my source told me, but they are called "*kosilay*." Throughout Muyuw there was a tendency to refer to all of them as either *kakam* or *kausilay*, and then differentiate them. People may have referred to the trees this way because they understood the others as kinds of these two, but sometimes because they were speaking inside of what they imagined I knew. In either case, I begin this account with *kakam* and I bring it to a close with *kausilay*, using the southeastern Muyuw pronunciation.

As implied in the table and as should be clear from my description, the trees' ecologies are distinctive. This phenomenon reflects what I learned first from my informants and the observations to which they directed me. My reports counter the systematic literature. In 1974 Stevens begins summarizing what he knows of species limits as he introduces an ecological understanding: "The majority of species are clear-cut, but the limited amount of variation and the lack of obvious relationships and ecological preferences ... have aggravated the uncertainties caused by poor collecting" (1974: 354). By 1980 the situation had not changed very much (Stevens 1980: 158–61). As a herbarium expert can distinguish among the species, many of them can be found across diverse environments, although how that accords with notorious intraspecies variation remains an open question. I suspect, however, that few collectors had guides as knowledgeable about these trees as me. As I gradually learned, the people of this region pay exact attention to these trees.

Kakam

Reyava or *leyava* from Iwa to the west, *kakam* from Gawa to Muyuw, and the obvious variant, *kwakwam* on Koyagaugau and Ole are the names for the most widely distributed of the *Calophyllum* species, *C. inophyllum*. In Iwa I was told that its flowers are called *kakam*, so I investigated a possibility of some formal meaning as the name changes across these islands. Does an Iwa-Trobriand part, the flower, become a whole in Muyuw? Nothing came from my inquiry.

The trees are common. Very occasionally they will be found back from the shoreline and growing, therefore, more or less vertically into large, spreading trees with modest boles; this was the case with the first one I collected, a tree perched five to ten meters above the shoreline on a coral ledge near Wabunun. It was about twenty meters tall, but its trunk was more than a meter in diameter at breast height. However, these trees are frequently found along the rocky or sandy shoreline sending enormous limbs first out over the water before secondary and tertiary branches bend upward to the sun. As noted earlier, people in this region claim that *kakam* maintain a beach's coherence, recognition of the effects waves and wind have on beaches and this tree's ability to mitigate their consequences.

A mature tree has craggily silverish thick bark and profuse white sap, which sometimes has a yellowish tint common to the genus and in the bark of young trees in Muyuw for all but one species. *Kakam* sap is sticky enough for Muyuw to use as glue. Underneath the bark the wood quickly becomes blood-red. Large obovate-shaped leaves have a dark semigloss top and a lighter, duller bottom, coloring common to the genus. Unlike the other trees in this set, I recall *kakam* flowering almost all of the time. The cymose flowers—characteristic of the genus although some are umbel-late—have brilliant white petals surrounding an equally brilliant golden corolla. Although the genus's seeds, "stones," vary by size, the relatively large *kakam* seeds tend to be round, three to four centimeters in diameter. A mature tree produces hundreds of them, some falling and sprouting beneath it, while waves and tides carry others away. *Kakam* seeds are common in the flotsam all over these islands. Muyuw people believe the seeds can float away and sprout elsewhere, unlike other local *Calophyllum*, which they presume had to be planted in their locales or carried by birds.[16]

My recollection of the near-continuous fruiting and flowering behavior of this tree is not consistent with what some people think. When I sailed from Ole to Muyuw in July 2002 the captain of our boat kept pointing to the flowering *kakam*. To him, that flowering meant that the southeast wind should increasingly be blowing regularly. It was not; instead, the hard south wind blew. His planned short trip to Muyuw ended up being more than a two-month excursion because the winds refused to conform to a pattern that he tied to flowering *C. inophyllum*.

Iwa and Gawa people spoke of these trees as if individual persons possessed them, or some of them, and as such they are passed down from one person to the next. An Iwa informant told me he performed kula magic under his tree. And when I arrived on Gawa in early February 1996, my host was building an outrigger for Iwa friends almost directly beneath a *kakam*, which he claimed and which had been denuded for the work. Such claims are not generally the case in Muyuw, although I would not be surprised if somebody asserted prerogatives over some of them. Nevertheless, in Muyuw, as elsewhere, individual *kakam* are critical sources for various tools.

The trees are singled out because of their specific strength, ease in carving, and, when dry, lightness. The "interlocked" (*kasiliu*) wood produces the strength. This quality then becomes coupled with the way that so many of the tree's large and medium-sized branches arc up to the sun, making the wood best for products that have complex bends, forms that in many cases operate in stressed circumstances. From the point of view of their eventual human uses, contingencies of growth are turned into well-defined structures. With a characteristic "7" shape, adzes, *igayoy*, or *tawalu*, those once fitted with Sulog's stone heads as well as many contemporary ones made with steel points, were and are almost always made with *kakam*. The longer handle derives from a branch going one way; the sharp point is fit on the angle running more or less perpendicular to the handle. The change of forces on these tools and the delicacy with which they have to be used for shaving thin slices entails complex action and subtler control than that necessary in the more modern axe handles or for the arced handles used for sago pounders, neither of which are made from *kakam*.

Large primary *kakam* limbs, perhaps more accurately boles, are occasionally appropriated as keels for the *anageg* class of outrigger. The outrigger canoe I sailed in 2002, built in Yalab, was fabricated from a *kakam* keel. The boat's name, *Lavanay*, was derived from the name of the shoreline location from where the tree was cut.[17] All Budibud *anageg* begin with *kakam*. My Wabunun informants told me that Budibud people guided their trees by tying them from an early age to create the desired arc, what sailors call the craft's "rock." I have no reports of other people training these trees, though they certainly know that would be possible. Most *anageg* keels and strakes are built from *kausilay*, and these are often situated to curve as they grow. But like *kausilay*, large arcing *kakam* limbs or boles must be big enough so that the keel can be cut from the tree's curving meat-red heartwood, the only part considered suitable. All boat parts are taken only from a tree's heartwood. This might seem to be common sense, but tree and boat dynamics vary throughout the Pacific. To make the traditional craft of New Zealand and the Philippines, heartwood was rotted, burned, or chipped out of the trees that formed hulls, *Podacarpus totara* and *Heritiera litorles* respectively (see Tone 1903; Scott 1982).

Smaller *kakam* branches are used for making *geil* or *gulumom*, what are called the "ribs" of a boat in many European traditions. These pieces form the sides of the largest two outrigger canoes. Shaped by obtuse angles, their short sides rest across and are tied to the inside top of the keel. Longer sides are cut for and rise along the angle of the boat's side. Two together form a pair, tied at their bottoms to each other but going up each side of the boat on slightly different angles. Strakes are tied to these pieces. Details of a well-formed system, the knots and knot names for these ties are discussed in the next chapter.

The effect of waves on the outrigger float creates different dynamics for each side. The side next to the outrigger float should have a more obtuse angle than the side opposite it. This is because wave action sucks the outrigger float, and hence the boat, down. The broader angle of the outrigger side then acts like a lever that forces the boat back up. These ribs then effect a continuous rising action after the waves lower the float, a rocking more or less perpendicular to the movement of the boat through the water. A structural consequence of these forces follows in that the strakes tied to the "ribs" should vary in thickness with the forces they meet: those on the outrigger side taking the full force of waves should be slightly thicker than those on the opposite side.

Holes are drilled into the ribs to facilitate tying of the strakes; another hole is drilled into the ribs through which run a pole from a special tree whose property facilitates a supple strength that runs the length of the boat. There will be more on these gunwale-like forms later. As for the *kakam* pieces, the natural angles and the interlocked grains provide the necessary strength for the part's function, forming the strakes to the curvature of the keel along with other dynamics that will be reviewed in chapter 6. With a wood of a different quality—the cross-grained character is the critical variable—the various holes and angles would weaken the craft to an unacceptable degree.

The angles and sizes of each pair of ribs are unique, a feature that is a major theme for these complicated structures. These boats are finely engineered to the forces they constrain; beginning with wood selection, their forms follow finely reasoned dynamics.

The interior space in the keel should be noted. For various reasons these boats constantly take on seawater, so bailers (*yelum*) are specifically designed to fit this U-shape. Prototypically bailers are cut from *kakam* trees, usually from enormous thigh-sized roots found curling over the limestone rocks at the water's edge. The bailers are oval shaped so that water can be scooped up from inside a section of the boat kept free for bailing purposes. Opposite the instrument's handle, parallel to and the same level as the flat top of the tool, the end is shaped so it fits easily into the keel's trough. It is easily grasped, held by a clenched fist. One examined in 2006 was about forty-seven centimeters long, seventeen centimeters at its widest point, ten centimeters deep from the inside, less than thirteen centimeters from the outside, and with a handle about twenty centimeters long. Although they appear hefty, they are very light.

Kakam is the source for two other intricately carved pieces, one of which is called the *kuk*.[18] Designed to represent the head of a rooster, it is fashioned so that it fits perpendicularly into the top of a boat's mast by a mortise joint coupled by several ropes that pull it simultaneously up and down. A hole is drilled into it that functions like a pulley for hoisting the sail with a rope called *yawasay*, a halyard. I return to this structure and its imagery in chapter 6.

Far and away the most important part constructed from *kakam* is a ladle-shaped piece usually cut from a single branch. It does not just tie the middle of the keel to the middle of the outrigger float, it synthesizes the boat's experiences of the forces of wind and water. Fixed perpendicular to the keel and float, it will be three to four meters long,[19] approaching a third of the length of a boat. Its "cup," *kunusop*, is placed in the bottom center of the boat where it holds the mast as part of a very sophisticated mast mount, to which I will return. Its extension, *duwadul*, curves up along the outrigger side of the boat out over the outrigger platform.[20] Tied to it are four heavy-duty pieces of wood (*taniwag*), spars that angle down and are pounded into the outrigger float. These pieces are cut from the tree called *kaboum* (*M. fasciculata* [Warb.] H. J. Lam), Boagis village's ritual firewood. Forty more pieces, *watot*, cut from the same tree also connect the outrigger float to the main part of the craft. But the four angled off the *duwadul*, likened to a leg's hamstring muscles, are considered the prime pieces holding the outrigger float in place out from keel. People say all the *watot* might break, but these four must not.

Outrigger floats are always into the wind and waves and so are subject to violent motions. Much of the wind's force is pumped into the mast. Hence the combined piece—*duwadul* and *kunusop*—mediates the two strongest and potentially most violent forces. Shortly after learning the significance of the structure, I asked a close friend what would happen if there was no *kunusop*. He struggled with the question for a moment then recalled a person who quickly rigged a makeshift mast into a boat to sail the mile or so from Waviay to Wabunun. Figuring the trip was too short to worry about the lack of the *kunusop* function, he proceeded only to find the force of the wind and sail on the mast so strong that it drove the mast through the boat's keel, splitting it and swamping the boat. After pausing to make sense out of my question, he gave me an answer that described how the form is intrinsic to the calculation of forces in the boat. The *duwadul/kunusop* structure absorbs the force of the waves on the float while it modulates the forces moving the boat forward, helping convert destructive forces into propellants.

Completely fitting the *kunusop/duwadul* form entails the most complex tying on these boats, and that means in the culture. I was first confronted with the significance of tying while pondering the emphasis put on "inter-locked grains," and so this is an issue that bears on *kakam*. But too many other things have to be described before those relations can be brought to light.

Kakam are not critical just because of the complex shapes people con-struct from their growth patterns. *Kakam* leaves provide models for the top and bottom curves of the sail type that customarily goes along with the *an-ageg* class of boat. Describing what is at issue in this conceptual operation helps situate the significance of these craft, so I end this discussion with an initial orientation of sail types.

The generic term for sail is *mweg*, and Muyuw now distinguish three different kinds. The type that is becoming dominant in the region now, and so far as I know is completely new and spreading from Panaeati, is called *seylau*, a term now used for any outrigger craft employing the sail type. The shape is a trapezoid. Another type Muyuw call *yabuloud*, partly named after the *Pandanus* type used to make it, *loud*. Its shape resembles an isosceles triangle. Its "base" (*wowun*) goes into the prow or stern of the outrigger canoe depending on the course of the boat relative to the wind. This is the sail form for the western half of the Kula Ring, the outrigger canoes eastern Kula Ring people call *tadob*, Malinowski's *masawa*. They are built and sailed from Iwa through the Trobriands to Dobu. It is also the form that used to be employed for the finest middle-sized eastern Muyuw outriggers called *kaybwag*.[21] In 2002 I raised the question of the crab claw sail type of the *lakatoi* double-hulled craft, a sail common to Polynesia but almost nonexistent in Melanesia except for Papua New Guinea's south-east coast. Maylu people used to sail into the Kula Ring in general and Muyuw in particular looking for conus shells, which they used to make their own shell valuables. I expected my main informant, Dibolel, to give it a different name and sense. To my surprise he denied that it was a dif-ferent sail type, and instead insisted that it was a simple transformation of the *yabuloud* form. Using his fingers, he showed me how the sail is moved from the forward part of the craft, with one of its sides parallel to the wa-ter, to the middle of the boat, with the two sides of the isosceles triangle pointing to the sky in a V-like shape (Horridge 1987: 142–43). A form from outside the Kula region is internal to its own dynamics.[22] I return to how the respective canoe forms are transformations of one another in the dis-cussion of the *Calophyllum* tree Muyuw call *apul*.

The *anageg* sail type is called *aydinidin*, the making of which I discuss in the next chapter. It is rectangular in shape with curved ends. Upward of seven to nine meters, the long sides are bound by two spars or yards called *kaley* and *kunay*, which are cut from trees called *yals* or *ayniyan*[23]; because the latter of these is also *Calophyllum* and used for *anageg* masts,

I will return to it later in this chapter. The former is never used as a mast for an *anageg*, but it might be appropriate in a smaller and different kind of boat. Conceived to be shaped like the two pieces that make a spring to hold the *kunusop*, they are tapered, having "slender" (*igiligil*) ends and "thicker" (*patupwat*) centers. The reasons offered for this tapering are the same for the curvature of the sail ends. The curvature is referred to as the sail's *gudugud*. If you do not get the curved top and bottom correct, people laugh at you. Knowing how much people enjoy looking at these boats in full motion, I suggested the issue was just a matter of appearance. I was immediately and emphatically told that the issue was how wind billows out of the top of the sail, not just how it looked. The tapered *kaley* and *kunay* contribute to this action. The aesthetics are functional. And both the slender *kunay* and *kaley* and sail curvature help the boat veer toward the outrigger side when it is underway. If the curvature is not correct, then the wind would not billow out of the sail in the way it is supposed to and the boat might go opposite the outrigger side. Shape and appropriate motion are intrinsically related to this form, and wind dynamics are related to water dynamics.

Kakam leaves model the curvature of *anageg*, *aydinidin*, sails. Young boys are instructed to rig their tiny model sailboats with a *kakam* leaf. Usually the mast is nothing more than a stick fastened to a hull-like piece of wood. The other end goes through the "sail," the rectangular-shaped leaf, positioned more or less top to bottom. By building these model boats, boys experience aerodynamics appropriate to these craft long before they sail or are responsible for how a boat works.

After explaining *kakam* leaf usage, people said another tree's leaf could be used as well, one of the island's *Ficus*, *tatoug aulekel*; the type that grows inland differs enough to be inappropriate. Shortly thereafter I learned that Budibud people model their sails after the rounded snout of a shark, not *kakam*. In 2012 I finally learned that the *kakam*/shark distinction leads to a slightly different shape for the sails. The *kakam* form is asymmetrical, the shark form more symmetrical. So the rounded ends of the sail vary, with a more pronounced bulge toward the *kunay* side of the sail on the *kakam* form relative to the Budibud shark model. The Budibud informant explained that the *kakam* form was much stronger (*tautoun*), meaning it is faster. This has to do with how the wind spills out of the sails when the boat is underway.

The functional significance to which *kakam*-derived parts are put is elemental to the received understanding of this tree. From the way it is conceived to protect beaches to the forms its unique shapes and qualities allow people to imagine and construct, *kakam* is vital to a way of life. Through this tree people experience a transformation of happenstance into structure. From each branch seemingly arcing haphazardly toward the sun, stunning pieces of engineering are fashioned for every *anageg*. Women are not credited with being sailing experts on these islands, yet I learned many subtle facts about boat design first from them. Virtually everybody in the region knows about these facts and, therefore, never sees a *kakam* without thinking about its pivotal relations. The principle here runs through the culture, not just in boats. The creation of people is understood in the same way. How a woman becomes pregnant was formerly unknown. But as in the past, so now, once a woman becomes pregnant, the making and shaping of the child is conceived to be planned and deliberate action celebrated in ritual from the moment of birth until the final dissolution of a marriage upon the deaths of a couple's children (Damon 1983c, 1989b, 1990: chap. 4).

Apul *and* Siptupwat

By my second week back in Muyuw in 1995, I was beginning to piece together the constituents of "the group" of trees that, after my informants, I was calling *kakam* or *kosilay*, still unaware, among other things, that most southeastern Muyuw people called it *kausilay*. In short order I had a list of what seemed to be "interior *kakam*," an inventory consisting of *kosilay*, *apul*, and *dan*. Soon I understood that they were ranked by the degree to which their grains were interlocked, *kasileu*, though it would be a long time before I understood the force of that classification.

Several kilometers north and west of Wabunun on the road to the island's igneous center and the old mining/European center Kulumadau, I was shown an enormous "old" *kosilay*. With a gleam of anticipation in his eye, logger Rolly Christensen would eventually talk to me about it. He drove by it every time he went to the airstrip on the island's southeastern peninsula or the store in Wabunun, by the mid-1990s the largest and best stocked on the island. According to him, it was a "true *Calophyllum*," the several other types not really "true." This manner of speaking, common among the Europeans exposed to the island, was one of the facts that eventually made people ask me why there was only one English name—the genus name—for this set of trees. In any case, that road led me to Kaulay in north-central Muyuw where the beginnings of the mysteries of this set were gradually posed. For there the man, Talibonas, who eventually became my new north-central guide told me about a large concentration of "*kosilay*" in the "wet," *siposep*, region. This region was off the considerably dryer and higher limestone platform that forms Muyuw's southeast sector. The "true *Calophyllum*" on the road from Wabunun was the beginning of that concentration. I later learned that it was situated well into the dry area, but it indeed marked a beginning, although its placement was something of an anomaly. In any case, Talibonas told me I should ask Sipum to show me the "wet *kosilay*."

The return from Kaulay left me in good position to hear from more than one person that three of these types of trees were regularly distributed across dryer, wetter, and swampier land. I already understood that *kakam* was a beach tree, but this new realization was the beginning of the formal articulation of recognized differences among the set. No longer did I see them as kinds of *kakam* or "*kosilay*." So on July 30, 1995, Sipum and I walked to the beginning of the "wet region," the sago orchard called Upwason, to find good examples of *apul* and *dan*. There I was to collect my third and fourth *Calophyllum*, having earlier obtained samples of *kakam* and *kausilay*. I shall talk about *dan* next. In this section I discuss what I have learned about the tree widely understood to be called *apul*.

Altogether I have gathered seven sets of *apul*; I've seen dozens.[24] Included in these specimens were seeds and sprouting seeds, though no flowers, which Muyuw say look just like *kakam* flowers. The first of these was identified at Lae and said to be *C. peekelii* Laut.[25] Stevens never saw that one. It was the one collected with Sipum on July 30, 1995. Eventually others were also collected in that region, though not from the same tree. From 1996, Stevens consistently identified the specimens as a new species.[26] He has named it after the recognized Muyuw name, hence *C. apul* P. F. Stevens, a fact that greatly pleased Muyuw friends. In any case, Stevens distinguished *apul* from *C. peekelii* Laut, because the two look quite different when dried. He seemed sure of this because he thought he found one of the latter in my collections. The first collection I made in 1996 was from Iwa, obviously of a *Calophyllum*. An Iwa person had planted it on the upper flat oval in Iwa's interior for a future outrigger canoe. I knew the tree grew in the "wild" so I checked the box for that fact on the tree's voucher specimen sheet; however, I also checked the "cultivated" box for, in keeping with the rest of Iwa's landscape, the tree's existence followed a human design. I presumed it was *apul* because on both Kitava and Iwa I had already verbally confirmed knowledge of that type. However, my Iwa guide said the tree was called *siptupwat*, and Stevens identified it as *C. peekelii* Laut. Stevens found its existence interesting and surprising because, while represented in New Britain, New Ireland, and along the Solomon chain (Stevens 1980: 586), it had so far not been seen in southeastern Papua New Guinea.[27] Unfortunately, I never showed that specimen to my instructors in Wabunun, but I told them about it. They told me, however, that *siptupwat* means "big leaves"; *apul* have big leaves like *kakam*, and since Iwa people do not have enough trees to know what they are talking about, I should not have listened to them—it was *apul*. When I returned in

1998, I told them that Stevens had identified two different species, the new one, *C. apul P. F. Stevens,* and the Iwa tree *siptuwpwat* as a different species, *C. peekelii Laut.* They told me Stevens was wrong. This was the beginning of their mediated interaction with somebody who knew almost as much about their trees as they did. They insisted there was only one kind of tree with these dimensions, and it was *apul.*

Determination difficulties aside, the tree I call *apul* grows to be one of the largest on these islands. It is well-known, both by the local culture and any European interest that has cut trees for money. It becomes a canopy top tree, approaching forty to forty-five meters in height. It has a long bole but spreads out at the top. Its leaves are big, like *kakam,* and with the same rectangular obovate form. There is a difference among them beyond my appreciation, for whatever the similarity in size and shape between them, informants said *apul* was not appropriate as a model sail.

As noted, *apul* was Rolly Christensen's favorite tree because it was often so big. The only other *Calophyllum* that approximates its size is *kausilay.* The size issue is significant because many of the other trees that grow large often become diseased. Neither *apul* nor *kausilay* have that tendency, although the latter seldom reaches the size of the former before it dies and is replaced by a different suite of trees. Sooner or later, however, the same thing happens with *apul*; it is not a late successional tree.

Christensen started a plantation of *C. apul* below and south of Kulumadau near the road leading to the wharf from where he stored and shipped out his logs. In 1996 the plantation was doing very well, unlike another one he had started for ebony. Although I never saw the remains, I was told that the plantation burned up during the severe 1997–98 El Niño. During that season a great many fires spread out of control around the island among those places that had been logged during the previous decade.

Through 1996 I only paid attention to *apul* because it was part of the set of six *Calophyllum* trees. Muyuw are rather dismissive of it. They do not use it in their boats, though they told me that Gawa and Kweywata people make *anageg* masts from them. Although Gawa people told me otherwise—they said *kou'silay* (*kausilay*)—Muyuw never look closely at whatever mast comes with a new boat because they know they couldn't be as good as theirs—more on this later. Although the tree grows very straight, apul is considered moderately interlocked and heavy. These two features together make it unsuitable for a mast. Because of its apparent insignificance, aside from learning basics and identification properties, this tree fell out of my interests.

Nevertheless, I recorded a critical fact about the tree. Although neither Muyuw people nor the original producers of *anageg* regularly use the tree, it is the tree of choice for the *tadob* boat class, Malinowski's *masawa*. This form is the largest and most finely fashioned sailing craft plying the Kula Ring's western half. For the most part, positive knowledge about this tree and the form associated with it will have to wait investigation where it is the prototype.

Yet the boat form associated with *apul* makes it of exceptional interest, one that returns to the sense of transformations introduced with the discussion of sail types. My prime 1999 informant, an experienced maker and sailor of *anageg* from Yemga, told me that he refused to sail in *tadob/masawa* because he was terrified of their dynamics. They represent for him an unstable form. He designated this by pointing to the gap between the outrigger float and the hull, an area Muyuw call *miliyout*. In an *anageg* it is often near two meters, approximately the width called *ovatan*, the unit between the outstretched hands of a human.[28]

The *miliyout* area is a nontrivial space. First, it creates a useful platform. Although this platform is only good for supporting paddles and poles on Muyuw's two lowest classes of boats, for *anageg* it is a place where people regularly stand, where cooking occurs (fireplaces are made on it, formerly with sand deposited on woven coconut fronds, now often on sheets of metal), where the person steering the boat perches at the aft section of the structure, where the toilet facilities are located, and where equipment, including the sail, and pigs are stored. Second, given its size, it is the place where fine adjustments are made to the mast and sail by various lines tied from the mast's top and sail's corners and midsections. Dynamical forces of the wind on the craft are partly controlled in terms of this space; I return to this later. Third, it creates a structure used for navigational sightings. So, if you are sailing from Nasikwabw to an island called Panamut, you do so by making sure the former "sinks" within the *miliyout*. Finally, and most crucially for my Yemga informant's perspective, although the *miliyout* length is proportionate to the keel's length, minor variation in its width affects the speed and stability of the boat. If the distance between the outrigger and keel is lessened, the boat goes faster; yet this adjustment makes it less stable. Increasing the width makes it more stable but slower. Thus critical hydrodynamics orient the structure.

In contrast to the *anageg miliyout*, the distance between the outrigger and the hull in *tadob* is much narrower. I never measured this width but I would guess it is about a meter. Undoubtedly this form leads to efficiencies for paddling. *Tadob* are frequently paddled for significant distances; in Kitava in 1996 some of the Trobriand people there for kula purposes easily paddled some thirty kilometers back to Vakuta; in 1974 several Iwa *tadob* paddled to Wabuun. By contrast, for all intents and purposes *anageg* cannot be paddled. I have been in them twice with no wind and each time it took us a day or more to move less than thirty kilometers, even when we could pole the craft. *Tadob* also take a different sailing rig—the fore or aft set *yabuloud* form—and with that rig they are undoubtedly efficient for their purposes. But to my Yemga informant they represented a transformation beyond the bounds of reason.[29] *Apul* is not just a large, more or less useless tree that Muyuw people observe yet all but ignore. The type also points to a regional transformation, inscribing different ideas about appropriate order. In the axis of transformations between Muyuw and the Trobriands, it is of interest that the beginning point for the production and use of *tadob,* Iwa Island, is also the beginning point of the Trobriand calendar and the practice of pruning yams.

Although I have no idea what proportion of *tadob* are made from this tree, on Kitava, Iwa, and Gawa I was told that the tree is the paradigmatic choice for the type. The structural reason for choosing it is that *tadob/ masawa* are long boats built from a dugout shape. The most significant and largest part of the boat derives from a single tree. For this form *apul's* girth and its straight bole are perfect. This contrasts with the eastern Kula Ring *anageg* whose base is an arcing keel upon which three planks are tied. *Kausilay* provides the prototype for that form, and I return to the contrast later when I discuss the *kausilay* species.

Unfortunately I never pursued the implications of this contrast by an intensive inquiry of Muyuw ideas about *apul* as a form for *tadob*. Although I am not convinced there is a lot more to be learned from Muyuw people on this matter, there is more to be learned about the western side of the Kula's understanding of this tree and the proportion *apul : kausilay :: tadob : anageg ::* western half of Kula Ring :: eastern half of Kula Ring; and the dynamic entailed by the Mailu *lakatoi* form sitting astride these two.

What did fascinate me about *apul* was its peculiar distribution. I thought I knew the facts by the time I left the island in 1996, but they did not make sense until I returned in 1998. I've already had the occasion to repeat the Muyuw stereotype. The trees are found in soils that Muyuw class as *siposep,* wet. One could easily draw a topographic map of the island, point to all the low, but not necessarily swampy, places, and, given one further condition, expect to find them there. The rare, odd one is found on the dryer limestone platforms, the one several kilometers north and west of Wabunun being a good example. Others I have seen on the high ridges just above their "natural" place on the north side of the southeastern rise. But inside the region where they are supposed to be, trees are found in grove-like conditions—sometimes. In 2002 I went north of the southeastern ridgeline heading toward the Sinkwalay River into a wet zone appropriate for the tree. I had already seen several that were up on that area's ridgetop, just outside of their environment. But once we got into their region, I spotted nary a tree. By then, however, I had already deduced the further contributing factor that explains their occurrence.

Where the road to Kulumadau falls off the higher limestone platform forming southeastern Muyuw, the trees are very common. There is the occasional big one near the road, but most of those were cut for timber when the road was built in the early 1980s. However, alongside the road dozens of younger, shorter ones appear. Saplings and medium-sized ones also rim, and occasionally are found inside, the meadows I discussed in the previous chapter. And leading up to those meadows, though I did not appreciate this until my return in 1998, the biggest of them appear. These are trees of tremendous girth—often near a meter in diameter—with crowns rising to forty-five meters. This situation finds its parallel in north-central Muyuw where large *apul* are found interspersed in the region's *tasim.* Two enormous ones rose out of Kaulay village's best known megalithic ruin, Bunmuyuw, their elaborate root structures part of the process leading to the building's dissolution. These trees go along with aforementioned trees *mamina* and *bitok* as signs of *siposep kaynen,* wet-soil trees.

This pattern was startling to see as I entered the Upwason area with Sipum on the appointed day in 1995. We headed from the slightly raised roadside bordered by many young *apul* into a swampier area, the distribution of trees changing as we moved. By the time we were walking in mangrove swamp–like conditions,[30] *apul* disappeared and were replaced by *dan.* There was a transition region in which small hillocks rose a meter or two up from the swamp, and often large *apul* would be perched on these; *dan* often surrounded them. In the previous chapter I noted the hill *pwan likilok,* "trochus shell butt hill," which, because of its occasional use by Unmatan people, had a suite of trees remarkably like two garden fallow regimes. But it was also notable because *apul* grew on the hill while *dan* thrived in the swamps, their respective canopies intermingling. I found this relationship repeated around the meadow called Bungalau. Standing inside it and looking at its surrounding trees apul and dan appear as if they both ring it. But when I went to the edges it was clear that the *dan* were back in the water whereas *apul* perched on rising ground.

The dramatic differences in the locations of these trees became prime facts for a possibility that Muyuw did not just carefully distinguish the regular locations of these *Calophyllum* trees but instead used these trees for talking about ecological zones—that the trees were ways of talking about spaces.

A further peculiarity became evident with regard to *apul,* though not *dan. Dan* trees of all heights and diameters could be found, but this was not true of *apul.* Along the new road in the wet areas running to Kulumadau and ringing meadows, *apul* were quite common. In high forests, only canopy-top trees were evident. Often these would be surrounded by hundreds of sprouting *apul* seeds that never became saplings. A hypothesis about this peculiar distribution presented itself toward the end of my 1996 research period during another trek along the northern rim of the rise that forms the southeastern sector of the island. Along the western end of that rim no *apul* (or *kausilay*) were found. But both began to turn up as I approached the region more or less directly north of Wabunun. This, as it turned out, was the recent western limit of the human occupation of that ridgetop, the area beyond called Kweyakwoya becoming increasingly less settled from the late nineteenth century (see Damon 1983b). Hence it appears that the condition for the tree's existence is major modification or leveling of a forest canopy, most often done on Muyuw by humans. I had seen several *apul* growing on Gawa and was told many were there, as well as on the small island north and west of Gawa and Kweywata, Digumelu. Iwa people had told me that people had recently gone there to make four *tadob* from *apul* (in the conversation referred to as *siptupwat*). These small islands are totally anthropogenic landscapes, so the existence of the tree there fits the hypothesis perfectly. When I proposed to Rolly Christensen that he was in fact cutting his "true *Calophyllum*" in regions regularly occupied or modified by humans over the last 150 years or so, he could do little but consent to the probability. Since his area of most active cutting surrounded Kulumadau, he was operating in regions from which Muyuw had been dying since the arrival of European diseases and in which miners had been traipsing since at least 1895.

While this hypothesis explained the presence of many *apul* saplings along the road Christensen had cut through the center of the island, it did not explain those within a kilometer or less of the meadows I was frequenting, especially Bungalau. Sipum told me that new *apul* would grow up in the gaps created by fallen older ones. This idea squares with ecological models of some tropical trees: seedlings wait until a gap appears with the collapse of a mature tree, then a fortunate tree quickly shoots the gap to the top.[31] However, I came across a few enormous downed *apul* and saw no emerging young trees. (The same situation prevails for *kausilay*.) The only places where I saw smaller rising trees were on the edges of human disturbances, recent roads, edges of occupied or garden areas, and the El Niño–created meadows. And it was my return to those areas in 1998, just after out-of-control fires ravaged parts of the island during the severe 1997–98 El Niño that enabled me to formulate a plausible model for the appearance of the trees: when Muyuw regularly burn off the meadows during the droughts, they cannot control the reaches of the fires. And once that became clear I noted that many of the large *apul* I saw while moving toward Bungalau or its sister meadow, Salayai, were within sight of those meadows. Hence this hypothesis: the occurrence of *apul*—and therefore one of the island's most valuable products for the Western-inspired international timber market—is a direct or indirect consequence of human action.

This hypothesis is consistent with what seems to be the case on Manus Island with respect to the related tree species *Calophyllum euryphyllum.* Following extensive logging in the 1980s of forest dominated by this species in the southwestern Manus, archaeologist Jean Kennedy and a class of archaeology students from the University of Papua New Guinea began site surveys in the area. Apart from the completely unexpected discovery of a large number of prehistoric upland village sites on ridges and hilltops throughout the logged area, Kennedy found two things surprising about the forest. First, as an earlier botanical survey by Karl Kerenga and James Croft had recorded, much of the forest was dominated by mature *Calophyllum euryphyllum,* of which there were sprouting seeds but no saplings. They also recorded this rare *Calophyllum*-dominated forest type at other locations in Manus. Second, the forests overlaid and therefore postdated the archaeological sites, which cover an uncertain age range from a minimum of a few hundred to at least a few thousand years ago. Thus, human disturbance is likely to have some role in the forest's establishment.[32] This situation parallels that concerning Muyuw's *apul* (and *kausilay*). If not over village sites, the *apul* (and *kausilay*) distribution is associated with human disturbance (see Bayliss-Smith et.al 2003).

Dan

In the evening after my July 30, 1995, visit to *apul* and *dan* locations, Dibolel quickly remarked of the latter that they were "*singaya igogeu.*" *Singaya* translates to "very," *gogeu* to something like "wobbly." The context of this evaluation had to do with the comparative assessment of the *Calophyllum* set given their overwhelming use in outrigger canoe construction, something I did not yet appreciate. What I took away from the description, however, was a dismissive attitude, one similar to my attitude about *apul,* and as with *apul,* one that stayed with me until 2002. However, as far back as 1974 I was accumulating experiences that might have marked the tree as worthy of inquiry.

Dan are distinctive trees. Although occasionally standing alone, and although *apul* and *kausilay* often grow in groves, *dan* are remarkable because people speak of them as a "*dan* forest." People will talk about going to the "*dan*," or finding a pig "in the *dan*." These forests are certainly distinctive. They reminded me of mangrove swamps because the aerial roots were so prominent they made passing through them problematic; as in a mangrove forest, one usually wades through water and mud, occasionally finding or losing one's balance on protruding roots. In these circumstances the trees come in all sizes, shorter slender ones amid larger taller ones. *Dan* also have the smallest girth of the group—"like a coconut tree" is often how people would describe the largest of them. One cut down in 2002 was about fifty centimeters in diameter at breast height, the largest I ever saw. Yet if they are small in girth, they grow to the canopy top, or even above. That 2002 downed tree was at least thirty-five meters tall; we took one down in 1995 that was upward of forty meters, although it was only about twenty-five centimeters in diameter at breast height. While I had qualms about felling these trees to obtain voucher specimens, my informants felt no compunction whatsoever. This is because, according to them, *dan* readily grow back from their stumps, one of their diacritical features.

Dan bark is often brownish red, and if blacker, it is clearly marked by streaks of red visible in its creases. Its sap tends to be clear and is not nearly as profuse as other *Calophyllum*, especially from leaves and twigs. White wood right underneath the bark is streaked, *sidasad,* and unlike the smoother—*sipusap*—wood found underneath the bark of, for example, *kausilay.* And the enveloping inner wood is less pronouncedly red as the others. The tops of the trees are also very different. Both *apul* and *kausilay* tend to spread from long boles. *Dan* have more of a Christmas-tree shape. I took a picture of one from underneath the Creator's sago tree inside of Bungalau because Sipum had pointed it out as close to his stereotype. And in fact this straight, tall, slender form is part of what Dibolel invoked when he described the tree as very wobbly.

Stevens has had difficulty specifying the species name for the voucher specimens I collected in 1996, 1998, and 2002. He thought the 1996 and 2002 specimens were *C. vexans.* In 1998 I also collected sprouting seeds over which we carried out a discussion, wherein Stevens said that the seed looked like an unidentified tree from the Louisiade on the one hand, or a species called *C. acutiputamen* on the other.[33]

From his 1974 and 1980 descriptions of *C. acutiputamaen,* I could not construct a likeness given my observations and Muyuw reactions. Stevens, moreover, doubted my suggestion that walking through a *dan* forest was like walking in a mangrove swamp. Never confident of my own observations, I took those doubts back to Muyuw in 1998 where they were greeted with laughter because Muyuw thought my characterization was correct. When we went to the *dan* forest we made a point of taking a picture with Sipum holding on to one of the aerial mangrove-like roots. Botanists call these structures pneumatophores; they are breathing roots for swamp-dwelling trees to make up for the fact that there is little oxygen readily available in the soil. Stilt-like lower trunks probably are a related phenomenon, and were very evident in the *dan* I examined in 2002. These characteristics, moreover, fit another species Stevens has described, C. piluliferum, which "grows in swampy places subject to floods of short duration. ... The field label of Pullen 7531, which has galled stems, mentions 'butt surrounded by peneumatophores" (1974: 388). Moreover, he describes the tree with a red underbark (1980: 613), a fact that particularly stands out in the Muyuw *dan* since, as noted elsewhere, all the other Calophyllum tend to have the distinctive yellowish cast to them. Although *C. piluliferum* shares some of the features of the *dan* I have seen and described, the Muyuw trees have much larger leaves and fruit. A final determination for this tree is in the future.[34]

These are technical matters for *Calophyllum* systematics and at first sight beyond the interest of the current inquiry. However, for the deep geological history of this region and, perhaps, for its much shorter human history, these are matters of interest. From the geological point of view, a spreading zone separates Muyuw from the Louisades. Some million or more years ago these were continuous landforms; Muyuw has frogs that could have only come from the Misima/Louisades region.[35] Although Muyuw believe *dan* is unique to their environment, it would be interesting to know about similar species across these islands. Yet that natural history enjoins the region's human history for one critical reason: the *dan* forests are part of Muyuw's "wild pig" culture. These pigs eat *dan* nuts and, according to Muyuw, the nuts of no other *Calophyllum* tree. People know this for several reasons. First, unlike with any other *Calophyllum*, it is difficult to find *dan* seeds. Pigs devour them.[36] Second, wild pig excrement reportedly changes color when the pigs consume the seeds. This is something Muyuw observe because the flavor of the meat also changes, and for the better. Consequently, while people often frequent *dan* forests looking for pigs, they make a special effort to do so when they know the seeds are falling, evidence for which they obtain either from the nuts themselves or black pig excrement. We experienced this in 1974. I knew nothing then about these trees except that they were near sago orchards and that pig flavor changed from the animals eating their nuts—I was reminded of baking turkey the first time I experienced it. Although Muyuw do not consider any *Calophyllum* seeds food, Sipum tasted *kakam* and *apul* seeds. He had an idea that they tasted sweet, and knowing that pigs loved *dan* seeds, he tried them with that expectation. To his surprise, he found them unbearably "bitter," *yayan*. He suspects that this characteristic is related to a final observation, and that is when the pigs eat *dan* seeds, round worms are never found in their intestines. The fruit is a vermifuge.

When I asked people if they planted *dan* forests, they answered "no," that the phenomenon are *misiken*, "fixed," perhaps "intrinsic," to their locales. It was the same with the sago orchards found in the area. Yet all this asserts, really, is that origins are not critical for their understanding of relationships. For by the end of my 2002 research experience it was clear that the *dan* forests were as much a part of a well-formed environment as were the *sinasops*, meadows like Bungalau of exactly the same region whose existence people reproduce through El Niño events. Moreover, the height of *dan* trees are thought to be critical for guarding the underlying sago orchards—the tall trees protect the lower, less stable trees from wind gusts—their towering "wobbliness" facilitates this. Although positive about these interrelations in south-central Muyuw, I was unsure of their existence in the north-central part of the island that effectively extends into the range of Kavatan's resource base covering the eastern half. But during my 2006–7 visit, an elder from the north-central region assured me that identical ties between meadows, sago orchards, *dan* forests, and wild pigs exist there. As noted in the previous chapter, the meadows are designed partly as wild pig refuges. *Dan* forests contribute to this set of associations. Like sago trunks whose tops and bottoms are left intact for pigs, they are elements in a network of detailed, intertwined phenomenon, known, experienced, and used in well-formed ways.

Ayniyan/Aynikoy

With the exception of *dan*, all Muyuw *Calophyllum* have a peculiar yellow cast to their bark and, especially when shorter, sometimes their sap. The trees I first got to know as *ayniyan* preserve this feature into their maturity. Sipum, who first examined a tree's bark whenever I asked him its name, told me he picked *ayniyan* out of the forest by this coloration. *Tigitag* is the Muyuw term used for the hue, a category that includes what we distinguish by yellow and green. Once the quality became familiar, their recognition became automatic.

Although not as distinct, this coloration is preserved through *ayniyan*'s variant, *aynikoy*, which I gradually learned about. This way of naming the two trees is unlike any other set of similarly named or related trees. People would often say one was the "real" *ayniyan*, and so *ayniyan*, while the other *ayniyan* was, instead, *aynikoy*, a kind of marked instance of the former, a locution probably motivated.

Both are said to be Sulog trees, meaning they are found in the region surrounding the island's higher igneous center. It is said that the trees rot relatively quickly, so they are precluded, among other reasons, from being used as keels or boat planks or strakes. But both are the prototypes for *anageg* masts, as reported by Wabunun and many other people. The tie between these trees and masts is deep. Recall the leaf "test" in chapter 2.

One person who took the test couldn't think of the name *ayniyan* but instead said "*vayiel*," the word for "mast." The "straight"-grained wood that makes these trees ideal for masts also makes them the tree of choice for poles (*altowatan*) for pushing outrigger canoes. Consider a pole vault: the pole has to be able to bend a great deal but then straighten, providing locomotion in the process. That quality is sought for poling a canoe. *Dan* "wobbliness" precludes them from that function.

The use of these trees as poling devices led me to question the Muyuw understanding of their ecology—the Sulog location. I eventually gathered all of my voucher specimens for the two in the Sulog region. Although I did not count examples of the tree in 2002 when I accompanied people looking for a replacement mast, *aynikoy* was clearly one of the numerically dominant species on the upper slopes where we were walking—there were scores of them. I never saw a concentration of these trees anywhere else like that. However, my informants occasionally found one far onto the limestone platforms, sometimes in mature forests close to Wabunun. And if these were upward of four meters tall or more, people frequently cut them for an outrigger canoe pole. Reportedly "birds" distribute the seeds beyond the Sulog area. But it may be that the trees are more common in Sulog's area not only for its geological or geomorphological conditions but also because elsewhere they are avidly exploited. Whatever the case, Muyuw see the trees as a characteristic of the Sulog Mountains and areas rising to it, one of the resources making the area central to the regional culture.

And it is not just Muyuw who understand *aynikoy* as the mast proto-type. In 2002 Koyagaugau and Ole people reported that they could find the same species closer to hand, but they felt that those near Sulog were the best. I overheard Ole/Koyagaugau people talking about these facts unprompted by me. The issue came up naturally because my 2002 voyage to Muyuw from Ole was made with a mast constructed from the species Muyuw call *kausilay*. Once I contracted the voyage to Muyuw, the boat owner's additional motive became finding a suitable mast. The mast he found snapped in the sector between Panamut and Nasikwabw called "insides fly," and he replaced it in Nasikwabw. However, Nasikwabw has neither *ayniyan* version, so the replacement had to be cut from *kausilay* again. By the time its *aynikoy* replacement was cut and trimmed, the re-cently cut *kausilay* mast was splitting—*sasal* is the term, a quality not seen in either of the "*ayniyan*."

Both *ayniyan* are desired as masts because they have the straightest grained wood in this set, the prime quality people specified for the masts. This condition allows the mast to bend toward its top while the lower end of the piece remains firm. The vibrating mast facilitates wind regularly spilling out of the top of the sail, as it must do for it to work properly. Other members of the *Calophyllum* group, presuming they are straight enough, are considered either too heavy or too "interlocked" to bend appropriately. Although I do not understand the physics of this issue, the physicists, aeronautical engineers, and sailing enthusiasts with whom I have discussed the issue say the Muyuw reckoning is correct. And it is not casually acted upon. As noted, when our Nasikwabw replacement mast was cut, it had to be from *kausilay*. When it neared completion,[37] Duweyala tested it. One man stood on its "base" next to the ground. Its middle was raised a meter by placing it on anchored crossed sticks. Another man, positioned at its "tip," lifted it up and pressed it down. Duweyala stood to the side gauging how it moved with varying degrees of pressure. He directed more trimming near its base. At one point the diameters of the old and new masts were measured with a string about two meters above their bases. The point here wasn't to make them identical since they were the same inappropriate tree. But a studied judgment was manufactured. When he thought it was finally done, Duweyala set the piece in the testing position, went to its top, and on his own pushed it up and down to feel how it moved. When it was finished, the mast measured 8.83 meters long, replacing one that was 8.84 meters.[38] Just before it tapers quickly to fit into its *kunusop* socket, it was about 37 centimeters in circumference, the original 34.6. At its midway point it was 36 centimeters in circumference, and near its top, at the 8.4-meter point, where it is rigged to hold the *kuk*—the contraption that holds the *balau* ("shroud" or "stay"), the line used for pulling the mast toward the front and back of the boat—it was 18 centimeters in circumference, the original about 25.

Several days later, only a few hours into our journey to Waviay, Duweyala became dissatisfied with the mast's performance. It was too soft, he said, bending inappropriately because it was the "wrong tree" (*tobwag kay*). Experienced models govern the selection of *ayniyan* or *aynikoy* as appropriate for this function.

With some boat forms there is a certain way of holding one's body to illustrate how the part is supposed to be and function. The sail is particularly important in this regard, its curved shape not only modeled by kakam leaves but also one's curled tongue. The degree to which the sail is curled controls how the wind is caught and helps determine how and where the boat moves in the water. Sometimes when I asked people to tell me how they would sail between two certain points, expecting to hear star courses, they would instead say it was easy; they'd lock their bodies into a distinctive shape and hold up a cupped hand to exemplify how, together with their body, they would set the sail to that place. I tried photographing one of these poses, but the form is too distinctive to catch with one camera shot. As it turns out, another such pose is used to exemplify how masts are to work, and, explicitly, why both *ayniyan* are the trees of choice. People hold up one hand, often the left, in a semi-cupped form. The other would be fit into it with the index finger extending up, surrounded by the enclosing thumb and index finger of the first hand. They would then wiggle the top of the straightened index finger back and forth, keeping the encompassed bottom motionless. This is the kind of movement the mast is supposed to have, bending at the top, motionless at the bottom. Every time Duweyala illustrated this for me he burst out laughing, something I never experienced with eastern Muyuw who had performed the same act for me. Duweyala laughed because the motion is thought to be like sexual intercourse. The mast is a penis, the supporting cup, the *kakam kunusop*, a vagina. Often leaves from two trees and a vine are placed in the *kunusop* where the mast rests, and if the boat is really sailing well, water will be forced up from the impact of the mast on the leaves in the *kunusop*. So far as I know, everybody imagines that this relationship—mast/mast mount—models sexual intercourse, though for my Muyuw hosts the analogy wasn't taken so literally—it was an expression of complementarity, for which male/female contrasts in general and sexual intercourse in particular are the focal models. I shall return to the issue of body metaphors and boat parts in chapter 6. For now, having said this, I must report that nobody *in Muyuw* makes the next step to say that *ayniyan/aynikoy* are male while *kakam* are female. *Ayniyan/aynikoy* are trees, *kay*.

By 1998 I was far enough along in my understanding of these forms that two of my best informants laid out a neat paradigm describing the differences between *ayniyan* and *aynikoy*. Table 4.2 reproduces their model. The features that Muyuw use to differentiate the two *ayniyan* become operational in their primary uses as masts. *Ayniyan* is lighter, *aynikoy* heavier, presumably because it is denser. Consequently the bending characteristics

Table 4.2. *Ayniyan/Aynikoy* Comparison

ayniyan	quality	*Aynikoy*
gagab (light)	weight	*momovit* (heavy)
kapakopw (fat/rounded)	leaf shape[39]	*igiligil* (thin/pointed)
tigitag (yellowish)	bark color	*bwabwel* (red)
mameu (flexible)	flexibility	*mamatuw* (stiffer)
a little red	heartwood color	very red

of each tree type vary slightly. According to Muyuw reckoning, masts made from *ayniyan* can be larger in diameter, while *anikoy* are smaller. This factor, along with density, introduces variability in frequency of vibration. When Sipum and I were looking for examples of both in 1998, he had a hard time finding *aynikoy*. There were only very small trees, saplings a meter or two high. Since the leaves of immature trees are often poor guides to what mature trees look like, he could not reliably identify the trees by their leaves. So he started swaying them back and forth. The heavier *aynikoy* exhibits more resistance than the lighter *ayniyan*. He was trying to distinguish them on the basis of how easily 1.5-meter-high saplings moved back and forth when swayed.

Wabunun sources refused to say if one type was better than the other for a mast. According to them, one prefers to have a "light" (*gagab*) mast. However, this desire becomes modified by the necessity of balancing between the qualities described by the terms *mameu*, pliability or flexibility, and *mamatuw* or *kalamatuw*, hard or stiff. The bottom of the mast should be the latter, the top the former. The trees present different options, so the choice of one over the other depends on the specific ways a mast's operating qualities fit the other characteristics of a given boat. Watching Sipum in 1998 make judgments about the different types was a lesson for me, seeing in action a principle feature of these boats: that there is a ratio-like understanding to their component parts bound complexly with the differential qualities of the trees used to make them. One is better than the other depending on how their qualities relate to a boat's other dynamics.

Whenever a new *anageg* is acquired, one of the first things its new owners do is head to Sulog to replace the existing mast. Although I suspect *aynikoy* is the first one usually selected, to be consistent I will just note that one or the other is first used, and then the person responsible for the boat decides how it sails with that particular mast. If it does not feel right there are always more trees. The critical issue is how the mast vibrates, which is what Duweyal was examining as he tested his Nasikwabw replacement mast.

Not everyone was ambivalent about which tree should be used. In 1998 my prime Nasikwabw informant said he always cut *ayniyan* and never considered *aynikoy*; I was never in a position to go back to him when my understandings were transformed by the 2002 experiences. Yet, in 2002 when the Koyagaugau/Ole crew went to Sulog to select a new mast, Dibolel, who with his younger brother Ogis is a stickler for minor details, gave the group instructions to cut *aynikoy* but not too big a tree. The option was firmly in his mind, and he was putting it forward to the guests to govern their selection. He directed another Wabunun man—Gumiya—to accompany us to help find the appropriate tree. Gumiya led us straight through where we might find *ayniyan* toward those where *aynikoy* were expected to be abundant. Later attempts to have Dibolel explain his directions were unproductive: he made a judgment about what that boat required and directed us to it.

This point brings us to several ecological features of these trees. Unlike *kakam, kausilay,* and *apul,* these trees grow up underneath a high canopy. Although I found plenty of *kausilay* and *apul* seedlings beneath some high canopies, smaller trees and saplings were rare. This is not the case with *ayniyan* or *aynikoy.* Where they are found, trees of all sizes are common, and the mature trees go to the canopy top. In the Sulog region that was at least forty meters. This brings us to the other differentiating feature between the two: even though we were finding really short saplings in 1998, Sipum was becoming frustrated by our inability to find what he was presuming would be the appropriate voucher specimen. So he stopped, thought for a moment, and then immediately found a rise to a hilltop. We started climbing, and in short order we found a canopy-top tree that he thought was *aynikoy.* The differentiating sound between the two names is the morpheme *koy,* which by itself means hill or mountain. *Ayniyan* are more common in the lower, wetter grounds in the Sulog vicinity, *aynikoy* on the hills. And to check himself, Sipum slashed some of the thick bark off the tree, examined the first layer of wood inside the bark, and found what he called *katuson,* apparent furrows running vertically up and down the tree's bole just inside the bark, forms not found, according to him, on *ayniyan.* This condition is referred to as *sidasad,* and contrasts with "smooth," *sipusap,* which are *ayniyan.*[40]

These acute discriminations should be contrasted with Peter Stevens's identifications. Although he noted that species variation makes determination conditional, based on my 1996 voucher specimens he was fairly certain of one of the "Sulog" *Calophyllum,* calling two of the specimens *C. goniocarpum* P. F. Stevens. But he did not think my lone *aynikoy* from 1996 was complete enough to do anything more than distinguish it from the others. That seemed to be rectified by the voucher specimen Sipum found on the hilltop in 1998. Stevens was relieved to be able to label that one *C. soulattri.*[41] The clarity of this situation, however, disappeared with the 2002 specimens, not only by its shape but also by its location. We had in fact walked straight though the lower, wetter areas near the place where we landed toward ecological conditions appropriate for *aynikoy.* Once on the rise near the mountaintop, we found dozens of them. The first tree Duweyala cut for a mast was readily classed as *aynikoy.* However, Stevens decided that the voucher specimen I made from it was *C. goniocarpum,* and the *ayniyan* I collected in 2002 was *C. soulattri.* This reversed what I expected. Although these trees are similar enough that from standard voucher specimens distinguishing them can be tricky, Muyuw expect the two trees to be found in very distinct ecological settings. Was Stevens wrong?

When I went with Duweyala to Suloga to find the replacement tree for his mast, we followed Dibolel's instructions and landed at a bay called Ulgekeks on Sulog's western side then climbed the east-facing rises. The canopy top there is forty meters high or more. But breaks are common, giving the region a mottled appearance. The phenomenon is recognized, the openings referred to as *kamnat*, which means "clear." Significantly, these clearings were invariably filled with *legis*, the prime species of *Pandanus* used to make *anageg* sails. Like the mast trees, they are noted to grow especially in the Sulog region, constituting one of its major resources. On the other side of the Sulog harbor there is a point, Shoute, that has well-named landings where Nasikwabw and other sailing village peoples regularly pull up their boats when pursuing these Sulog resources. *Legis* availability became another lesson about "patches" for me.

It didn't take us long to find many *aynikoy*, and we quickly cut two small trees, one for poling purposes, the other for the *kunay*, one of the sides of the triangle-shaped sail called *selau*. Toward the top of the ridge, when we started to find larger "mother" *aynikoy*, Duweyala eventually found a tree of the size he thought appropriate. He cut it down, stripped it of its bark, and set it against another tree to examine it, finally rejecting it because he thought it was too skinny. He continued his hunt and found what he wanted forty meters down the haphazard trail. But I took a voucher specimen from this first one as well as a leaf for Christine Notis's DNA work and noted the *katuson* in the wood inside the bark. I also found the new stump interesting. I was reminded of pins sticking up from a pincushion: was this was a phenomenon of the tree's reputed straight grains? I then noted small stumps as well as faint trails throughout this area. Although apparently in a climax forest, we were following well-worn tracks.

Exactitude is required in choosing these masts because critical sailing conditions—and therefore life expectations—follow from their characteristics. A few notes should make clear the intentions and conditions that guide the selection of an individual tree among the many that are present.

People speak about sailing with respect to three wind positions. The specifications follow from the wind's direction relative to a boat's intended course. For reasons of balance, these boats always sail with the outrigger float into the wind so that it counters the force of the wind on the sail. The wind positions, therefore, are figured with respect to the outrigger side of the boat, often referenced as *walam* ("to" or "at" the outrigger float) as opposed to *katan* or *watan*, the side opposite the float; actually going in these directions is *yiwatin* or *buluk*. The wind direction terms are *gitamatan*, *duwadul*, and *lilimuiy*.

Gitamatan refers to sailing close to the wind. The wind comes from the front outrigger side of the boat. In Western sailing terminology this is "beating."

The wind at *duwadul* is a right-angle wind, "abeam," more or less perpendicular to the boat's course. The Muyuw name derives from the *kakam* boat part whose bottom cup holds the mast, its extension over the side of the boat tied into the outrigger float. Sailors consider a *duwadul* wind the best sailing wind. I left Muyuw in 2002 taking a diesel boat heading south by southwest from Boagis to Yemga with a stiff but manageable southeast wind. Fellow Muyuw passengers started talking about how much fun it would be to be racing *anageg* with that *duwadul* wind.

Lilimuiy refers to winds coming from behind the boat, more or less pushing the boat in the direction that it is sailing, "running." These winds are the most dangerous and difficult for sailing, for, among other things, they so intensify the force of water on the *kavavis* that it becomes hard to manipulate. These were the winds with which we sailed northeast from Panamut to Nasikwabw when our mast snapped in 2002.

Mast and sail adjustments follow from how the sail "catches" (*kon*) the wind, only a little with the *gitimatan* position, a lot with the other two. The force of the sail on the wind pushes the boat "windward," *yiwatin*, or *walam*, "to the outrigger float." Among the adjustments made, two are with pieces that are inserted or withdrawn by degrees. One of these consists of two wedge-shaped pieces called *enam*, which help regulate the angle of the mast from straight up to leaning toward the outrigger float; the second is *kavavis*—"rudder."[42] The *kavavis* is only raised and lowered, not turned. The more it goes into the water, the more it intensifies the movement of the boat "leeward," *buluk*, or *watan*. This counteracts the wind's effect. A *lilimuiy* wind drives the boat more toward the outrigger float, so the rudder will be inserted deeper into the water. Traveling north from Panamout in 2002 the wind was so strong that Duweyala ordered the use of two rudders. In the following discussion I concentrate on what is done with the mast since these understandings flow directly into the choice of which tree will be selected for it and how that tree is trimmed once selected.

Wind conditions generate three alterations to the mast/sail coupling. The first, irrespective of the wind direction, concerns the strength of the wind, and hence pressure, torque, exerted downward by the sail on the mast. The greater the wind, the lower the sail on the mast, thus the toque is lessened. If there is not much wind the lowest corner of the sail may be tied to the mast a meter or two above the point where the mast rests in its mount, the sail's long sides running nearly parallel to the mast. With a strong wind the sail is lowered as much as possible. In 1998 I sailed on an *anageg* darting across a lagoon with a squall coming in behind us. The sail was pulled down so low that its lowest corner was tied toward the front end of the outrigger platform, the angle of the long sides of the sail cutting something like a forty-five-degree angle with respect to the mast. The wind was *lilimuiy,* from our rear, and even though the sail was low the craft was nearly out of control. The man handling the "rudder" shouted that he could barely manage the craft's dynamics: this reaction is important to keep in mind for the next and last tree I describe.

The second alteration concerns the pitch of the mast from a more or less vertical position with respect to the front and back of the boat. This is accomplished by altering lines, functioning as back- and forestays, called *balau.* (People speak of the *balau* as one strand but usually there are two, in a "mother/child" relation.) About fifteen centimeters below the top of the mast, a hole is drilled into which is inserted a small piece of *kaboum* whose ends protrude out of the mast enough for the *balau* to be wrapped firmly around them. The two ends of the *balau* are tied forward and aft on the outrigger platform. Exactly where and how they are tied governs the pitch of the mast. The base of the mast remains stationary, but because the mast mount is a socket joint, the mast easily pivots as it is pulled. Pulling the *balau* lines even keeps the mast in the center; tightening the forestay end, *takavatay,* pulls the top of the mast "forward," *wanhogw.* This is done with the wind called *gitimatan,* that from the front end of the boat. With this wind the top of the mast leans forward. With the other two types of wind it moves back toward the center.

Also by means of *balau* adjustments the top of the mast is bent or pulled (*takayamen*) toward the outrigger float so that it arcs more or less. The mast is thus said to be *kankako.* This alteration is partly effected with the *balau* ropes, partly with the pieces called *enam;* they may be completely removed with a *gitimatan* wind (from the front), but doubled with a *lilimuiy* wind (from the rear). So with a *gitimatan* wind the top of the mast arcs toward the outrigger float; it does so much less with a *lilimuiy.* Whatever the relatively fixed positions the mast is pulled to, it is also expected to sway at its top, movement that facilitates wind spilling out of the sail. Pictures accompanying this chapter show this.

To sum up: the mast is angled forward or centered; it will be pulled, at least a little and maybe a lot, from the top toward the outrigger float; and, depending on the strength of the wind, the downward force on the mast from the torque generated by the sail is kept within the limits of the materials and structures at hand by raising or lowering the sail on the mast. These positions help vary the vibrations the mast experiences. Choosing either *ayniyan* or *aynikoy* configures the tensions generated by these positions. Selecting one or the other, and thus manipulating the density and diameter of the mast, amounts to varying the internal boundary conditions of the structure that is the boat.

These terms and relations refer to ideal circumstances; every situation is, of course, more complicated than such referencing can convey. One dimension of this I cover in the next chapter devoted to the place of tying in the culture. But whichever direction the wind is coming from, given the intended course, there are continuous eddies off its main thrusts. And it is the responsibility of the man working the boat's *kavavis* to watch for these eddies. Both the front and rear of the boat and the sail are decked out with streamers, telltales, *bis*, now as often pieces of plastic as the more traditional folded *Pandanus* leaves. People avidly watch and greatly appreciate the fluttering of these streamers. Waves spray up, over, and around boat structures as the craft plow through swells, creating visual effects that add to the streamers dancing in the wind. People rightly think these scenes are beautiful, and every passing or approaching *anageg* draws a crowd to marvel at its visual dynamics. However, those on board study the quivering streamers for slight changes brought about by gusts or eddies that might instantaneously turn beauty into disaster. While sailing, the boat is in a delicate balancing act with the effect of the wind on the sail, hitting its inside, counterbalanced by the weight of the outrigger float. Should the wind suddenly come from the opposite direction, conjoined rather than opposed outrigger float weight and wind might capsize the boat. We experienced two situations like this when we were sailing from Panamut to Nasikwabw. Running on a course about 40 degrees, the first time we suddenly switched to about 120 degrees, the second time to 60–70 degrees, and both times because the wind started to hit the "backside" (*tapwan*) of the sail rather than its *nuwan*, inside. Experienced as reversals or major shifts in the wind, these eddies are called *kubub* (*kubut* in the Ole/Koyagaugau language). When I finally understood them, Ogis started pointing them out to me as we looked out on the lagoon waters off Muyuw's southeastern shoreline, which, though somewhat protected, were full of one- to two-meter breakers. I couldn't see them, but they were obvious to those with the experience needed to recognize them. At night—the preferred time for initiating long voyages—when the "streamers" are invisible the man operating the *kavavis* must not wear anything above his waist so that he can monitor these currents by the feel of the wind on his shoulders.

The density and diameter of the mast is part of the internal boundary structure intrinsic to a boat's form. Mast vibrations must be able to withstand the winds' forces. The winds and eddies form half of their external boundaries, ocean dynamics the other half, and we come to these with the next tree.

Harnessing and controlling the unpredictable powers of the wind constitutes one of the deep themes of this cultural system, and this may be partially illustrated here—I return to this issue in chapter 6—by a further consideration of the streamers tied to various boat parts. For not only are sails and fore and aft structures decked out in *bis* (*Pandanus* streamers) but "big men" (*guyau*) also wore them when, for example, they dressed up for kula contests (and, I was told, when they went to war). *Bis* were tied to the ends of combs tucked into one's tufted hair as well as into armbands. Their fluttering was considered a sign of the wearer's power. The twinkling of stars is also said to result from the quivering of *bis* tied to them. Big men and stars are equated, one aspect of which is experienced by their respective deaths, for both release powers that are realized in squalls or periods of wind and rain.[43] As noted previously, for the major stars there is a regular pattern to their "deaths," their heliacal setting in the western sky at dusk. Good gardeners say they time their gardening activity to them. The deaths, and resulting disturbances, of big men are not so predictable. Following the demise of my old and great informant, Aisi, in late 1996, Muyuw was hit with a modest hurricane and then the 1997–98 El Niño, both of which some people attributed to his death.

Aniyan is the name of the power released by the setting of stars and deaths. I assumed the word for this power was identical to the tree name *ayniyan* and its close associate *aynikoy*. In late 2006 I asked Sipum about this equation. He denied it and distinguished the respective sounds, *aniyan* versus *ayniyan*. However, a few days later I queried his older brother, Ogis, who was far more experienced with these boats than Sipum. Ogis immediately affirmed the identity. He didn't focus on the slight differences in the pronunciations but instead emphasized the respective concerns with powers intimately related to the movement of air. Masts and their dynamics are focal points for that aspect of an *anageg*. These relations return us to one of the points in the meadow/sago section in the previous chapter. *Ayniyan* trees are not just significant because their dynamics afford ideal structures for *anageg* sailing dynamics; they are also, along with *aynikoy*, placed in the vicinity of sago orchards and as such are part of the forest structure conceived to protect those sago trees from uncontrolled outbursts of wind power.

A concluding point to this discussion: I closed the previous chapter by beginning to specify how relations and knowledge in this region are part of an intricate network. We can now see how this network is beginning to be realized in the structures made with these trees. With the *kakam Calophyllum* I showed how peculiarities of the tree's growth are turned into critical components—as actual pieces or conceptual models—of complex structures. With the discussion of *apul* I have shown how another *Calo-*

phyllum becomes the vehicle for the transformation of one form into another, *anageg* into *tadob* or *masawa,* the two forms effectively different phase states seen, by some at least, as mutually incompatible. Similarly, *dan* are part of an actual and conceived set of interrelationships in a resource set that generates and maintains sago orchards, meadows facilitated by El Niño droughts, "wild pigs," and tying materials for boats and, no less, the armbands into which big men fold their *bis.* We see this issue rising again emphatically with respect to the *Calophyllum* masts whose internal properties become the condition for propelling the craft. But the significance of these masts is greater than that, for they illustrate the composite nature of the sailing craft. *Anageg* are produced on islands where the best trees for masts are not available. Part of the understanding of these boats becomes the fact that when they are first moved to Muyuw, the initial inferior mast is replaced and will continue to be replaced by the superior trees found in the Sulog resource base. Other than the Trobriand term for their class of outriggers, I do not know what *masawa* means, if anything. But the term Muyuw—and nearby peoples—employ, *tadob,* translates to "cut [*ta*] thing from Dobu." People from the southeast corner of the Kula Ring, Koyagaugau and its vicinity, call *anageg* "*kemurua,*"which translates to "Muyuw [Murua] wooden thing [*ke*]." *Anageg* are composites of the differences that compose this set of places.

Kausilay

The Ellen quote headlining this chapter urges that ethnobioloigcal inquiries examine the "cognitive processes (means, agents, instruments) which help us first comprehend the world and then negotiate our way through it." The trees I have examined to this point are the means, agents, and instruments for social action throughout the eastern side of the Kula Ring. These facts achieve their summation with *kausilay.* Whenever I posed the question of why *kausilay* are so important, I was first told that it is because the tree's grains are so "interlocked," *kasileu.* They may be carved very thin yet remain strong. Carved house gables are perhaps the thinnest items made with these trees but they lack structural significance. The tree's primary use is for boats, for which they are turned into extremely sensitive forms.

There are three classes of outrigger craft: *anageg, kaybwag,* and *kuwu* (in descending order). The first two are based on keels with, respectively, three and two strakes tied to the keel; *kuwu* are simple dugouts with no additional strake. Where *anageg* and *kaybwag* craft are not predominant (e.g. Kavatan and Mwadau), the much simpler *kuwu* tend to be constructed from kausilay, as I found in 1995 and 1996. In villages with many *kaybwag,* constructed almost invariably from *kausilay* keels—I note another option later—*kuwu* are made from inferior trees.

As they do with the tree *kakam*, the people of the northeast corner of the Kula Ring turn idiosyncrasies of *kausilay* growth into an intricate structure. A complement of and opposed to *apul*, it is, like that tree, the centerpiece of a transformation into a phase state, a principled order. Like *dan*, it is part of a resource base by virtue of its juxtaposition to a fire-generated landscape. And while the two *ayniyan* provide the internal boundary conditions by means of the wind, *kausilay* establish the same set of relations through water. Seemingly summing all these relations, this species, *kausilay* (*Calophyllum leleanii P. F. Stevens*), is an immensely complicated social fact. It is a standard, and from its form other parts and their associated trees take their places like solutions to the problem this tree defines. But the tree is not just the standard: it is the condition of existence (for *anageg* and *kaybwag*). These boats are named, and each lasts until its keel wears out. Until then all other parts of the boat will be replaced, most especially but hardly exclusively the handsomely carved prow boards (*dabuiy* and *kunubwara*). Those are made from one of several *Ficus*, trees with wood easily carved but quick to rot. But as long as the keel is neither punctured nor too thin to have new strakes tied to it, the boat remains. And in fact they often remain after this time. Both Koyagaugau and Ole islands were noteworthy for old keels lying about them. Not a few (mostly *kaybwag* keels), far too old to be used, are tied to the undercarriages of Wabunun houses. To me they were a bit like bones left in the nooks and crannies of old Muyuw burial practices—a dispensed presence. At least one man in Koyagaugau told me he performed mortuary rituals over his *anageg* when the keel was no longer usable. Wabunun people know this Lolomon practice but do not do it themselves. Yet from its function as a standard and condition it perhaps follows that these relations are underlined by cultural understanding that stipulates that the tree is personified—it is considered a female person.

This concluding section outlines these relations. Although I focus on the personification problem at the end, the weight of the local understanding is intimately related to the tree's socially recognized and contrastive ecology.

On Discovering the Socioecology of the Tree: Events to Structures—Shape

During 1995 and 1996 I set out six one-hectare survey areas of varying shapes and five others of indeterminate area but fixed purpose. Three of the hectares were along the road heading to Kulumadau northwest of

Wabunun, one just beyond its gardening area (by 1998 it had been tran-scended); the other two were far removed from any village, one of these close to the western limit of the uplift that defines southeastern Muyuw. Two were in north-central Muyuw. One was toward Kaulay's northern lagoon in an area that had been timbered in the previous decade; the other was to the south beyond what I presumed was Kaulay's gardening region (although quite close to a megalith Simon Bickler eventually plotted). The final hectare surveyed was close to one of the lakes on the western half of the island, also in a region cut by Milne Bay Logging Company in the 1980s. Two of the indeterminately sized survey areas were near Kavatan. Another was west of Unmatan along the beach and designed to examine one of the *Syzygium* species (*kikiyau*) that grows in sandy areas just behind high tide marks, as well as a set of *Intsia bijuga* found in the same envi-ronment. A final survey region was a several-kilometer-long track cutting northeast by southwest across a peninsula marking the eastern side of Suloga Bay: I noted all the significant trees within five meters of the line we tried to follow. The rolling landscape went through several swampy areas; it recalled images of Humphrey Bogart pulling the *African Queen*.

I had two reasons for conducting these surveys. One was to provide a well-defined method for learning about Muyuw forests, hence the nota-tion of every plant that people could name in the initial surveys as well as several smaller ones focused on *tasim*. The other reason concerned a project organized with Dr. H. Hank Shugart's encouragement. Given that the island was being commercially timbered, we devised a plan to cre-ate a growth model for the island's commercially significant trees with the idea that if one could understand forest regeneration, it would be possible to responsibly use the island's resources. For this purpose I gen-erated estimates of tree height, measured tree circumferences (thus gen-erating diameter at breast height) and put expandable belts, more or less at breast height, on seven hundred trees, thirty-four different species alto-gether. The belts were designed to measure circumferential growth. I and my Muyuw associates have reexamined these trees subsequent to their original belting, for some twice in 1996, then again in 1998 and 2002. We learned much through these procedures, especially about species distribu-tion. Most important was seeing the differential growth patterns among the species, which was evident in the circumferential data, though not from height estimates whose margins of error were too great for the time intervals involved. Also of great interest was the reaction of my assis-tants. Although most of my helpers found this exercise unbearably boring, one, Gumiya, got so involved in the project that he kept me disciplined. In the survey areas in which he participated—three of the hectare-sized ones, two of the others—he watched me write down every measurement.

Gumiya was a detail person, and once I'd inscribed the circumferential growth numbers of about twenty-five trees in the first survey area, he saw the results and said, "I thought so," demonstrating that he already knew trees like *akuluiy* (*Mastixiodendron smithii* M. & P.) and *kaga* (*Pomentia pinnata*), even though they are found only in secondary or tertiary successional stages and grow very fast while others, like *abanay* (*Podocarpus sp.*), grow slowly.[44] The growth data are interesting, but more important is the fact that proportional differences are part of the acquired cultural repertoire.

In all of these areas I found one *apul*—far too big to belt—but no live *kausilay*. The ecology of *kausilay* is similar to that of *apul*. In the high forests where both mature trees might be found, intermediate-sized trees and saplings are nearly nonexistent. Such forests are not particularly old; in areas that are older neither tree is found. In the first survey unit, ten meters wide and one hundred meters long, one enormous *kausilay* had been growing there. (Later I learned it was on a line that roughly bisected the survey unit.) However, it had died and recently fallen out of the survey unit. Although now rotting, its crown's gap remained evident. This survey area started some hundred or so meters back from the road, so chosen because I thought it was well out of the range of any recent human activity. In the high forested areas several hundred meters east and south from here, potshards, clam shells, and flakes of igneous rock from Sulog stone tools were evident—this area was easily examined because it was being gardened in 1995 and 1996. With the exception of one small broken stone adze head, there was no evidence of human presence inside the survey site. It seemed clear that we were beyond the range of any settlement, a deduction more easily made in the two hectares surveyed to the north and west along the same road where—both inside and outside the hectares—there was no evidence of human occupation whatsoever.[45]

After making continuous visits to the first transect area, in June 1996 I noticed that as we walked to its beginning we passed under several enormous *kausilay*. By this time I realized that *kausilay* tended to appear in groves, one of several species violating what I thought was a principle of tropical rainforests: great distances usually separated species of the same type. So I asked a young associate, Gumiya's son Danny, to see if this was a grove and if so exactly what its limits were. Within a few minutes we plotted seventeen mature intervisible *kausilay* close to the size of the one that had fallen in the survey hectare (approximately fifty centimeters in diameter at breast height). About the time we were finished with this appraisal, my associate stopped all of a sudden, looked around, examined the terrain through which we had been slogging, then announced that we were in a *sasek*.

Although all of Muyuw except its igneous center is an uplifted coral platform, here and there one finds more significant concentrations of coral. In these places there is often very little soil; instead, one finds sharply edged coral clumps or boulders. Called *sasek*, only fools go into them lest they leave them bleeding profusely. They can be found everywhere, but even if they tend to be lines reflecting old ocean levels or island elevations, they take different shapes. Those I'm familiar with in southeastern Muyuw tend to be smaller, fractions of hectares; in far-eastern Muyuw, just west of Kavatan, there is an enormous one that, reportedly, might run a kilometer or more. In these regions, given another condition, people expect to find *kausilay*. I walked the eastern edge of the big *sasek* near Kavatan, and it was edged by mature *kausilay* (an area close to where Kavatan, or its gardens, would have been before it moved to the shoreline; there were potshards and other bits of evidence of human occupation). When Danny explained *sasek* structures to me I told him I wanted to find a bunch of smaller *kausilay* that I could belt. He said that would be simple. By then I had spent many hours traipsing through high forests looking for small *kausilay*. To be told that finding *kausilay* I could belt would be easy was disconcerting. But the very statement was rounding out what was then becoming part of the tree's mystery. The next day Danny easily found about fifteen youngish *kausilay*, all arranged around several *sasek* well within Wabunun gardening areas that had probably been farmed for 1,500 years. With that came an echo of a comment from Wayavat village men, meaning men *east* of Wabunun toward the area that has, also, been intensively and continuously occupied for a long time. I knew *kausilay* were prized for the international timber market as well as for traditional outrigger canoes when I started the tree-growth project. Consequently I told people I wanted to belt them, among many other trees. The Wayavat men told me I could find many smaller *kausilay* in their region. Not knowing enough to make sense out of their statements, I reproduced errors of the Oxford botanists looking for the endemic cuscus, going where people had not been—hence the survey sites to Wabunun's *northwest*.

I did not understand the tree's ecology when I visited Gawa in early 1996, but when I returned to Muyuw in 1998 I asked some Gawa men in Wabunun where they found the tree. Along with people on Kweywata Island, Gawans are the prime builders of *anageg*, so they had an immediate answer to this question: "Everywhere," the *tawan*, mature, high forests below the high oval platform on top surrounding the island, and *wakoy*, "on the top" or hill. Like the steep declines with their ledges, Gawa's top is also a patchy environment. The entire region is extremely mottled, an effect easily witnessed in Google Earth. This mottled condition is precisely *kausilay*'s environment. The Gawan men added that when one finds a growing tree and thinks it will be good for either a keel or set of strakes—the difference depends upon its curvature—one reserves it for one's *kadas* or *tibus*, sister's children or alternate generation descendants.

Muyuw say that *kausilay* grow at *matantalas*, places often proximate to *sasek*. The word derives from *matan*, "eye," "end," "edge," or "endpoint," and *talas*, "axe." It means the place where people stop chopping forests for the purpose of making gardens and, therefore, the beginning of *tasim*. This is not the place to stop cutting one year with the presumption that one will or could start cutting there again the next. It is a more fixed transitional zone, part of the relatively permanent landscape design that generates the spotty interrelations of gardens and *tasim* across the island. Chapter 3 brought up the related concept *matan opweypwey*, as the location of a certain kind of plant, as if the plant and location were fixed. *Kausilay* are the same, point-of-transition trees.

Transition as a concept is recognized and significant in the culture. The terminology for tides also illustrates the form. In Muyuw there are somewhat peculiar tides, named *potan*, which seem to be equinoxal tides.[46] They occur at what is called the *matan yevagam*, an expression analogous to the conceived microenvironment for *kausilay*. But this transition demarcates the time when the high tide no longer comes during the day and is thus beginning to come at night (and vice versa). This is significant because it means that the island's underground "water"—*yevagam*—lens is no longer forced close to the surface of the soil when the sun is hottest. People circumscribe this time because it combines the best conditions for growing, the sun's heat plus water. In short, the tie between *kausilay* and the liminal interface, "end-point-of-the-axe," is a recognized pattern.

Danny easily found a bunch of young *kausilay* because he operated with the *matantalas* concept. And when we were examining the grove in the very high forest he deduced not only that the trees were growing in a sasek but also that it too had once been at the edge of what somebody had cut as their garden areas. (In all likelihood the *kausilay* line extending from the one that had fallen inside the survey area was another one, evidence of some limit of anthropogenic activity some hundred years ago or more.) During a trip to explore the flora transformations leaving Unmatan heading north and a bit west toward its sago orchards, my guide and I came across a band of *kausilay* that extended for a considerable distance east and west, but was quite narrow north and south; it undoubtedly represented another former limit of a larger population and its gardening endeavors. Over my returns to Muyuw from 1995 to 2002 the island's population was growing to what it probably was had been in the mid-nineteenth century. So all over southeastern Muyuw people were gradually cutting into forests pockmarked by mature, canopy-top *kausilay*: this fact was a matter of idle commentary and gossip in 2002 because many great trees ideal for canoe-making were going for naught. By 2006, a region proximate to where Danny and I found the *sasek* was being called *"kausilay"* as the then-major garden area of a significant proportion of Wabunun's population; it has replaced as a destination site *"waukw"* of the mid- and later 1990s.

By late June 1996 I was quite sure of the tree's ecology. And so, toward the end of this, my longest research period, I returned to Kaulay in north central Muyuw, partly to identify, but mainly to measure the height and girth of as many trees as I could. We found no *kausilay* in the survey areas I had set out. Close to my departure time from Kaulay and out of desperation I asked people if they knew where I could find the trees. Immediately they referred me to a well-known place called Ulgulag'; it turned out to be the strangest location I had come across. It is a named patch of ground just up from the shoreline on the north shore of the island west of Kaulay's lagoon. It's the kind of region Muyuw call *umon*, meters removed from the highest line of wave action but clearly a deposit from sea activity. The grove-like conditions of the area resembled others where I had found *kausilay*, though there were several other species in the patch as well. But nowhere else did I find this tree so close to the sea, and nowhere else did I find it growing in what appeared to be the mixed sand and clayey materials that compose most *umon*.[47] I had no idea why they were there, and at the time, unfortunately, I never thought I should ask. Were they planted there by people from Gawa who would want a ready supply of replacement timbers if one of their canoes was in trouble when it sailed to Kaulay? Or by Muyuw for any number of other islanders from the whole eastern side of the Kula Ring who customarily frequented the region to replace other boat parts? Since they tended to be straighter than most *kausilay*, they would have been ideal for *anageg* strakes.

I returned to this place repeatedly during my 2009 visit, and its mystery remains. It is exploited for the lowest class of dugout canoe by Kaulay people, and it is well-known by people from Gawa, Kweywata, and Nasikwabw. The Gawans I spoke to in 2009 denied that they used the tree for replacing *anageg* strakes; however, Kaulay people told me that Budibud people, who build a kind of *anageg* form but don't have *kausilay*, did use the place to find and make strakes. One Nasikwabw man told me he knew of it and also considered it an ecological anomaly. I had just finished a conversation with him and Sipum's widow during which they both explained to me that kausilay were found in sasek—which this place was not. They guessed that the coral limestone typical of sasek were invisible there but probably located under the sand, a possibility that didn't seem unlikely. The Nasikwabw man went on to point out that Nasikwabw "had one too," that is one place where a patch of *kausilay* grew in similarly anomalous conditions. The anomaly is a marked category.

Because the ecology seemed so peculiar, I realized I should collect a voucher specimen of those trees to in fact prove that they were of the same species. The tree we collected[48] just before I left Kaulay in 1996 was a sapling, so the leaves were large. After a previous visit to Kaulay I had shown my specimens to my Wabunun instructors in southeastern Muyuw, so I did the same with this batch. Again I was startled by the exactitude by which some of the plants were known. Without any hesitation whatsoever—which was not always the case—Dibolel identified the leaves as both from a sapling and from *kausilay*. He prefaced the identification by the age of the tree because the leaves approached the size of those customarily found on both *kakam* and *apul*. The differentiating criteria he used were the "sharp point" of the leaf's end as well as the sharper shape of the whole leaf; both *kakam* and *apul* have rounder ends and a more oval shape.

Before continuing with these critical social facts, however, I need to fill out the evident understanding of the tree's ecology. The *sasek* made it difficult to move around the *kausilay* grove first explored with Danny. (The next one we examined was part of a *tasim* amid Wabunun's long-standing gardening areas. Danny wisely walked around its edges while I barreled straight in. When finished I walked back into Wabunun to the shock of people who saw my slashed legs and bloody feet.) One of the things that became evident in that examination, and was confirmed by the younger *kausilay* we found and belted in the next few days, was the tree type's root structure. The roots were almost as gnarly and evident on the surface as the beach-dwelling *kakam* (*C. inophyllum*). They circled around, over, and through the outcropped coral forming the *sasek*.

This brings us to Nasikwabw, famous for large numbers of *kausilay*, which, indeed, I saw in 2002. Nasikwabw is an enormous *sasek*. Geologically it is part of a barrier reef, encircling Muyuw in the way that coral shelves often encircle volcanic atoll structures. However, in this region classic atoll structures are altered by a north-south slope. The northern side is uplifted, while the southern side is usually below the surface. Virtually all of Muyuw's southern side is below the surface. Nasikwabw is an exception. A geologist told me it is a small portion of reef uplifted and carried south by a series of secondary faults. And, in fact, the human communities on the island have been confined to its western end, the only place where there is soil. As I explored portions of the island in 2002, it became obvious that the further east one traveled, the more impassible the terrain became because of the protruding coral. (The same conditions, i.e. human impassability, make the island ideal for one of the region's ebony species.) While Nasikwabw people do not live or garden toward the east, they traverse the area. Old and new trails testified to this, as did rotting *kausilay* stumps, piles of wood chips from the fashioning of *kausilay* boles into keels or strakes, and a reposed bole waiting for its owners to turn it into a canoe. In a few days I saw many such trees—the southeast wind was about to strengthen making travel difficult so perfect for constructing the next set of outriggers or repairing those already in use; underneath many Nasikwabw houses there were sets of boat ribs (*geil, gulumom*) waiting to be mounted on some craft. Nasikwabw's *sasek*-like structure provides one of the ideal ecological conditions for the tree, and, hence, the sailing communities built around them.

Nasikwabw's structure eventually led to an explanation of two other locations on which I found the trees, places that surprised me because they were clearly in much older forests that humans rarely used. These places were on the sides of two hills, one of which I found several kilometers east of Kaulay when I was looking for small ebony (*gav*) trees to belt; I discovered the other near the western limit of the southeastern uplift, somewhere north and west of Unmatan in a region otherwise typical of where *apul* grow. Both of these locations had steep hillsides, and on both several mature *kausilay* were growing, conforming to the tree's arcing patterns. What was interesting about these hillsides were the other trees. They would grow to a certain size and then collapse because their root structures failed to hold them to the earth. *Kausilay*'s roots enabled them to hold on to the treacherous landscape. The fallen trees created hillsides of partial light similar to the interstitial conditions found at *matantalas*. I presume a similar context prevails on Gawa's *tawan*. And these are the conditions that generate the other feature of the tree gradually becoming evident to me and critical for indigenous purpose: *kausilay* rarely grow straight, unlike *apul*, which are generated in nearly the same microenvironment; instead, they bend or arc. Not all of them do, of course. Of the dozens I've seen, I recall one, a "mother tree," located a kilometer or so behind Wabunun, that not only grew straight, but spread quite quickly and so had a modest bole. But most of them arc in significant ways. So prominent is this feature that it became my chief criteria for picking mature *kausilay* out of the high forests. My eyes were not good enough to recognize distinctive color and leaf shape out of the upper reaches of the canopy—thirty to forty meters above. But I quickly recognized the arcing boles; finding fallen leaves beneath them was then easy. Nowhere on Muyuw did some other tree's shape ever fool me, for as noted earlier few trees grow to *kausilay*'s size.[49]

Nasikwabw's situation is a bit different, for it has another tree that can be coupled with *kausilay* in size and setting, as well as initially in the appearance of the bark, alligator-like and darkish gray. Called *akuyak*, it is considered by Nasikwabw people to be *kausilay*'s cross-cousin, *nubien*, though it is not personified. The tree does not grow in eastern Muyuw, and the name "*akuyak*" is used with a different tree. I did not take a voucher specimen of Nasikwabw's *akuyak*, but it is not a *Calophyllum*. Although an observer like me would initially confuse the massive boles of these two tree types, after a few moments I picked up their prime differentiating feature: *akuyak* boles tend to be straight.

All small *kausilay* I came across in the vicinity of *tasim* in 1996 were beginning a tilt that would cause them to become like the large ones I came to recognize in higher forests. Muyuw—and Gawa (see Munn 1986)— people pick up these features from early in the tree's life. And when reasonable, the trees are marked for future individual usage. The tree then becomes known as this or that person's *vayas*, replacement. Many middle- and large-sized canoes have names that are passed through time from one boat to its next realization.[50] Hence a *kausilay* designated ahead of time to be this or that person's *vayas* is very likely designed to fill the slot of a particular named boat.

Kausilay arc as they grow in microenvironments created by a locally devised sense of order. I shall return to aspects of this arcing later, but for *anageg* two terms delimit the variation: *amwagan* and *esol*. The former refers to a keel that is "arced" (*kaydodo*), the latter to one that is "straight" (*idumwal*). The latter, straighter, shape is appreciated because when the front (*dabwen*) and back (*wowun*) ends of the boat crash into waves they create great splashes of water, likened to "smoke" (*museo*). This visual effect is aided by dozens of cowry shells tied to the boat's ends. The former has its prow and bow pieces much higher so they tend not to go as deeply into the waves. This contrast was explained to me in 1998; when I first heard it I presumed the straighter version was preferred. However, the boat I examined and sailed on in 2002 was considered more toward the arced end of the continuum, and I was told it was "very good." As in a number of other cases where variation was expressed in terms of binary contrasts, the point was describing a ratio of differences rather than stipulating a preference, morally (i.e. aesthetically) or technically determined. Nevertheless, different properties emerge from these forms. One informant stipulated that if it is short, a boat that is more arced tends to weave, and is therefore harder to control. And if it handles well with a beating wind direction (*gitamatan*), it performs more poorly with a wind from the rear (*lelimuiy*). While I presumed these distinctions led to favoring the straighter form, I was told instead that such craft tend to more easily sink when heavily loaded, the front and back of the boat tending to "hit" the water and impede movement.

These arcs are critical because they become the basis for the curves, the rocks, of all the outrigger canoes made on these islands. Generated from a combination of its distinctive properties, each tree and the keel produced from it becomes the condition for a new boat. Every other part becomes sensitive to the initial form fashioned from the tree's curve. The tree's ecological setting, defined by intentional human activity, defines a craft's first variables.

Configuring a Transformation

Both *apul* and *kausilay* require sunlight to progress from seedling and sapling stages to more mature trees. Neither makes this transition under a well-developed canopy. So *tasim* provide one of the ideal circumstances for both trees. But there are differentiating features as well. *Apul* tend to require wetter soils while *kausilay* thrive on those that are better drained. Consequently, while *kausilay* are readily found amid southeastern *tasim,* *apul* are witnessed in those in north-central Muyuw.[51] Two other differences are apparent. *Apul* grow somewhat larger and straighter than *kausilay*. These two facts become initial conditions for making the Kula Ring's two largest and most complex outrigger canoes: *masawa* or *tadob* in the case of *apul,* a longer, straighter craft which can be both sailed and paddled efficiently; and the *anageg* or *kemurua* built up from a keel with three strakes and arced far more than the former. People use the different forms to describe a phase change between two points, and this is conceived as moving from a manageable order, in the case of *kausilay* and the *anageg* form, to *tadob*'s unmanageable form.

Sensitivity to initial conditions exhibited in a keel's curvature also appears with its length. The Australian National University biologist and Oceanic sailing vessel expert Adrian Horridge made me aware that some Pacific sailing traditions have precise numerical terms for their boats. In 1999 I questioned my Yemga informant very carefully on this matter. What goes by way of a numerical standard begins with the keel, and everything else fit proportionally to it. The Yemga man said *ovatan* are used to figure keel lengths. This standard unit of measurement means the distance between outstretched arms, from the fingertips of one hand to the other.[52] Almost all parts of a boat are analogized with parts of the human body, so it is not surprising that a unit taken from the human body becomes the vehicle for expressing a craft's fundamental dimensions.

Anageg come in three sizes: "three plus one" ("very small"), "four plus one," and "five plus one" ("very big"). As noted earlier, the distance between the keel and the lam is proportionate to the former's length. But outrigger floats, masts, and sails are roughly calculated to fit to these dimensions; a sail, for example, runs the distance between the boat's two prow boards, so the keel is the determinant. These are rough approximations, however. One of the *anageg* I measured had these dimensions:

Table 4.3. *Lavanay Anageg/Kemurua* (2002) Measurements

keel length, straight, end for end	10.7 m
keel length, following the arc	11.85 m
distance between two prow boards	8.37 m
length of sail	8.5 m
mast length	8.74 m
outrigger float length	11.32 m

While my measurements here are descriptively informative, this mode of expression should not be confused with the Western reliance on numbers as models for reality. When I questioned the Yemga man as to this expression, he told me that the "plus one" unit did not mean another full length of one's outstretched arms. Instead it referred to an indeterminate addition: so "five" and some additional part of outstretched arms. These boats become exact in the process of production and sailing, not in the initial conception. So, long after a keel is first cut, two pieces that function as shock absorbers for the mast and mast mount will be measured once they are appropriate for that boat. When they break, new ones will be cut to the specifications of the old ones, then reworked as the properties of the new pieces of wood are realized in experience. The point here is not that a keel is cut to a size. Rather, once the keel is fashioned, other parts become sized to it; it is the standard. Not measuring units, but the experiences of operating boats become the standard in this region.

As my discussion with the Yemga man continued, I asked, "Why not 'six plus one'?" The response was immediate: that was impossible because the boat would be too hard to rig. The expression is *"angineol singaya kikay,"* where *singaya* means "very" or "extremely," *kikay* translates to "hard" or "difficult," and *angineol* is "its 'tying' or 'rigging.'" In general this expression refers to all of the tying that goes into the boat. But in particular it pertains to the specific ways in which the sail is braced with respect to given wind conditions and directions.

The form "five plus one" thus expresses a mathematical—that is relational—limit to which these boats can be made, a limit defined by the combination of human strength and available materials given the wind and water conditions in which the boats operate. Such limits are, of course, experienced. In 2002 I participated in one when our mast snapped. In 1998 I experienced another when we sailed across a lagoon from southeastern

Muyuw to Waviay. A squall was coming up behind us, the wind *lilimuiy.* The wind was strong, the craft on the brink of disintegration. The sail was pulled down as far as it could go; the steersman yelled that he couldn't control the boat. Fortunately we reached our destination before the most intense part of the squall hit us.

In the context of my discussion of keel sizes with the Yemga man, I asked him about the western Kula Ring boat form, *tadob* or *masawa.* His reaction was similar to that for the six-plus-one expression: he would not do it. The reasons for this I reviewed earlier in the section devoted to *apul.* The critical issue concerns the dynamics generated by the distance between the outrigger float and the keel, a distance that in the *anageg* model balances speed versus stability. Bringing the float closer to the keel affords greater speed but less stability. Relative to an *anageg,* the distance between the float and the keel in a *tadob* is so small that the craft is too unstable to sail, according to my very experienced Yemga source. His perspective is obviously not shared by the people who sail *tadob.* But his expression rivets our attention to the fact that these boat forms are complex models of intricate dynamics. Inside there is order, outside disaster.

Element in a Resource System

Chapter 3 showed how *sinasop* were near inverted instances of *tasim,* the former fire-induced patches of low vegetation amid higher forests, the latter patches of high forest not burned in the processes that made the surrounding gardens and fallows. The tree-type *dan* first appears of coincidental interest because it provides food for wild pigs. But once that relation is understood as part of an anthropogenic environment containing a set of resources, the apparent coincidence becomes part of a structured set of relations. For boats, *dan* help define what is appropriate because they are too "wobbly" to be used for anything. However, recall their correlation with the two trees appropriate for masts, *ayniyan* and *aynikoy,* by *dan*'s proximity to them. As a set, these trees stand over, "watch over," "wait on"—*kayamat*—sago orchards, protecting them from gusting winds. By contrast, *kausilay* as "*tasim* trees" block the sun from crops and regenerating forests, which is said to encourage faster growth. *Kausilay* are obviously much more of a resource than *dan* because they become valuable for other human-generated forms, yet they maintain a close parallel with *dan* in the *sinasop/tasim* contrast.

Another parallelism in these opposed yet correlated forms concerns the relationship between *tasim* and the cuscus, *kwadoy,* the small animals that inhabit these islands. *Dan* facilitate will pig production in the meadow/sago orchard relations. *Kausilay* participate in cuscus environment because they are significant components of the high flora of *tasim*. These nocturnal animals forage among the early successional forests generated by Muyuw gardening practices but retire to proximate high forest to sleep during the day. Muyuw people hunt and eat them when winds make the seas too rough for reef resources; boys and young men pursue the animals by climbing up certain trees, searching amid epiphytes growing in the canopy top. They slowly inch along a branch until they can either knock the animal to the ground to a waiting bunch of men or grab it and immediately tie it up; the next chapter describes a special knot called *sipkwadoy* precisely for this purpose. When I asked people what trees they regularly climbed for *kwadoy*, *kausilay* were prominent on the list.[53]

Independent of their primary use as a raw material for sailing craft, *kausilay*, as live trees facilitated by a designed landscape, like *dan*, are elements in a set of relations that Muyuw use to sustain their life needs. They provide shelter rather than food for protein substitutes.

Boundary Marker

Kausilay grow successfully at the edge of two microenvironments, the "end of the axe." They are boundary markers, and coincident with that they define the boundaries of sailing craft (and in a way houses through gables, though only aesthetically rather than structurally). They do in water what *ayniyan/aynikoy* as masts do for air. This correlation and opposition runs through the relations that define these trees. If the *ayniyan/aynikoy* set derives from the least anthropogenic environment of the *Calophyllum* set, *kausilay* is a product of the most human-defined environment. *Ayniyan/aynikoy* grow straight and have straight grains, while *kausilay* arc and have extremely interlocked grains. Young *kausilay*, of the size appropriate for selecting masts, have relatively little red heartwood in them, while *ayniyan/aynikoy* have considerably more. Hence a *kausilay* mast is likely to be too "rubbery," *woweu*; *ayniyan/aynikoy* are much less so. This difference relates to another contrast, *awoyan/vat*, light, spacey, impressionable, or soft versus dense or hard. Fingernail impressions can be left on wood that is really *awoyan*, I was told; really *vat* wood is so dense that it might sink. When young, *kausilay* tend to be *awoyan*, one of the qualities that make them relatively easy to carve and shape; and carved and shaped they are. As the wood ages—both as the tree matures while it is alive and after it is cut—it becomes increasingly *vat*, dense. Although *ayniyan* is a bit like *kausilay* in this contrast, both *ayniyan* and *aynikoy* tend to be more on the *vat*, denser side of the scale, and this is one of the qualities that leads to the wood's superior performance as a mast. A mast is never really carved; rather, it is selected near its desired length and diameter then trimmed. *Kausilay* (keels, strakes, and gables) are carved out of the center, redder wood of the tree and fashioned into well-formed designs. As elaborated in chapter 6, a bird image is always carved onto the two ends of the keel; each end is gouged to hold one of the front boards called *dabuiy* and grooved so that the strakes can be pulled up and held by the form. The resulting structure situates all of a boat's critical dynamics.

Kausilay boles arc, those that arc more are desired for keels, those less for the three strakes forming the craft's hull. As noted before the contrasting terms *amwagan* and *esol* express the nature of this arcing, the former more arced on a vertical plane. We must now go into this relationship in more detail. Note Figure 4.1.

Although there are aesthetic dimensions to this contrast, more than one informant told me that the more the keel arcs, the boat will have less tendency to go *yiwatin*, which means toward the wind and outrigger float side of the craft. Consequently there will have to be less forcing the boat in the opposite direction. This forcing is done with the "rudder." Most *anageg* keels also have much subtler horizontal arcs so that each end veers slightly toward the outrigger float. This design feature should nudge the boat toward the outrigger side—which is said to be good.

Figure 4.1 charts some of these relationships. Arrows D to C and D to E describe the contrasting effects of the wind and the rudder (*kavavis*) on the boat's direction. While the small veering of F to E goes along with the direction that the rudder effects, the much more pronounced horizontal

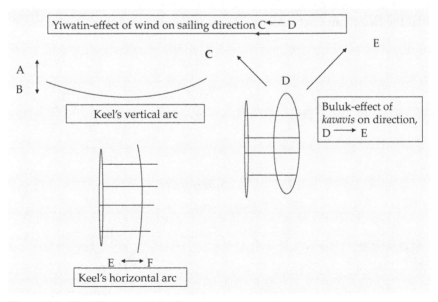

Figure 4.1. Keel Design and Its Consequences

arcing, B to A, is supposed to minimize the D to C vector, thereby reducing the use of the *kavavis*. These forces are unleashed by and follow from the keel's carved shape.

The more or less V-shaped sides of the keel are called *bab*. How they are cut organizes variation among the boat types. Figure 4.2 outlines ideal contrasts among Muyuw's three classes of outrigger and then *anageg* models the two basic variants within the *anageg* class. The *kaybwag* diagram is exaggerated to make a point that the sides of those boats, when constructed correctly, supply a lift to the craft not intended with *anageg*. *Anageg* do not hydroplane, whereas *kaybwag* do. The contrast between Gawa and Kweywata keels shows two dimensions and initiates a discussion to which I return. First, there should be a difference in the side angles of the *bab*/keel that becomes reflected in the strakes eventually tied onto the piece. I have illustrated the point by the imaginary perpendicular line running vertically from each keel. Hence the angle a-b is greater than the angle b-c. The strakes should be cut and tied to continue this difference. This relationship is not found in *kaybwag* because, unlike *anageg*, they are not designed for the large wave dynamics.

Second, stereotypically Gawan keels are "thicker" than "Kweywata's." The Gawan model features a longer a-c than the Kweywata a-c length. The actual thickness of the strakes follows this distinction so that those in a Gawan craft will be thicker than those in a Kweywata rig. When I learned about this difference in 2002, I also learned that, unlike many other subtle design features, not everyone knew these two. But it is widely understood that the two main boat-making islands make two slightly different boats.

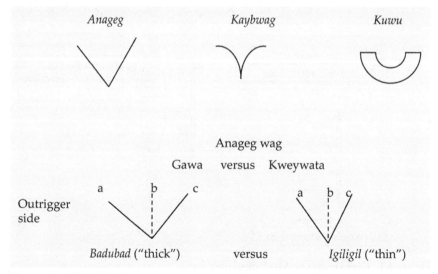

Figure 4.2. *Wag/Bab* Shape

If a boat is made in Yalab, people assimilate it to one of the two types; the 2002 boat I sailed in was made in Yalab and was classed as the Kweywata type. Although there is not really a third type of boat, people told me, once I got into these distinctions, that Budibud *anageg* were thicker than Gawan boats. However, people did not consider that as a separate type because Budibud *anageg* are not made to be exchanged. They are made by and for Budibud people, and their primary purpose was shunting Budibud products to other islands in exchange for sago and other vegetable food. Although Budibud people do most of their commerce with Muyuw, occasionally they sail to Misima for sago. The prime quality sought in Budibud boats is the ability to carry a significant load, and the wider the keel, and consequent width between the strake, the more the craft can hold. This reason brings us to the Gawa/Kweywata contrast. The difference between the two exchanges carrying capacity for speed. The Gawan type is "thicker" so it holds more; the Kweywatan type "thinner" so it travels faster.

These distinctions are interesting for two reasons. First, they illustrate a principle of social organization for the larger region: communities organize by institutionalizing minor differences that create useful functional relationships that people act on (see Mahias 1993: 170–72). Second, almost as soon as I was told about this distinction, I was told another that concerns the spring and mast-mount structure running inside the keel and supporting the mast. To counter their different dynamics, the "Gawan" structure is made thinner, the Kweywatan one "thicker." This difference reverses the contrast between the two initial keel and strake forms.

Although I do not understand all of their dynamics, that they are designed to traverse the flows of the sea is clear. A story I was told underlines the point.

This episode was recounted in July 1996 as I was becoming drawn to boat dynamics. To understand the story it is important to realize that from the coming of the European order in the 1840s, the West has taught that the "natives" are uneducated and in one way or another undeveloped. Their creations are trite, primitive. This point of view is internal to contemporary Muyuw culture today; people pretty much hate what they are (see Sahlins 1992). In this context, of course, my patient interest in Muyuw lifeways appeared as an idiosyncratic and sometimes comical novelty. However, as I was becoming informed of these craft, people were seeing these complexities from a different vantage point.

So, one evening they told me a true story. One of their local pastors, and in fact a Wabunun man well tutored in these forms, decided to sail a boatload of people to Nasikwabw, taking one of their finely crafted *kaybwag*. There was a convenient wind so they were sailing well, the pastor at the boat's stern managing the "rudder." A *kaybwag kavavis* is a slightly enlarged paddle that is shaped much differently than an *anageg* rudder.

It is used only on the side opposite the outrigger float where it is pressed up against the side of the boat. By contrast the *anageg kavavis* are on the opposite side nudged close to the outrigger float itself where they are raised and lowered on the outrigger platform. The difference between these positions cannot be more than a couple of meters. In any case, by the time the craft was perhaps five kilometers into the voyage, the considerable pressure exerted on the *kaybwag kavavis* made the pastor's arms tire. Although he knew these forms well, he also knew that the prime authority for their order was phrased as *Geliu*, the Muyuw Creator. And since he followed the Christian God, there was little reason for him to pay attention to those prescriptions and proscriptions. So he thought he could rest his arm muscles if he just switched the rudder from the prescribed *kaybwag* to the proscribed *anageg* side. He did. And the boat blew apart. Since it was close to a small island along the reef, the crew easily swam to it with the boat's parts, and eventually the boat was reconstructed. Nobody got hurt in this adventure, and my friends thought it was humorous. But they were not just laughing. As I was becoming informed about the complexities of these creations, they were seeing them again in a new light. They were becoming conscious that their significance was not because the forms descended from Geliu. That notion was obviously a cover for a swirl of technical details about how these forms bind the boats in water.

Kausilay *as a Person*

When I first became aware of the tie-in between *kausilay* and the land category *sasek*, I also dimly recognized that seeds and seedlings were evident under some of those seventeen intervisible trees; under others they were not. When I saw Nasikwabw *kausilay* in 2002, where this fact was equally striking, I raised an issue with my prime informants, then men from Ole and Koyagaugau (the *anageg* crew) as well as Wabunun. I suggested that there were probably male and female *kausilay*. This was not an uninformed probing on my part. Although my experience on Iwa and Gawa was limited, I left with a distinct impression that practically every tree was easily classified as male or female; an Iwan man, for example, told me that *kausilay, apul (siptupwat)*, and *kakam*, the three *Calophyllum* that can be found on Iwa, were all female trees, a sorting I could not sustain in Muyuw. Muyuw people see that there are male and female papaya (*musumweis, Carica papaya*), and they class other trees as male and female, such as their two *Canarium* species. But they do not readily class trees by gender. *Kausilay*, however, presents a special case. Yet to my question, some of my friends replied that they had observed the same thing, a spotty distribution of seeds under groves of the trees. One of the Ole/

Koyagaugau crew, a man who had lived on Muyuw for perhaps a decade and who was a great source of "traditional" information, both lore and technical matters, looked hard at me and delivered a homily on the birds and bees. "There were male and female of everything," he scolded. As a matter of Western botanical information, knowledge of sex forms in the genus *Calophyllum* remains poor although there is evidence of male and female types within some species (see Stevens 1974: 351). However, the indigenous recognition of male and female *kausilay* is important. For at a remove from this level of empirical observation, all of the people in that encounter knew and had talked with me about the fact that *kausilay* are understood to be not only female but human beings.

This encounter became another moment when I realized that organized knowledge existed on more than one level. Various aspects of these trees are known very precisely; at another level, some of their features speak in uniquely fashioned ways to model life's conditions.

Although I do not know the spatial limits to beliefs about the person-ification of this tree, the idea is widespread. First reported by Munn, the standard myth is that men were attempting to make a boat out of rocks. A woman saw what they were doing, wiped her vagina with her hand, then wiped its blood on a tree. That tree became *kausilay*, the standard for keels, and a female person. This part-to-whole transformation, effected by a woman, clearly prefigures a massive set of subsequent wholes into fash-ioned parts and then into new unique wholes, for the most part effected by men. As boats these wholes become the communicative means for effecting other transformations.[54]

I have heard versions of that story on Iwa, Gawa, in western Muyuw, and from my Ole/Koyagaugau informants. The myth is one of the things regularly invoked if people are asked why they consider the tree to be a female person. Wabunun's story is different. There women were regularly leaving the garden to urinate at the same place—the privacy afforded by *tasim*—until a *kausilay* grew up where they were urinating. When I was in far-eastern Muyuw, Kavatan, I asked if people knew the *kausilay* tree myth and if the tree was a female person. My informants either did not know or refused to tell me what they believed. One, however, said that the tree was the "basis of blood," *wowun buyav.*

The association of the tree with blood is part of its identification with people. Freshly cut *kausilay* is dramatically red in color. And while the keel and strakes made from *kausilay* gradually become a bleached white, people observe a continuous effect that they liken to bleeding. Rainwater that settles in the keel of a boat looks reddish. And tied to this is a com-plicated reference to "the blood of *kausilay*." In this expression the tree references a boat and, more importantly, actions specified in relation to it. The blood refers to sexual fluids. When men travel from one island to another, especially on kula voyages, it is assumed that youths will have sex with women in these other places while their wives are back at home. (Elder men shouldn't do this because such fluids/activities interfere with their kula relations, making the valuables too greasy to hang on to.) And, typically, an elder's crew is composed of his daughters' husbands, men who assume their affinal responsibilities by boat work. Yet these are also precisely the people who are likely to be engaged in sexual relations else-where. When the daughter or wife discovers that her husband has been sleeping with other women, she becomes enraged. It is then incumbent upon the elder, her father, and the father-in-law who was using his son-in-law's boat labor, to mollify her by saying it is "only *kausilay* blood." This is supposed to excuse activities from afar that would be inexcusable in their village. If an initial reference to blood indexes femaleness, another invokes male activities, and the apparent ambiguity—or is it intertwining?—is matched by the conflicting alignments of persons managing the craft.

The idea that this tree is a person is well-formed. Another tree Muyuw people say is nearly as good as *kausilay* for constructing *kaybwag* keels is called *aunutau*.[55] Gawa people knew the tree but they do not use it for *anageg*. Its significance, therefore, may be restricted to Muyuw. And while much of my data about the tree comes from Wabunun, it is well-known in western Muyuw also.

The tree's contrastive ecology and properties throw into dramatic relief the saliency of *kausilay* personification. Recall chapter 3's discussion of the distinction between trees that are either *kalow* or *man ankayau*. *Kalow*, which literally translates to "fall down" or "drop," refers to those trees propagated by one or more "mother" trees whose seeds drop near it and sprout. These are the ones that generate grove-like conditions. "*Man ankayau*" refers to trees distributed by the errant and unpredictable arrival of (usually) bird (*man*) droppings.

There are two issues relevant to this contrast. First, *kausilay* is the prototype for trees that are *kalow*, the condition thus providing an indigenous explanation for their grove-like appearance. And people pass directly from this observation to the female personification issue. Just as people of the same subclan follow from a single woman birthing, so are many *kausilay* found near one another. By contrast, *aunutau* is virtually the prototype for *man ankayau* trees. It might be found anywhere, and as a standalone tree. Sipum told me he discovered them by the unique silver underside of their leaves. As he passed through the bush, he would see the leaves on the forest floor and realize they were from *aunutau*, and he would search for the tree until he found it. While I have come across them in the high forest this way, the tree I used for a voucher specimen was located in a *digadag,* the early fallows forest/gardening area proximate to Wabunun.[56] The kind could be anywhere. Human males are thought to be the same since they move around. Three other contrasts relate to this one. First, the tree is quite "greasy." When it is fashioned into a keel, the tools become covered in "grease," a euphemism for sexual ejaculate (*pwak*). Second, the wood is very "white" (*pwakau;* in Muyuw *pwak* translates to semen), contrasting with the blood-red *kausilay.* Finally, the wood does not readily crack when exposed to the sun, so it requires far less care than *kausilay,* which must always be covered with woven coconut fronds when beached and once or twice a day doused with seawater so that the wood does not dry out and crack. Like women, *kausilay* require continuous attention; like males (*tau*) *aunutau* do not.

These associations finally came together in 1996 when I was with Mwadau people in western Muyuw. I suggested to them that, while *kausilay* were female persons, it obviously followed that *aunutau* were men. The men said they had never thought of that but the deduction must be correct (mostly because I had said it; across the island I am known as *singaya tasinap*—"a very smart person"). When I returned to Wabunun with this new insight, the people who know me much better than those in the west belittled the suggestion. Although the logic is incontestable, they considered the deduction incorrect. *Aunutau* is not a person, male or female, it is a tree with masculine-like properties.

In 2002 I had a similar encounter concerning the tree (*vayoun*) used to make outrigger floats, *lam.* Somebody suggested to me that this tree was a male person to the female *kausilay.* When I checked this with my Wabunun informants, they said the logic was correct because like women, a *kausilay* keel sinks because it is heavy. Men, by contrast, are light and buoyant, and so resemble the "float." The colors also fit this mode: the outrigger float tree is white in contrast to the red *kausilay.* Yet this similarity in features didn't make the tree a person to them. The personification of *kausilay* is a selection among a number of other possibilities, but it is arguably a matter of reasoning rather than counterintuitive attribution (see Boyer 1996 and Rival 1998). A consideration of the tree used to make strakes for the *kaybwag* class of outrigger argues for this conclusion.

The tree is *antunat* (*Palaquium sp.*). As noted in chapter 3, although they are an *ulakay,* older, high-forest tree, they tend to be found closer to the shoreline and probably result from landscape changes brought about by humans. Their inner wood is as red as that in the *Calophyllum* species and so could be invoked in the same way the red color of *kausilay* is. But when I pointed this out the response was immediate. Upon cutting, this tree will grow back from the stump, whereas *kausilay* dies and rots right away. Many trees actually sprout from their stumps, but *kausilay* never does. And, I was told, this is a feature that distinguishes it from most or all of the other trees in this "group." To paraphrase my informants, just as a human dies when its head is lopped off, so *kausilay* dies and rots when it is cut down.

The boats constructed from *kausilay* are meticulous objects that require painstaking care. There is a widely known story about a boat owner/ captain who pays no attention to his boat and another who obsessively douses his with water every day while constantly checking its lashings— the boat controls his actions by making him take compulsive care of it. Then both men go on a trip; the former is never heard from again. And so people say that one must take care of one's wife, husband, or children just as one takes care of a boat. The qualities that derive from these trees and how those qualities translate when they are fashioned into sailing craft become a mechanism for reasoning and reminding people how they have to treat each other, as products of and means for other products or relations. A tree conceived to be a person becomes, by means of human intentions, a thing useful for contemplating how a person should be.

Kausilay exist amid a network of empirical facts and conceptual relations. While providing a base for outrigger sailing craft, they also contain within their properties, real and conceived, all the facets of the other trees in this group. Conceived to be derived from human action, they are also located at the limit of human activity, the end of a gardening area, *matantalas*. And like the human agents who make those spaces, who also come into existence by means of gardening activity, they may be fashioned, once they exist, into new instruments of relating and of talking about relations—as are human beings (and kula valuables, which also derive from the edges of human action). If this personification is a reification, it is one that functions to specify the relations among the categories it concretizes. The form shows what is at the base of the reasoning realized in the networks of these relationships—transformations, like a child from a womb who becomes the reason for a lifetime of obligations; and the odd coincidence at the created conjunction between a garden and a *tasim* that becomes the vehicle for sequences of conversions. This tree is not mystification, a covering up of relations; rather, it is the very sign of intentional, ordered sequence.

Notes

This chapter draws from, corrects and transforms accounts previously published or presented in Damon 1997, 1998, and 2008a and b.

1. Our mutual interest intensified before my 2002 return when Stevens connected me to Christine Notis, who was then conducting DNA research on *Calophyllum*'s family. See Notis 2004.

2. Created by Peter F Stevens, see the Angiosperm Phylogeny Website, Stevens, P. F. (2001 onwards). Angiosperm Phylogeny Website.

3. The Sulog tree is *Eucalyptopsis alauda* Craven, now understood to be a new species only found there and in the Moluccas. See Craven 1990. Lyn Craven, personal communication, July 2005.

4. Personal communication, July 1996; see also Kennedy et al. 1991.

5. Table 6-2-a Major Tropical Log Species Exported by ITTO Producers, 1995; 231,000 m^3 after Homalium foetidum which was 326,000 m^3; after Calophyllum was a Planchonella at 195,000 m^3.

6. Bryan Finegan, e-mail correspondence with author, May 25, 1998. For additional Central American data see Coe and Anderson 1996.

7. Work on the biochemical aspects of the genus *Calophyllum* appears from many sources. See Mohd et al. 1998. When I tell a colleague of *Calophyllum*'s medicinal uses he responds, "Yes, I know of the medicinal connection and can tell you that a number of people … have fantasies about making money by cultivating it."

8. Polynesians or the Spanish brought seeds to the New World? According to Stevens (1980) four species have been reported for the Americas. Colleagues from Duke University—Charles H. Cannon, Donald E. Stone, and Gary S. Hartshorn—have referred me to *Calophyllum brasiliensis, C. soulattri* and *C. longifolium.*

9. This correspondence went from 1997 to 1998. From Sather on June 1, 1998: "Note, the terms chinaga and penaga refer to different trees in Iban. You are right, apparently, in Iban too, penaga refers to Mesua spp. I am not certain the terms are reconstructable to the same root form. And, yes, indeed, Iban is a Malayic language with lots and lots of Sanskrit, particularly in the bardic language." See a parallel set of speculations comparing practices in India to those in Melanesia (Damon 2007).

10. See Pearson and Brown 1931: 47–49; also Stevens 1980: 235–38, 220–26. Stevens reclassifies *C. tomentosum* to *C. polyanthum.*

11. The Pacific literature suggests that many places use or used *Calophyllum.* Among others, see Howard (1995: 125) in Feinberg (1995).

12. Stevens' response: "It must be a bad key."

13. Linus Digim'Rina notes that *leyava* should be *reyava.* "L" and "R" are not discriminating sounds throughout much of the northern Massim. On Kitava and Iwa, I heard "L" and did not think to see if I would be understood had I said "reyav."

14. According to Digim'Rina (personal communication, 1998) the Trobriand name is *kweisira.*

15. Stevens's initial reaction: "Represents an undescribed species."

16. In 2002 I learned that there is much knowledge about different birds' ability to move seeds, the details of which are beyond my competence.

17. This is a common but not prescribed way of naming boats. The 1998 vessel I examined was from Kweywata and named after the plot of ground from which its keel (from a *kausilay*) was cut, *kaiyelu.*

18. Responding to descriptions of my work sent to him, Pierre-Yves Manguin writes, "One thing I noticed recently during archaeological work in Vietnam

is that the oldest Southeast Asian statues we have (Buddhist), carbon dated to ca. 3th—5th AD are made of Calophyllum inophyllum Linn; from the little reading I did on the matter, it seems that statues in the Pacific were also made of the same tree. None of the boat timbers of the 1st millennium AD I found in archaeological context in the same area (western Southeast Asia) uses Calophyllum, though. I also found a resin made of Calophyllum inophyllum nuts, mixed with shark oil, being used for caulking in the Maldives" (personal communication, August 16, 2000).

19. On the 10.7-meter-long *anageg* Lawanay, the *duwadul* was 3.82 meters long.

20. Other than a reference to sailing directions, the term *duwadul* has no meaning that I know of. *Kunusop* does. Before floats from industrial-scale fishing nets regularly came ashore, freshwater was stored in hollowed-out coconut shells. These were then stored in a *kunasop*, a basket (*kaynad*) made from woven coconut fronds. "*Kunusop*," I was told, applies to this outrigger piece because when the mast is not in place, water gathers in it; when sailing the mast pushes the water out.

21. For my elder informants from the 1990s on, this was the sail type they first experienced and think of as their own. Most were woven from *Pandanus* leaves when they were young, in the 1940s and 1950s. When I was first in Muyuw, Wabunun had two finely built *kaybwag* that sometimes shared a single sail of this shape made from canvas material preserved from the US Army and World War II.

22. Maylu were forbidden to sail into Muyuw, if not the whole Kula Ring, in the 1960s; my elder informants miss the coveted materials they brought.

23. The sail type *yabuloud* has a *kunay* and *kaley*, the former the top of the sail, the latter lower, closer to the outrigger platform, and often longer. The trees used for these parts differ from the same named parts on an *anageg*. They could be either *abanay* (*Podocarpus sp.*) or *busibuluk* (Ebenaceae).

24. Lae Herbarium (no. 47) or P. F. Stevens determinations in quotation marks: Damon 47, *apul*, Guttiferae C. *peekelii* Laut; Damon 48, *siptupwat*, "Clusiaceae *Calophyllum* C. *peekelii* Laut."; Damon 112, *apul*, Clusiaceae *Calophyllum sp.*; Damon 232, *apul*, "Clusiaceae *Calophyllum* (Stevens says = 112, correct)"; Damon 306, *apul*, "*Calophyllum apul p. f. stevens*"; Damon 307, *apul*, "*Calophyllum apul* p. f. stevens"; Damon 323, *apul*, "Clusiaceae *Calophyllum sp. Nov.*"

25. Lae systematists also identified the first *kakam* I collected as C. *peekelii Laut*. There is a small possibility that they are correct. However, I am presuming that it is C. *inophyllum*. My interaction with the systematists became part of the ethnography of my research. Outside observers of "science" presume a kind of reliability and consistency that is different from the experience of insiders. If at the genus level *Calophyllum* is immediately recognizable, its speciation is very complex. My experience with Stevens trying to determine what I collected was consistent with his warning about the complexity of his classificatory task. The reader should not forget the episode recounted in chapter 3, where Sipum swore at me for trying to make him identify the well-recognized *Calophyllum* species on the basis of their leaves.

26. Notis's (2004) DNA and Morphological analyses consistently distinguished C. *leleanii p.f. stevens* from *apul* which she designates as C. *sp nov.*

27. The possibility that this tree is really *C. peekelii* Laut raises the question of its exact relation to *C. apul* and possible human commerce across the Solomon Sea, in the distant past (see Tochilin et al. 2012) or the present. One man I met on Iwa was on mining furlough, first trained in Bougainville's Panguna Mine.

28. On one boat, I measured the distance between the *milyout*'s outer edge and the arcing hull using the ten *kiyads* as the frame of reference. The extreme ends were 194 and 202 centimeters; the three shortest lengths were 166 centimeters.

29. Fearing that this judgment was based on a single informant, I queried several people about it in 2009 and had the perspective confirmed.

30. Neither *dan* nor *apul* are found in true mangrove swamps, nor do the island's various mangrove trees grow in freshwater swamps.

31. I thank Lyn Craven, principal research scientist at the Australian National Herbarium, CSIRO Plant Industry for discussing these matters.

32. Jean Kennedy personal communication, July 1996 and August 2005.

33. Personal communication, August 4, 1998.

34. Informed of my suggestion that the tree might be *C. piluliferum*, Stevens felt it was unlikely (personal communication, 2006). I examined the holotype of *C. piluliferum* on June 27, 2006, at the Harvard Herbarium and found its leaves and fruits to be much, much smaller than those I've observed and collected.

35. Allen Allison, personal communication, September 9, 2009.

36. Although we found seeds in 1998, in 2002 we were unable to locate any. How this fact relates to the continuous regeneration of the trees and the nearly novel fact that trees of all different sizes are readily found in the *dan* forests remains a puzzle. In 2009 I found seeds in leaf crevices in low-to-the-ground *Pandanus* trees.

37. A difference between *kausilay* and both the *ayniyan* is that relatively small *ayniyan* suitable for masts have the desired red heartwood closer to the outer edges.

38. A mast on a Nasikwabw *anageg* examined in 1998 was 7.82 meters long.

39. As I shall note later, Peter Stevens eventually settled on a determination for *ayniyan* as *C. goniocarpum,* and *aynikoy* as *C. soulattri.* In Australia's National Herbarium in Canberra, the one quality that stood out between them was this noticeable difference in leaf shape.

40. A third way the first layer of wood can appear is "knotty," *patpatuk,* not relevant to local *Calophyllum*.

41. *C. soulattri* is a widely dispersed species (see Stevens 1980: 277–92), the category itself, Stevens told me, perhaps a dumping ground waiting for the next evaluation.

42. For most readers "rudder" best describes what this piece does. For sailing experts it is like a "leeboard" (Geoffrey Irwin and Pierre-Yves Manguin, personal communications, 2000).

43. See Damon (1990: 36–37) for the star-human relation and their parallel powers.

44. He was an interesting character; someone warned me that he was "not very smart," so I should be careful about what he might tell me. Indifferent to tree names, he probably misled me with a flower for one voucher specimen. Unlike Sipum who only got intrigued with this survey when we came upon

things he did not know, he nevertheless followed my recordings and immediately picked up patterns. Probably in his late forties or early fifties during the belting, so roughly my age, he had at most a few years of schooling and could barely read. Yet he is an accomplished carver, fascinated by the intricacies of the craft. He is indeed a detail person.

45. Simon Bickler examined these plots more carefully than me. By 2002 I figured out how gardening and fallowing wore down coral outcrops and so determined again that these areas had never been continuously manipulated by humans.

46. In my experience they occur in April or May. *Tan* best translates to high tide, the *po* almost an emphatic marker meaning exceptionally high. Most high tides are at the second (full) and fourth (new) moons. In 1996 in what seemed bizarre at the time they came during the first and third moons.

47. The proximate location to the sea and the likelihood that underground water would be salty make the underlying ecology of the place much more suitable for *kakam, C. Inophyllum,* than *kausilay.* The condition is a puzzle. Back a few hundred meters from this spot, however, is a freshwater swamp. Perhaps it drains through the region, thus providing freshwater approximating the dryer limestone shelves on which most *kausilay* are found.

48. Voucher specimen Damon 273.

49. *Apul* grow larger but are always straight and usually in a different ecological setting; *kaga (Pomentia pinnata)* might grow larger but are usually in an older successional stage. Unlike *kausilay* they have buttresses, and their bark is red and smooth, the bole straighter. A final tree, *ameyleu (Syzygium sp.),* grows larger but in a different ecological setting. An emergent tree, it grows straight and up through the canopy, remaining solitary.

50. Up to 2002 Dibolel had been responsible for acquiring four *anageg,* three from Gawa, one from Kweywata. Two of them had names conceived to be traditional for their builders and passed from one boat to its replacement.

51. Over the course of our interaction, Linus Digim'Rina and I came to one tentative conclusion: that the near-constant availability of an underground water source in the Trobriands and its relative absence in Muyuw was a major differentiating criteria between the two. Water has also loomed as a major focus of Mosko's Trobriand research (personal communication)—pursuing its relationship to *apul* might be very revealing.

52. *Ovatan* is a noun classifier. Two units is *ovey,* three *oveytoun,* four *oveyvas,* and so on.

53. Included were the *Ficus akisi, bwibuiy, abuluk,* and others.

54. On the transformation of metaphors into metonymies see Fernandez 1974, 1998, who draws from Lévi-Strauss 1966. I have worked this out for the kula in Damon 2002.

55. See chapter 3, note 9

56. It was a large tree for that class of forest, nearly thirty-two centimeters in diameter at breast height, about eighteen meters tall.

PART III

Synthesizing Models

It is the processes of adjustment that are taken to be fundamental... (Winterhalder 1994: 33)

As with any great work of art, one acquires by study a deeper awareness of the work's inner relationships. (Wills 2002: 85)

CHAPTER 5

Vatul

A Life Form and a Form for Life

> Indeed, ... the problem rendered spatially in the knot is not a problem that can be conceived experientially at all other than on the basis of binding itself; that is, it is unlikely that the knot is the result of a projection of pre-existing concepts derived from social-culture experience, but that it evidences a complex relational and transformational field. (Küchler 2001: 71)

My first leap of insight about the *kausilay* tree concerned its "cross grains" paralleling a quality about gardens. As cross grains give the tree its strength, intertwined male and female yam vines make a garden productive and beautiful. As other facts came to light, however, the *kausilay/* garden insight seemed superficial—because I paid no attention to tying. Yet long before the centrality of string, chords, and ropes came into focus, people were talking about their qualities, slowly creating an argument I realized I had to follow.

In 1995 Sipum reported that the strangler fig *akisi* had a lot of "power," *tautoun*. The prototype for four trees noted in chapter 2, *akisi* is a well-known *tasim* example often observed on the edges of garden areas or along paths or roads. Sometimes their downward-growing roots braid together in a rope-like fashion, creating a seamed bole with a hollow center—from the disintegrated host—running twenty to forty meters up to the canopy top. Photographs accompanying this chapter include a young *akisi* just starting its winding course and a mature one.

Although they are thought to be bad for gardens, *akisi* leaves facilitate love magic. They can insert the tree's entangling powers into a lover to entwine a couple through death. During a final mortuary rite, a widow is supposed to be an incarnation of death, and wears one of death's signs, a long, black, braided cord, hanging from her neck. A couple remains married until rites performed over their deceased children end one cycle and enable a new sequence of transformations (see Damon 1989b). Moving from being unconnected, entwined, and then released to start over again are fractal sequences of great cultural significance.

After talking about *akisi* leaves, however, Sipum said that while some trees have substantial power, vines, *vatul*, generally have more. A small unobtrusive one called *diday* is used in garden magic. It grows quickly, and the magic is thought to transfer that property to the male yam type (*kuv*), increasing its already fast growth; they then pass that speed to associated female yams. The very end of a growing shoot should be inserted under the garden's centering rocks (*ananun*), the stones defining its intersections. From them garden efficacy derived from *diday* tips, spreads. At least one other vine is employed similarly.

As soon as I arrived back in Wabunun in February 1996, Sipum and his companion Mayal stunned me with another episode. As always I appeared with gifts, but since I passed through the Trobriands, Kitava, Iwa, Gawa, and western Muyuw, they had to be small. I had purchased for these two in Port Moresby an artificial lure, a Finnish Rapala, similar to those I used when young. What would they make of it? The lure didn't interest them. What grabbed their attention was the knot pictured on the lure's package. They had never seen that knot before and doubted it would work. The two studied the instructions, finally pronouncing it OK. Then they made it. This fifteen-minute event astonishes me to this day. These men rapidly identified a knot they had never seen before, imagined its coursing, conceptually determined it sufficient, and reproduced it. Here I witnessed something fundamental about organized thinking for these people, undoubtedly in the Kula Ring and perhaps throughout the Austronesian world.

Several years later as the issues of string and tying were coming into focus, I stood by an *anageg* with Sipum and, pointing to all the different bindings, suggested that string and tying were important. Struggling with the naïve comment, he patiently pointed out how important cordage was to boats; next he said they were critical to forests because vines bind trees, giving forests their strength; then he moved on to gardens because intertwined yam vines, "*vatul*," gave them their strength and beauty; and finally, *vatul* are fundamental to persons because—he said while flexing his muscles to show his bulging arm veins—these "string/*vatul*" enabled life.[1] String and things bound by them are everywhere.

Thinking through that encounter brought into focus this 1999 episode in Alotau, Milne Bay Province's capital city where I had extensive discussions about *anageg* there. As my time there was coming to a close, a Yemga man agreed to make a model sail. After seeing string I purchased for him at one of Alotau's stores, he insisted that we both hike down to the store where, at a corner sewing section, I waited with impatience and bewilderment as he carefully felt dozens of different types of material. He was searching for something with his fingertips. Neither he nor I liked the color of the one he selected—green and very much unlike anything that he would have found at home—but it matched qualities he thought necessary. The feel and strength characteristics he sought in the material derive from the inner bark of *ayovay* (*Hibiscus tiliaceus*). Villagers often plant this tree within or proximate to village boundaries because it is useful for many different purposes, foremost cordage.[2]

Moments like these built a point of view. For this research I collected 329 voucher specimens, obtained from higher parts of the tree. I started watching how my assistants casually seemed to draw a flimsy plant stalk or pull the bark off a tree, fluidly tie the ends together to create a circle called *kaywat*, slip it around their ankles, and then shimmy up a tree. Yet there was nothing casual about this. They selected materials the exact tensile strength of which they understood. They knew exactly which trees could—usually Sterculiaceae[3]—and could not provide bark useful for the purpose for which they were chosen. They had observed and practiced tying since they were infants and had climbed many trees. This was craftsman's expertise on display far removed from my idle appreciation.

Stimulated by Küchler (2001), this chapter changes that appreciation. Küchler suggests knots' symbolic and conceptual efficacy may result from the fact that they tend to "exist apart" (2001: 71).[4] I argue instead that string and cords and knots are a fundamental experience and logic purposely fixed in the consciousness of these actors. In a social system predicated on relating, they are an initial experience of linkage, which is to say transformation. They are constitutive of the region's fundamental conceptions. This chapter shows how this is so.

Küchler lends credibility to the idea that string, strands, and cordage are platforms for life. Drawing from the mathematician Papert and Lévi-Strauss, she invokes the latter's critical insight: "We ... think with objects ... by using them as image in thought, and maintain in thinking the properties that these objects would demonstrate in the physical world" (63). Knots are "evidences a [of] complex relational and transformational field[s] which can be discovered simply by doing [them] and looking at [them]" (70). I add to this discussion tying material and items produced with them.

Putting this sensibility in the center of my inquiries is critical for understanding the qualities of judgment that enable life in the world described here.[5] These are, after all, lives predicated on the active transformation of one thing into another; a forest class into a vegetable resource *and back* to that forest again; a village devoted to one set of activities relating to another village's activities; the transient and unpredictable arc of a branch or bole composed by specific kinds of grains becoming a finely hewn structure; the wind from one direction moving a boat in another direction.

These transformations come from a world of string regularly fashioned together for culturally determined purposes. Sipum's 1995 message about vines being stronger than trees was just the tip; we search for the base.

This chapter first works through an inventory of vines and a discussion of their names, replicating the structure of chapter 2. Reproducing chapter 3's logic, I then discuss tying, knots, and the forms they enable, ultimately modeling a way of life. Both these sections share chapter 4's concern, the problem of organization *Calophyllum* understandings present. At one level *Calophyllum* and string are physical objects, varieties of trees and cords and materials into which they are transformed. At another level they present conditions of knowing and thinking. If the qualities amid the *Calophyllum* set were not created by long-term human interaction and selection, they were discovered and matched for a set of designs that, whatever their physical necessity, are human inventions of extraordinary scope. The category "*vatul*" contains identical phenomena, astonishing relational sets. The chapter concludes with string figures and their places in this cultural system. They formulate a geometry of motion. This facilitates a passage to chapter 6's discussion of the *anageg* form. *Vatul* anchors and energizes devices, and everything proceeds from it.

Vatul as a Life Form, and Names

Table 5.1 provides an alphabetical list of the forty-four vines collected or described in my notes. Far from exhaustive, the list excludes yam varieties, the "Creator's string (*vatul*)." As with trees, people do not name all the vines they know. Numbers 42 and 43 are the two large vines noted in chapter 2. Although the vines are well-known by their properties, none of my informants had names for them.

As noted in chapter 2, the Muyuw classificatory system organizes eight major life forms. "Vine," *vatul*, is one of these; perhaps best translated as "string" or "cord," the term is generic for vine and an extremely important concept. *Vatul* carries its own noun classifier, *ul-*. So *ulitan* translates to "that vine/string," *ulita* "one vine," *uli* "two vines," *ulitoun* three, *ulivas*, four, and *ulinim* five. Just as many tree names begin with the tree classifier *kay*, *ke*, or *a*, many recorded vine names begin with the *ul* sound, for example, *ul*buniwan, *ul*agubaguba, *ul*akaykay.

While "*vatul*" specifies a kind of plant, it also refers to materials used to tie things. And not all of the latter are the former. The best tying materials for boats derive from trees, *ukw* (*Sterculia sp.*) and *ayovay*. People extract *im*, another important *vatul*, from the inner bark of aerial roots of the shoreline "tree" *loud*, one of at least twelve *Pandanus*. *Im* is used to make

Table 5.1. Vines

	Muyuw Name	Scientific Name / ID Voucher Specimen #	Place	Uses / Miscellaneous
1	*adilaboub*			3 large leaves, long barbs; female
2	*amlumwala*			slender leaves and slender stem
3	*Avled*		*Demiavek*	
4	*bwadabwad*	Melastomatiaceae *Medinilla* / VS297	forest tops	old source of caulking material for middle-sized outriggers
5	*Dadow*	Asclepiadaceae *Hoya* / VS298	freshwater swamp areas	flowers, and their perfume qualities, used as *antanon*—corsage-like—to capture women or men
6	*Diday*			transfers its fast-growing properties to the *kuv* yam type
7	*Dodiv*	Solanaceae *Solanum* / VS162	early succession forests and gardens	considered female because it has barbs; moderately sweet fruit
8, 9, 10	"*domwet*"	Piperaceae *Piper* sp.	various	At least three kinds. One is the standard for betel chewing. Another grows in central Muyuw and is known as *kakawowa*—it is not used for betel chewing. Eastern Muyuw people recognize it as different. Another piper species, very red in color, is called *kweybok* after its location. Although some people chew it, Malas clan people may not. It is said to be from one of their clan mates; it is called *buyavis tau*, "blood of a man," "because of its red color."

(continued)

	Muyuw Name	Scientific Name / ID Voucher Specimen #	Place	Uses / Miscellaneous
11	*Dud*	Gnetaceae *Gentum sp.* / VS281	high forest canopy	Source of dark-colored string for sewing sleeping mats (*sags*). Using it adds a blended appearance to coconut leaf skirts (*dobs*). It is also used to separate teeth in combs, among other uses.
12	*igenapwey*	*Ipomoea sp.* / VS202	beach	considered male, used with a weed (*auyow*) with a long taproot, called *pwankulas*, for preparing boats for fishing and washing yam seeds
13	*Kokoyit*	*Gelichinia* / VS292	meadows, e.g. Bungalau	stem contains larger and smaller thread-like material, both of which are used to make black woven armbands, calfbands, and waist bands
14	*Koliu*	P. F. Stevens said it was a fern.	freshwater swamps	slender leaves; leaves eaten in Wamwan; roots dried and smoked, then soaked to make pliable for tying strakes to keels
15	*mamad "tatonen"* (="real")	*Cynanchum ovalifolium* Wright / VS103	oldest class of forest, *ulakay*	one of 3 vines used to tie highest class of outrigger canoe; put in water for a week until bark and inner bark rot off
16	*mamad puvakau*	Apocynaceae	oldest class of forest	substitute for *mamad tatonen*
17	*mamad kumkum kamey*	Malpigiaceae	oldest class of forest	substitute for *mamad tatonen*
18	*mamayak*		climbs *tabnayiyuw*—oldest class of forest	has hooks and slender leaves, *kalokol won*—i.e. barbed or hooked

19	*namon sigeg*	*Passiflora sp.* / VS92	early succession forests and gardens	female; used in magic; fruit and leaves rubbed into fishing nets; irritates flesh and has barbs
20	*Niwoun*	Arecaceae *Calamus* / VS299	oldest class of forest grows high	like *weled*, but very thick, cord-like; very strong—used to edge sails
21	*Pwaplow*	*Passiflora foetida* / VS67	early succession forests and gardens	seeds eaten raw, leaves used for skin rashes, vines very occasionally used for tying *domwet*
22	*Sekita*	Arecaceae / VS155	oldest class of forest	Old fallow-forest, barbed. Coupled with *agavagav*, a bush serving as a distance marker living along the shoreline, the two are very powerful. They make the body stand up, be strong, and shine. Good for kula and women. Its fruit is rubbed into nets. Irritates skin.
23	*Seypwan*			fruit used as beads; worn for its pleasant smell
24	*Siyam*			three round leaves, short barbs, and female (because of barbs); sweet fruit.
25	*ulagubaguba*			leaves used as corsage to attract women or men Name is explained from verb, *gub*, which means to climb hand over hand. Western Muyuw use in male/female exchanges (*takon*).
26	*ulikaykay*	(Convolvulaceae) *Erycibe* / VS231	everywhere, but only large in *ulakay*	Grows big and is used for hauling large logs/*wag* from forests to villages. In Nasikwabw, Gawa, and Iwa it used in place of *yit/it*, which doesn't grow there. But it has to be cut or divided into many strands; *yit* is customarily divided into two strands.

(continued)

	Muyuw Name	Scientific Name / ID Voucher Specimen #	Place	Uses / Miscellaneous
27	*ulaweydon*	Caesalpinia (Fabac.) / VS76	edge	leaves crushed and used for birth control by women; salty, so associated with death and infertility—blocks blood; plant has lots of needles on it, and is dangerous to go near; leaves are rubbed into dogs so they can catch pigs
28	*ulbunibwan*	*Uncaria* (Rubiac) / VS62	toward sulogSulog	named for eagle because it grabs; also used as birth control device
29	*Uliyan*			substitute for *ulakaykay*, for it is very strong
30	*ulmwadeyo*		sago swamps	medium large leaves; similar to *yakawom*
31	*Ulsiyan*	*Cissus* (Vitac.) *spp.* / VS66	early succession forests and gardens and beyond	minor tying of betel pepper bundles, firewood; flowers enhance singing and change a woman's mind; fruit rubbed into nets for magic; irritates skin, barbed.
32	*ulubutobot*	*Parthenocissus?* (Vitac.) / VS95	early succession forests and gardens	named after seeds put on *veiguns*, which are from a plant that grows in the Alotau region; like *ayolal* tree, burns skin; used in net magic.
33	*ulyaktakwit*			small double leaves
34	*Vayu*			substitute betel pepper obtained from seed, which is similar to *kwayen* seed only hard
35	*Weled*	*Flagellaria* / VS72	early succession forests and gardens	flowers with nectar similar to *gwed* (*Rhus taitensis*)—cuscus drink it; used to tie houses, fences, but not outriggers; fruit rubbed into fishing nets as magic; irritates skin; leaf ends barbed

36	*Weybit*			similar to *yokoluta*, barbed, *"kolokon won"*
37	*Yakawon*	kudzu?		Leaves are used as large leaf covers in earth ovens and for washing yams. Its white milky sap can blind for days.
38	*yagug* (yams)			distinctive feature of yams, which are *vatul*; one kind is male, the other female
39	*yawelloks*		*siposep* (soggy soil) vine	very broad leaves; the vine may be used for wrapping sago, especially the large-sized *ibsisay*
40	*Yokoluta*	*Simlax* / VS88	many environments to canopy tops	The tendrils (*udag*) of this plant are extracted and placed at garden intersections (*ananun*) with the idea that they can impart their qualities to the garden's yam vines. The same parts also used to be rubbed on a person's throat to help him or her sing better.
41	*yoyita/it*	*Araceae? Raphidophora?* / VS115	more mature forest	large-leafed vine from which *yit* is extracted; the major fastening cord used to tie stones to various wooden handles, garden fences, and all parts in houses, from posts to the sago leaves used to make roofing thatch
42	unnamed 1	Rubiaceae *Uncaria* / VS165	oldest class of forest	used for drinking when on trek; irritates skin
43	unnamed 2			used for drinking when on trek; does not irritate skin.
44	unnamed 3	Rubiaceae *Psychotria* / VS143	oldest class of forest	while not named, Kwadoy wrap themselves up in it to sleep

a relatively fine string for the kula valuable *veigun* (or *bagi*), parts of women's coconut-leaf skirts, the sewing material for sleeping mats and sails, fishing nets in north-central Muyuw, and is, everywhere, the ideal for string figures. String produced from this plant may be very thin and supple yet sufficiently coarse to hold a knot or place in a sequence of places.

The degree to which vines are grouped into genus-like *bod* like *Calophyllum* and *Dysoxylum* trees is uncertain. However, the two "yams," *kuv* and *parawog*, and a set of thick vines called *mamad* form genus-like groups. People are highly aware of their spatial contexts so that certain examples of them are found in the distinguished fallow types, or swamps, along beaches, or in other similar places.

Nearing the end of what I thought was going to be my last research time in late July 1996, and after a frustrating day with Sipum collecting few interesting plants, we walked down a beach close to Wabunun. Sipum started telling me that the beach was a great place because, as it was situated between the sea and the land, all kinds of remarkable and useful plants grew there. As if from *Purity and Danger*, the interstitial location seemed to explain the fact of the plants. One male vine grew from the land toward the sea; high tides frequently nipped it. Called *igenapwey*, a name that, so far as my sources know, means nothing, it complements another plant for garden and fishing magic. This other plant is considered the female in a set and some people know it by the name *pwankulas*, "excrement from the kulas fish," because the plant's sap looks like fish excrement. Its features, however, contrast with those of the male vine. It hardly goes anywhere, remaining a small clump of leaves on the sand. Yet its roots go deep. The magical effect is obtained by mixing together qualities oriented by horizontal (male) and vertical (female) dimensions. They then may be added to two different productive processes (which also combine of male and female qualities). Before planting, yam seeds should be "washed" with water mixed with these plants' crushed leaves. For fishing, one holds the crushed leaves and then swims around a canoe when it is in the water. The swimmer starts from the right side of the boat, opposite the outrigger float, and moves counterclockwise, ending up at the opposite back end of the craft whence the remainder of the leaves are set inside the boat alongside its "sternboard" (*matsibod*).

Igenapwey is considered a male vine. Table 5.1 shows that many vines are considered female. This is often because of their barbs, sometimes from their conspicuous entangling propensities. For now it is important to consider some of the names whose sense is readily conveyed, a minority of the names I have recorded.

Ulubutobot is interesting because its name was explained by the fact that its fruit, berries, look like berries found near Alotau, the capital for Milne

Bay Province some three hundred kilometers to the southwest and, from the Muyuw point of view, the southern extremity of Kula interests. Those seeds become part of the decoration for the Kula valuable called *veigun*. The naming principle here is com-

Table 5.2. Vine Names

Muyuw Vine	Translation
Ulubutobot	"another place's name"
Ulbunibwan	"eagle" vine
ulagubaguba	"*gub*" vine
Ulakaykay	"tree" vine
namonsigeg	sexual juices flow forever

mon, for the issue is resemblance to another item. But the interest is that the item is from far away, which is brought up close by the fact of naming. Perceived resemblance covers many of the other names on this list. *Ulbunibwan* derives from *bunibwan*, "eagle" and this plant is called "eagle's vine" because its barbs are said to be like those of an eagle's claws. Because of its piercing, grabbing capacities, this vine is to be avoided. Yet the same properties make its leaves useful, some believe, for aborting an unwanted pregnancy. The plant's properties supposedly tear a fetus away from the mother's insides. *Gubagub* was translated as "hand over hand," and refers to this often large and very woody vine found in high forests where it is used to climb trees, by putting hand over hand. In western Muyuw, whose customs approach those of the Trobriands, the vine—or one with the same name—is used to tie together large baskets of food that brothers annually give their sisters in prestations of significant display.[6] *Ulkaykay* translates to "tree-like vine," an appropriate name in that the vine's size approaches that of a tree. It is used to haul partially trimmed tree trunks out of the high forest for fashioning later into keels or strakes for boats.

Noted in chapter 2, the last vine here, *namonsigeg*, is in the *Passiflora* genus, whose sense I first misunderstood. The Western name invokes the Passion of Christ because flowers in the genus have a cross embedded in them; I thought the name referred to the plant's erotic appearance. My attempts to photograph the flower never captured its essence, but it is the most erotic, almost pornographic, flower I've seen. *Namonsigeg* has at least two related uses in Muyuw. One it shares with other vines for fishing nets. Four vines (table 6.1, 19, 22, 31, 35) and a couple of trees are used as magical additions to new net sections when they are spliced into an old net. All the vines, and at least one of the trees, are either barbed or generate a burning or irritating sensation when touched, the distinctive qualities. The *parawog* (Dioscorea *esculenta*) yam kind is often covered with sharp barbs; hands spent harvesting them often bleed. This prickly, barbed quality is among those properties understood to be feminine; young courting men proudly exhibit their gouged knees and thighs the next day. Although made by men, fishing nets are considered feminine; they are large, slow, and eventually concave-shaped when used. All the vines rubbed into the nets have feminine qualities, and the point is to add them to the structure to increase its efficacy. Adding male labor to their actual use, literally men or boys on a fishing excursion, makes these instruments more productive as well, thus complementing the addition of female properties expressed by the leaves.

Namonsigeg's other use, and the very name, draw on its capacity to "stick" men and women, not fish and nets, together. This understanding and the plant's name were outlined in chapter 2, where it was noted that the plant is bought into play for the most complicated moment in the life of young lovers, when they are deciding whether to become a couple. *Sigeg* translates to "forever," or "continuous." *Namon* derives from the possessive *na*, meaning "it/she/he"; and *mon*, literally translates to "fat" or "grease." The word connotes male and female sexual fluids. The sense of the plant's name then is something like, "May her sexual fluids flow forever." The plant is a joyous complement to the hoped-for effect of using the strangler fig *aksi's* leaves in love magic. Not only will a couple be entwined through to their deaths, but they will be ecstatically and passionately in love with one another forever. *Akisi* trees are commonly seen amid the terrain proximate to Muyuw gardens and villages; *namonsigeg* is more common as it twirls its way among the low-cut brush on paths leading from villages to gardens and across growing plots of small trees from recently used garden areas. Although not all of it is positive, Muyuw flora shouts out the messages people have recorded in its literal and figurative properties.

Tying and Its Terminology

This section moves from generic and particularizing names for tying procedures and knots through a sample of those things created by weaving and tying to illustrate a social order. There are names for parts of tying materials, patterns for things woven, different knots, and a tricky category that I translate as "methods of tying."

If initially like "table manners," low-level but important patterns for living, the cosmological import of this data quickly stands out. *Tapopau* is an example. It prescribes how any line should be bundled (*katipum*): a stationary left hand holds whatever the line is being wound around, the right hand doing the winding, going back to front, equivalent to left to right, clockwise. Another concerns the quality desired in boat cordage "hardness," in this case best understood for what it should not be—*dimwalikw*, a word best translated as "slippery." Friction sufficient to prevent unraveling is sought. Although store-bought line or beach flotsam now replaces many traditional tying materials, most lack what I translate as "hardness." The contemporary selection of inferior qualities is very much part of contemporary culture. Today much Muyuw culture rests uneasily in a complicated duality contrasting emerging realities with those passing from the scene. Intimations of a cosmology? Some people liken the movement of sago orchards, pigs, and people between clans—the Creator's garden clan model—to a tying process. An apparent movement from the minuscule to the encompassing runs through this material, a pattern evident in the common tying material *yoyit*.

The Vine Called Yoyit

The vine called *yoyit* (Araceae, probably *Raphidophora sp.*), found in older forests, provides material for tying or binding almost everything except boats. This item is an example of how a single plant implicates purposeful relations among different forms in forest growth, draws on the function and structure of body parts with correlative gender associations, and, through the position it plays in a set of possibilities, invokes differences between boats, houses, and technical forms across the region. Wamwan, central Muyuw people, used to employ this plant for darkening their teeth, transforming betel-nut chewing's motley effect to a brilliant black, once a feature that epitomized cosmetic elegance.

Called *yit*, sometimes *it*, the vine's pith is easily pulled from its centimeter-plus diameter. People use *yit* to stabilize all house posts, beams, and support structures. They also use it to sew sago leaflets onto sticks (*sinat*) upward of two meters long, which are derived from the long aerial roots of a *Pandanus* tree type called *dageliu*; *yit* is then used to fasten those sections to houses for roofing and siding. *Yit* holds garden fences erect. Young males search out supplies of *yit* before every fence-erecting task— the vine's availability is a minor consideration for placing every garden. Yam houses are tied together with *yit*. It is used for binding stone or metal cutting devices to sago pounders and adzes. But *yit* does not just surround people by virtue of objects raised with it. Metaphorically, the plant is considered to make humans stable as well. Turning from a vanilla color to a dirty white or gray when dry, *yit* is likened to Achilles tendons. Beyond similarity of appearance is an attribution of similar function. Running from the ground to the canopy top in older forests, the vine is credited with forest stability. People consider the Achilles tendon to run from a person's heel to his or her brain, because a person cannot walk without wobbling, cannot be upright, when it is severed, a quality attributed to a lack of connection between the brain and the foot's balancing mechanisms.

Yit may be used for more temporary but significant tying, bundling construction material obtained from the forest for transport to a village;

after being used for this, it is quickly discarded. I witnessed one attempt to use the material this way in a sago orchard in 2006; my companions commented on not finding any before they extracted bark from two saplings in the *Sterculia* genus (*silawow* and *ukw*). Although it can become scarce in areas close to villages and frequently visited sago orchards, *yoyit* grows in virtually every environment on the island with mature growth.

The pith is usually split in half, sometimes quartered, before being used. Very flexible (*manum*) at first, the quality most tying material should have, *yit* turns into strong, fast binding. When dry it becomes hard (*matuw*), tough, immobile, practically cast into place; in this condition the material cannot be used again, so if the item has to be retied people discard the old *yit*.

The contrast between fresh and older, stiffer *it* invokes a contrast that runs through processes and materials in this region. First of all, although they are analogized with one another, boat and house forms become differentiated by their relationship to this tying material. Its initial suppleness makes it very useful for houses while the hardness it eventually achieves corresponds to the relatively fixed forms houses take. But this change in quality makes it unlike much of the traditional material used for lashing boat parts that are reused until they wear out. Houses are hard and stiff, boats supple and fluid. The contrast, moreover, becomes part of a three-stage divergence in lashing materials that runs from flimsy to rigid. *Katipok won molay*, a term derived from woven coconut fronds (*molay*), the flappiest and least stable of Muyuw's roofing or siding materials, is the flimsiest. *Kaytol* is toward the most rigid or stiff end of the spectrum, the quality demanded of houses. Informants assured me that *kaytol* is related to *kaymwetol*, the term used when rigor mortis sets in. Both terms connote a tree-like stiffness; the words probably begin with the common Austronesian classifier for wooden things, *kay*. *Mameu* or *manum*, between *katipok won molay* and *kaiytol*, defines a persisting flexibility necessary for boats that is present in fresh *yit* but absent once the vine matures. *Mameu* becomes associated with the hands of women (and children) while men's hands and fingers, because of the nature of their work, become increasingly *kaytol*-like. This contrast also defines one of the main differences between the sail types; sails from the western half of the Kula Ring are more *mameu*, the eastern form more *kaiytol*.

With *yoyit* uses and meanings providing an analysis of variation, I now look more carefully at the range of ideas and objects associated with the phenomenon of "string" and knots.

"Soul" *as Method*

Soul is the word I translate as "method of tying." So, for example, boats and fishing nets have their "methods." The method for an *anageg* differs from that for a *kaybwag* or *kuwu*, the two smaller classes of outrigger. And some people will say that different parts of the same boat have unique methods. Interestingly houses and garden fences are not considered to have distinctive methods. People told me they tied any way they wanted, though the slightest visual inspection shows that houses are tied, and must be tied, with considerable care. And it is this tying which keeps all structures, boats, houses and fences, erect. Although other tied objects may not have their own methods, now and again specific patterns are used to tie this or that just as a matter of showing off, what Muyuw call *kalimwasau*. I specify the different "methods" when I correlate the knots to their usage, and as convenient I move from kinds of knots to illustrations of the uses to which they are put. This section leads to a description of sail-making processes, with traditional sails illustrating how social and technical relations are woven together. I close the section with a brief return to one of the last "methods" of tying, the one fixed in the Muyuw fishing net, a form that encompasses as it transcends some of the understandings internal to the beginning of any tying procedure.

The photographic essay that accompanies this chapter illustrates many issues I only outline here. It also juxtaposes photos of each knot kind with line drawings of each type.

Tying materials are understood in terms of common Austronesian distinctions, including the "tip," "top," or "point" (*matan*); the midsection (*tapwan*), the same word used for a body's back and trunk of a tree; and the "base" (*wowun*). The "tip" or "point" refers to the end points of any strand used for tying or sewing. Between them is the midsection, *tapwan*. When two or more things are tied together, the original point of anchoring the cord is called *atusip*. This becomes the "base" of the structure. If two pieces of material are spliced together, the "tips" are further distinguished between "left" (*kimau*) and "right" (*katay*). The former becomes the base, the latter leads. The entwining or wrapping part, now the single tip or top, becomes the *yawan*, a word that may function as a verb or noun meaning "tying." People deny there is a systematic gender distinction coincident with this contrast. However, once tying processes are put into motion, a female/male contrast governs the process. The left part of the structure, the "base," *atusip*, is "female" (*kudavin*) since it remains stationary; the right part becomes "male" (*kudataw*) because it moves, providing strength to the ensemble. It is widely understood that the correct tying direction, left to right, the natural movement of the yam vine called *kuv*, and the flow of armshells, *mwal*, are the same.

This paradigm may be selectively attributed to other facets of experience. Part of the "method" for tying all boats includes having the male part always move toward the "top" (*dabwen*) of the boat; this means that tying is conceived to enable motion. I have noted that Muyuw do not class trees into male and female types, yet they do ascribe this gender contrast to relevant situations involving trees. The three strangler *Ficus* (*akisi, bwibuiy, agigaway*) and *Fagraea* (*bwit*) are considered male trees with respect to whatever is their host tree, which then becomes female.[7] The experiential sense of this conception is important. Although the terminology is different, the resulting process of intertwined trees, denoted with the verb *bikwen,* is likened to what happens when two people, as the English expression goes, "fall in love"—*kalpwakit* in Muyuw.

Tying is understood by means of a set of distinctions. Sometimes cordage is used so that one end is fast, the other flutters. This is called *katsipus;* the object at issue may be the tying material itself or what it ties—streamers, telltales, attached to boat parts. The wrapping that closes many tying procedures, finally pulling the fastening tight, is called *tageg 'eyon. 'Eyon* comes from the word for throat, *kayon.* This procedure usually amounts to continual circling then tucking the end of the material inside a convenient loop. It is readily seen in houses and fences. At first glance the same thing might be said with some tying in outrigger canoes. However the difference is that nobody pays attention to the number of revolutions in houses or fences while these are prescribed when fastening canoe parts.

Attaching the outrigger crossbeam, *kiyad,*[8] illustrates this tying. The form runs from one side of a canoe through the other side to the outrigger float. On each interior side of the canoe it is tied to an extremely important structural part, *seisuiy,* discussed in detail in the next chapter. On its right end the *kiyad* is notched so that it fits right on the *seisuiy,* and the wrapping takes a slightly different name, *eyon tau* (literally "throat male"). These wrapping names are not the same as the "method" names for these two forms; my informants knew there were such names but could not recall them.

Although single strands were used, much tying was also accomplished with string or rope composed of individual strands of material, usually the inner bark of trees. The material used to make fishing nets (*wot*) is called *yawn,* and is made by working together (*-yawun*) several strands of one of two previously noted trees, *gudugud* or *gayas.* Although the term *yawn* is often heard for thicker tying materials, "rope" is technically referred to as *yawoywuwun.* It is made (*-wey*) by interlaying three inner strands of *ukw* or *ayovay.* This kind of material is used for the heavier duty lines on a boat, *balau, asan, alita,* and *yawosay.* The term *ikwatav* refers to a form of braiding. The form is said to be very strong and is used to hold pigs, and pull (*senau*) things together, including a belt-like structure called *kayovay*[9] that composes the interior waistband of a coconut-leaf skirt (*dob*). *Mwag,* the rope-like adornment women wear around their necks for mortuary rituals, made invariably of dark cloth or blackened material, follows this method. Although effective for its purposes, this is not the strongest weave. That honor goes to a form called *skoko yayun,* or "rat's tail." Unfortunately I do not have an example of this practice, but the method used for such things as anchor (*lon*) ropes, lines coming off sails, including the *otan, tabagon,* and *latagan,* and the ropes made for bent-sapling snare traps for pigs. *Tibwelon,* the generic term for arm (*sasi*), waist (*palit*), or leg (*kaykwas*) "bracelets" woven from the black inner strands of the vine *kokoyit,* come in two kinds, *kalamanag* and *kaylogwaw.* I have not seen the latter of these two, which is considered to be from Gawa—the name specifies that. Its different weave, called *man* (the word for "bird" or "animal"), gives it a "mixed" (*lamwelan*) appearance. Unlike the other materials just reviewed, *tibwelan* are not tied into knots or used for other fastening procedures. Yet all the peoples of the Kula region used to wear these bands, and those who still do are said to look good. They encircle body parts—upper arms, waists, calves—conceived to be composed by other strands and cords. I regret that I do not have data on the significance of these waning practices; they need to be integrated into the whole set of things organized by binding, bodies, houses, gardens, boats, and forests.

Sip Vinay *and* Sip Tawau

Two knots distinguish different ways of tying end pieces, either because the cords are too short for a purpose, or a single fiber is being used to wrap something up, like a bundle of firewood. These are *sip vinay* and *sip tawau,* "knot female" and "knot male" respectively. *Sip vinay* are said to be "ugly" (*kamgag*), made only by women, and they are only good for temporarily tying bundles of firewood, betel peppers, or leaves of cooking greens. When a woman walks to a house with an enormous bundle of firewood on her head, she will often just nod or tilt her head so that the bundle comes crashing down at the intended spot. The fastening usually snaps—which is fine. This knot form is discriminated against; it is the one I naturally make and I suffered slurs until I learned how to make the male kind.

Sipkibkeway *and* Sipkwadoy *and Boat "Methods"*

Sipkibkeway and *sipkwadoy* create a second contrast set. They are opposites because they anchor different boat "methods." But they are not the complementary opposites of the previous two. They are considered different, therefore they are not comparable. Nobody would say that *sipkwadoy* looks better than *sipkibkeway,* only that they serve different purposes.

Sipkwadoy literally means the "knot" used to tie together the feet of a captured "cuscus" (*kwadoy, Philanger lullulae*). In fact, this knot form is used to tie all captured animals, including pigs, and it is the basis of the *anageg* tying "method." With the exception of procedures named later, everything fastened in an *anageg* uses *sipkwadoy* for the original attachment, *atusip*. In the other two outrigger classes, the originating knot is *kibkeway*. Literally, *sipkibkeway* means "knot" for the lowest class of outrigger boat to Muyuw's south (Bwanabwana), called *Lolomon* or *Yelaman* by Muyuw. When I asked if the forms were equally strong, the first answer was yes, and I was told that one could use *kibkeway* with an *anageg* (although it would be considered incorrect). However, I was then told that the primary contrast between the two is that whenever possible *anageg sipkwadoy* fastenings are pounded tight. Every *anageg* sails with an igneous stone used for this purpose. The rock is called the boat's *tabu-*, "grandparent," and it is supposed to remain on the boat from its first sailing to its last. No such stones accompany the smaller classes of craft. When tied with *kibkeway*, the lashings are pulled as hard as possible. Pounding and pulling is the activity that transforms the "flexible" hands of male children into the harder, less delicate condition found among most men.

These two knot forms lead directly to the two "methods" for tying the different classes of outriggers: *kumis* for tying in an *anageg*, *kibkebway* for lower-classed outriggers. I was told that an elder might look at a boat, immediately judge that it had been tied incorrectly, and refuse to go on it or forbid it from sailing. This is a judgment similar to the one made of the lure knot I presented to the two men: people are trained to analyze fine details.

All craft should have their parts tied from back (*wowun*) to front (*dabwen*). This is supposed to generate a synchrony between production and use even though everyone knows that sometimes a boat's "bow," sometimes its "stern," leads depending upon the desired sailing and wind directions. As in many other parts of the Indo-Pacific, the outrigger is always into the wind, so sometimes one end of the craft leads, sometimes the other does.

A number of differences define the *kumis/kibkeway* "methods." The distinction was explained to me foremost in regards to the tying material used for connecting crossbeams (*kiyad*) to the outrigger float. Tied to the aforementioned *seisuiy*, these pieces run from the boat hull to a set of stakes, called *watot* (stanchions), pounded into the float. Although practically everywhere modern materials tie these parts now, people think of them in terms of the traditional materials employed for tying. *Anageg* require *mamad* vine both where the crossbeam connects to the main hull and where it connects to the outrigger float by means of the stanchions; one source estimated the length needed for each set was *ovey*, two units of the length measured between outstretched hands and arms, approximately four meters. This would mean about eighty meters for an *anageg*.

"*Mamad*" refers to three large vines found in mature forests that can be used for this function (see table 5.1). Although one is considered the "real" one, the other two substitutes, *mamad pwakau* and *mamad kumkumkamey* are acceptable. Smaller craft employ roots from the fern *koliu* for both these connecting points. Tying the crossbeams is sufficiently important for the distinctive processes to be generalized into a well-formed analogical set: "*anageg : kumis : mamad :: kaybwag* (or *kuwu*) *: kibkweway : koliu*".

While it specifies a plant whose roots furnish tying material, "*koliu*" also refers to specific boat parts that have to be tied and how they are tied. The process is called *kowag*. One instructor told me he could "do *koliu*" but he could not do the mast mount or the *duwadul*, the latter the *kakam*-produced piece that flows from the mast mount over the side of the boat to the outrigger float, partly described in the previous chapter; tying procedures associated with it are the most complex in the culture. My instructor in this case meant that he could tie strakes to the keel and to one another, but he would leave other chores for somebody more experienced (his younger brother). Strake ties have individual names, usually referencing fish, details of which I turn to later.

Materials aside, I describe the primary difference between these two "methods" beginning with *anageg*.

Ideally every *anageg* has ten crossbeams, *kiyad*. Two sets of two *watot* (stanchions) connect each crossbeam to the float. The top of the outside ("*walam*") watot of each pair points to the dabwen, the bottom, pounded into the float, points back to the wowun. The inside ("*katan*") watot of each pair reverses the outside direction: its bottom is pounded into the float on an angle towards the dabwen, the front of the boat, its top pointing towards the back. Each *kumis* tie begins and ends on the outside of the *wowun* side, the *walam watot*. It binds the two sets to each other and to the crossbeam.

A *kaybwag* is the opposite of an *anageg*. The inside, *katan watot*, of each pair leans to the *dabwen* and points back to the *wowun*. The outside pair leans back to the *wowun*, and points forward toward the *dabwen*. Using the *kibkeway* method, the *walam* pair's tie starts on the *katan* side and finishes on the *walam* side. The *katan* pair starts on the *walam watot* and points back to the *wowun*, stern. This direction explicitly mimics a paddling stroke. One paddles from the *katan* side of the boat, and when the stroke is finished the right hand is down and to the back, the left up and to the front. I first learned this while walking by one of these beached craft with a friend who knew I was trying to understand these details. He pointed to the stanchion configuration then made a sudden jerk with his two hands, jutting out the left fist, pulling back the right as if he had just finished a short, swift stroke—the Muyuw paddling motion—showing how tying modeled the stroke.

In short, the *anageg kumis* is designed to reflect conditions for sailing, and sailing only. By contrast, although the lower class of outriggers may sometimes be sailed, their primary mode of locomotion is paddling. *Kumis* and *kibkeway* forms instill the qualities of motion into these forms, a synchrony governing production and use.

Are these differences just a matter of aesthetics? Whether this order has significant engineering consequences or is just part of the no-less-important conceptual order that makes these craft the social phenomenon that they are is beyond my competence to judge ... but then there is the story I was told in 1996 in which a minister reversed the order instilled by the Creator, and the new forces blew the craft apart.

Although it is employed, the analogical formulation *anageg : kumis : mamad :: kaybwag* (or *kuwu*) *: kibkeway : koliu* is not absolute. Part of the "method" for an *anageg* includes the tying of its strakes and the mast-mount setting. These require *koliu* methodology. I provide more details about this structure in the next chapter, for it entails a complex spring considered the most important part of the boat, not incidentally the most encompassed part. But the pictures accompanying this section make some of *kumis* technicalities visible. With his left hand the tier holds down the dark tapered piece of wood while his right hand moves what is the moving male end. The fastening begins from the stern, left side of the construction and moves to the prow and right side. Both times I watched this process—on a model in 1999, on this real boat in 2002—the men who executed the ties told me that the prescribed revolutions and directions are counted.

Sipbalau *and Reversal*

The *Sipbalau* knot is commonly seen and probably entails the fundamental principle behind all tying, a reversal of forces so that accumulated friction provides the desired strength. *"Balau"* is the name of the rope (*yawoywu-wun*) used to control the arc and pitch of a mast; "stay" is the technical sailing term. In the illustrated cases the reversal is doubled or tripled, providing strength as well as ease of alteration. The latter is needed for the primary use of the knot and from which it takes its name. The form is used to encircle a peg, *amawuw*, at the top of the mast above the *kuk*, the pulley system. Each end is then pulled down and fastened with a *sipbalau* at the desired location on the outrigger platform—almost invariably to the third crossbeam in from each end—so that by degrees the mast leans more or less in the direction of movement and toward the outrigger float. The variables with respect to wind direction and strength were listed in the previous chapter's discussion of mast dynamics. Although *balau*/stay fastenings to the outrigger platform have to be strong, they also must be quickly changed; the knot's simple reversals facilitate both conditions.

Encompassed reversals are the dominant characteristic of these craft. In a sense reversals begin with the difference between the interlocked keel and the straight-grained mast. They extend to steering forms, which systematically counterbalance the effects of the wind on the sail to those of the water on boat's "rudder." They are found on each end of the *balau* from the *sipbalau* at the top to its attachment on the outrigger platform, a contrast that is also defined by the difference between a doubled or closed loop at the mast top versus the spread or open set of loops on the crossbeams. An encompassed reversal also defines what is considered the most important part of the boat, the spring under the mast mount. The most encompassed point of the boat, its structure generates upward forces reversing the downward force of the mast on the keel. This logic is also featured in cowry shell placements on the decorated prow boards and the carving in those boards. It may be stretching the terminology, but these relations *are* fractal moments of a pervasive structure, not surprising given that outrigger craft entail repeated balancing of opposed forces.

Lepwason *and Sail Structuring*

"*Pwason*," which translates to "navel," is the base of this next knot, *lepwason*. Trying to explain the name's meaning, my instructors stressed the point that a navel is the point of release between a mother and child, the intersection of combination and separation, a process not a source. The form was explained to me in relation to its critical use for positioning a sail, so I shall use it to enter into a detailed discussion of the sail type that remains intrinsically related to the *anageg* form. The immediate point for this knot, however, is simple. There are two ends—if you pull one, the knot tightens. If you pull the other it releases immediately. This boat fastening must be tight, but it must be instantaneously releasable—hence the quick-release form. Ogis explained this by noting that if the wind reverses and the outrigger float rises out of the water and threatens to capsize the boat, this knot can be quickly released. The knot form instantiates balanced teetering that ramifies through the culture.

Other means hold the lines to the sail while *lepwason* knots fastened to the boat position the sail in the wind. The Muyuw terms for these lines (approximate European sailing terms in parentheses) are *otan* (brace), *tabagon* (cuningham controller), *asan* (tack tackle), and *alita* (sheet). To understand the totality of these lines, the full sail structure must be made apparent.

The generic word for sail is *mweg*. As noted in the previous chapter Muyuw now distinguish three types: the *anageg* sail is called *aydinidin*, the other two are *yabuloud* and *selau*. *Yabuloud* is the Muyuw name for two-boom triangular sails rigged with the *masawa/tadob* outrigger form that dominates the western side of the Kula Ring. The Muyuw name derives from the shoreline *Pandanus* tree type, *loud*, which furnishes the leaves. Those leaves are relatively soft, pliable, short, fat, and strong. They are more suited to this form's more complex movements than *legis*, the name of the *Pandanus* tree from which *anageg* sails are made. Pliability is a recognized feature because not only are the leaves for the *aydinidin* firmer, the sail is constructed with slats to make it more rigid than the *yabuloud* style.

Selau is a trapezoid-shaped sail ushering in a new style of boat rapidly transforming the sailing culture of the eastern side of the Kula Ring. Not without European influence, the form is said to be spreading from Panaeati in the southeastern sector of the Kula Ring region to the north, a direction replicating the indigenous the history of the *anageg* form.[10] It is widely understood that the *selau* sail is easier to handle, and more efficient to use. It is made with plastic, so it is easier to acquire. In 2006 (and this was repeated to me in 2009 and 2012) Gawa boat builders were trying to adapt the *anageg* form to the *selau* sail rigging. If this experiment ends up working, it will do so because the *anageg* is more seaworthy with a load than the *selau* boat structure, which, while faster, easily ships water and is poorer for transporting building supplies, vegetable food, pigs, and people. The future on these matters will follow from the degree to which internal combustion engines, and thus a more thorough absorption into external capital structures, completely take over the movement of things and people.

The *anageg aydinidin* sail is rectangular and has rounded shorter ends, called *awomweg*, modeled after the shape of a *kakam* leaf (or shark's nose in Budibud). This shape is held to be critical for the way the wind moves over the structure, but it also plays a role in navigation. If sailors find themselves on the water at sunup, they set their course given the position of the sun on the horizon in reference to oncoming waves. The setting is then monitored throughout the day as the sun arcs over the sail's curve, *awomweg*. Exceedingly complex topographical forms come into play in these *calculations*, which are dependent on the time the year.

I measured the length of two different sails, one in 1998, the other in 2002. Both were calculated while the sails were rolled up and resting on the *anageg*'s outrigger platform racks. One sail was 9.3 meters long, the other 8.5. The yard and boom running the long sides of the 9.3-meter sail were 6.78 meters long.[11] So a sail's longest dimension will approach two meters greater than its sides. I never unfurled a sail to measure its width, but three model sails have been made for me, and their ratios of length to width are 2.4:1, 2.6:1, and 2.24:1. Applying these proportions to the sail lengths I have, the approximate widths would be 3.58/3.86/4.15 meters for the 9.3-meter sail and 3.27/3.54/3.6 meters for the 8.5-meter one.[12]

Not trusting information from earlier visits about ideal measurements for sails, among other parts, I again asked people about standards during the 2006–7 visit; I was again firmly told that there was no fixed scale or number for sale dimensions. The width of each end of the sail, however, is measured so that they are equal to each other. The only requirement for a sail is that it must fit between the front and back prow-boards, *kunubwara*; these tend to be slightly more than a meter in from the ends of the keel whose length and shape is the determining dimension for the boat. I asked if people sometimes had to trim sails once they had been made, and the reply was affirmative. A sail that is too large, called *kataninom*, tends to drive the boat into the water, slowing it down. While built to a specifiable design, the sail, like the boat itself, remains an individualized product continuously modified piece by piece as it is experienced. Although people sometimes measure boat parts after they are found to be particularly good, precision in these forms comes from the process of production and practice, not the originating design. New pieces are made of new materials so old calculations are only approximations.

A sail is defined by its *tapwan*, "back," and *nuwan*, "inside or stomach," and by its *wowun* or "base/bottom" and *mayien*, "tongue." The "tongue" image is literal. People curled their tongues, so I could understand the form. While the back/inside contrast is absolute and thus part of the internal structure of the sail, the base/tongue distinction is relative because if a boat has to tack so that the stern rather than prow leads (or vice versa), then adjustments have to be made, including tipping the sail so that what had been the "tongue" becomes the "base," and vice versa. Changing the leading part of the canoe and altering the sail end for end is called *yesag*. In the *yabuloud* style of sail, the base/tip contrast is fixed. The narrow point of the triangle form that is tucked into the front or aft end of the boat—depending on which direction it is sailing—is the "base," and the opposite broad, unbounded end of the sail is the *dabwen* or "top," not "tongue."

The wind should be "caught" (*kon*) in the sail's "inside," helping it form the curled tongue shape. It then spills out the top as the sail ripples in the wind. The expression *"bo katigusgub mweg"* is used to say that the wind and sail are working as they should, the process finally controlled by the lines connecting points on the sail to the boat by means of *lepwason* knots. For reasons of stability, the boat must be positioned so that the wind hits the sail's "inside"; if it hits the "back," the boat may instantly capsize, a very real possibility given the eddies that spin off wind currents. The back/inside distinction is another structure formed by means of an inversion. *Pandanus* leaves are sewn together so that the "back" side of a leaf always forms the "inside" of the sail.

The process of interweaving a sail's rigging is, of course, near the end-point of its production. Sail construction begins with the collection of materials, foremost the gathering of leaves from one of some twelve named kinds of *Pandanus*.[13] Budibud people make their sails from a species called *sigiyad* or *signinitun*. It used to be exclusively planted on their low-lying islands where it is still regularly turned into sleeping mats exchanged for clay pots and bundles of sago.[14] For most others, however, the preferred sail material comes from the *legis Pandanus* type found in the Sulog region. Although I never accompanied anyone when they collected it, through 2002 I always heard of forays to gather the material; the Nasikwabw crew of the *anageg* I studied extensively in 1998 gathered *legis* for a sail while they were in southeastern Muyuw during that year's ritual season. As noted in the previous chapter, *legis* is a gap-phenomenon plant also associated with soils classed as "wet." The plants certainly appear because of the human action there, though I do not know the degree to which it is the plant's governing condition.

Legis leaves are considered very long and slender in contrast to the shorter and fatter *loud* leaves.[15] I did not measure the width of any *loud* leaves, but the *legis* leaves were about ten centimeters wide. From pictures I estimate that they might approach two meters in length.

I saw Nasikwabw people in Wabunun make a real sail during my first research period (1973–75); unfortunately I did not pay attention to the process. I witnessed the model made for me in 1999 and some of the work on the second model made in 2007. I watched and photographed every step of the third model Ogis made for me in 2009. A first draft of the description below was completed before I witnessed the third iteration. My informants about this process were often animated and wistful. Although Ogis ridiculed the significance of the labor time that went into making sails in 2009, it was clearly a lengthy process, and because of that the passing of this technology is not completely regretted. Yet the wistfulness comes from the experience contained in the sail. The art of politics in this region is—to some extent was—turning necessity into enjoyable collective action, and those making a sail practiced that art of *communitas*. I was told that five or six people could make a sail in two or three days, whereas a whole village could do it in a day; the initial stitching I watched in the 1970s took a day when a large group spread out across one end of the village and worked on it together.

Making a sail is called "to burn a sail" (*gob mweg*), because once the barbs of the leaves are shaved off, the first significant step is to pass individual leaves through a fire. Two leaves are burned at a time, the leaf's "inside" directly exposed to the fire. The burner then places the leaves on the ground, lightly holds them there with his foot, and they are then pulled out from under it. They are then folded or rolled up. Individual leaves are often rolled up together in large bands then returned to a village. When they are ready, they are spoken of as "cooked" (*kamin*). Although dried to a degree, the leaves remain limber and take on a supple, leathery quality.

To start a new sail, two sets of two leaves are placed together. Each of these leaves has a "head" (*kunun*) and "tail" (*yayun*). The "head" is structurally equivalent to the *dabwen* or *matan*, top or end point; the "tail" is equivalent to the *wowun*, base, of a tree. The sewing always goes from "base" to "head." Since the base is slightly wider, the next series inverts this order. By coupling a head with a tail, that is a systematic reversal, the sail ends are made the same width.

Leaves are set so that their "insides" are face to face but slightly offset so that the edge of one is slightly below the edge of the other. The "backside" of each leaf thus faces out. The next two leaves are put over these two, reversing the order so that a new leaf's "backside" faces the backside of the previous set. One side of these four is then stitched together using *im*, inner bark strands from the *loud Pandanus*. These last two leaves, the new top and bottom, are then bent back and away from the first two, covering the stitching and reversing the side of the leaf that is exposed. The bottom one is bent back first. This is *kanogol*. Then the top one is bent back over it, *arpo*. These two words become unison calls to effect this action across the line of people so that they make the requisite flip together. The *kanogel/arpo* alteration creates a rhythm organizing collective action.

That action once left a residue of life, and people who experienced it now fear that the culture has lost an important quality. Whenever I would discuss sail-making with people, they would begin the *kanogel/arpo* cadence and break into smiles.

Sometimes the newly bent leaves are pinned together with veins extracted from coconut leaflets (called *senigeil*, which are also the material used to make brooms) until the next leaf set can be placed over it and sewn up.

The *kangol/arpo* sequence is then repeated over and over, moving from one long side of the sail to the other, establishing the sail's width. If only a few people are working, they move along the sail's length, a practice considered cumbersome. But if a large group works together, each person remains in the same position, their coordination generating the activity's effervescence. Once a sail is completed—a consideration that sometimes includes the encircling materials sewn around its edges—it is usually left do dry (and whiten) for several days before the yard, boom, and other lines are attached. It can then be used. When not attached to a craft, the sail is placed on racks on the outrigger platform where it is always covered with woven coconut fronds and what remains of the sail it has replaced, all to protect it from direct sun and rain.

When new leaves were first put down, the "inside" was up, exposed. Once the *kanogol/arpo* sequence is effected, the "backside" is up. The backside is considered the stronger surface, so it is open to the elements. Moreover, it covers the stitching, leaving a smooth surface to the sail. As I note later, the width of the sail is eventually crossed by many different lines (*latagau* and *taseu*). Their splices and the interior slats that run down the "inside" of the sail generate a surface of considerable friction, while the sail's "backside" remains smooth. Presumably this facilitates sailing closer to the wind (beating) so that the sail more readily functions like an airplane wing as the wind travels faster over the smoother arced backside of the sail. Speeds I recorded on a leg of my 2002 trip when we were sailing with a wind from the front side of the boat were only slightly slower than those recorded earlier when we sailed with a stronger wind coming from our rear.[16]

Once *Pandanus* leaves are sewn together, a sail is ringed with two different materials. Each is inserted inside exposed leaf sides or ends then covered with *nit*, the mesh-like material that protrudes from the top of a coconut tree. *Alawanay* is the name for the encased material, but it comes in four different pieces. This is a complex process that bears on the guiding lines coming off the sail; careful description is necessary.

Two strands of a relatively thick and sturdy "high forest vine" called *niwoun*, a rattan (Damon 299, Arecaceae, *Calamus*[17]), go along the long sides of the sail. The top of the sail, where the yard or spar, called *kunay*, will be tied, is strung first; the bottom, *kaley* (the boom), side second. The four ends of these two pieces of *niwoun* have "rope" (*yawoywuwun*) tied to them. *Yawoywuwun* is a weave composed of three pieces of *ukw* or *ayovay* inner bark. These strands run along the curving ends in the same way that the *niwoun* courses the sail's two sides. Although this "rope" is strong and firm, it makes the curving ends of the sail more supple than the sides. These two end pieces, forming the sail's curves, are carefully placed so that when they are finally completed they can be pulled taught, in part to make the end widths of the sail identical; measurements are taken to assure this. All of these pieces are conjoined, but in different ways so that differently functioning forms may be tied to them. A *sipkwadoy* knot first attaches the "rope" to the *niwoun*. The rope and the niwoun are twirled around each other for a certain distance before they are separated. When I first saw Ogis attaching the two ends I thought he was just winding the "rope" around the piece of rattan. Then I realized he was turning a simple revolution into a *sipbalau* knot by executing a series of reversals. When it is being made, this form protrudes from the corner of the sail waiting for other lines. Its knot will eventually be tied tight and large so that it holds the *kunay*, yard, close to the edge of the sail. Two other lines coming off the sail, the *otan* and *tabagon*, become attached to this intersection. From the other end, the boom/*kaley* side, an extension called *asan* is created.

Over this edging and so also encircling a sail is the material called *nit*. It encompasses the somewhat brittle leaf material and provides extra strength for fastening the yard and boom. It is likely to be doubled near the center of the *kunay* to withstand rubbing from the halyard. Pieces of it carefully overlap one another especially as the curved ends conjoin the long sides of the sail. Holes are carefully punctured in these joints so that the "rope" stitched inside the leaves can be pushed through them, knotted to hold everything in place, and have various lines tied to them. *Nit* is stitched (-*bes*) with *im*, the relatively fine material from *loud Pandanus* (or, I was told but have not seen, strands from the *ayovay* tree). *Im* is the *vatul*, string, of choice for this and other places because it is so flexible. The sewing encircles the sail edges to keep this material firmly in place. However, for at least part of the way around the curved "base"/"tongue," extra tying is needed to keep these ends "hard." These fastenings come after and encircle the *im* stitching that goes all around the sail and tightly encloses the *nit*-covered *niwoun* vine. *Ayovay* or *ukw* inner bark (or "European" material) is used here rather than *im* because it has to be stronger. These pieces, called *lawunay*, are not connected to each other; each tie is an independent construction. *Ayovay* or *ukw* bark is then also used to fasten the sail's yard and boom to the edge of the sail. Tying the yard and boom to the sail by means of wrapping that encircles the *niwoun*, *nit*, and the yard and boom is called *tulutoul*. When the sail is hoisted up the mast, the *kunay* is at the top, so around it is tied the *yawasay*, a halyard, the rope used to host the sail. Although nobody would specify the ideal length of the yard and boom, I was firmly told that their exact center must be determined for fitting the halyard.

A boat's original *kunay* and *kaleiy* are cut in Gawa or Kweywata usually from the *kausilay Calophyllum*. As with the original masts, they are almost immediately replaced with one of the Sulog *Calophyllum*, usually the lighter of the two, *ayniyan*.[18] Although the *kunay* should be slightly larger in diameter than the *kaleiy*, these two pieces have the same shape. They are thicker in the middle and thinner toward the ends, making the sail stiffer in the middle and more flexible toward the ends, the effective top and bottom of the sail. The shape of these pieces is said to be like the springs sitting below the mast mount, which are considered the most important part of the boat. With a good wind the two will bend as the sail catches the wind.

A quick turn to the two-boom triangular sail form called *yabuloud* is useful here because it calls attention to the ways properties and forms are explicitly tied together. Although, like the *aydinidin/anageg* model, the *yabuloud* form has a *kunay* and *kaleiy* constructed from very straight-grained trees, the species is different. Muyuw call it *busibuluk,* a tree only identified to be in the ebony family. Like the *Calophyllum* appropriate for masts and *kunay/kaleiy* pair, it may be used as a pole for propelling canoes, and it is also often found in the Sulog area. But a much longer tree is needed for the *yabuloud* form, and only *busibuluk* meets this quality. It never obtains the girth of the *Calophyllum* species, instead remaining quite thin though also strong.[19] Moreover, while the *aydinidin kunay* and *kaleiy* have specifiable "bases" and "tops," after trimming they look the same. By contrast, the natural tapering of the *yabuloud* forms is evident in the pieces and in fact facilitates the sail's more flexible structure. This difference redounds back to properties in the leaves used to make these sails. *Loud* and *pwakau,* Iwa's *kebwiba,* are considered much stronger than *legis*; the latter would shred if subjected to *yabuloud* forces.

The greater length of *busibulouk* makes it a favorite for another task, as a stick for pushing down into holes in the coral reef to scare fish into a waiting net.

Returning to the *aydinidin* sail form, paralleling the *kunay* and *kaleiy* are slats called *pamaloul* designed to give firmness to the interior part of the sail. I was told there should be seven or eight for each sail; there were seven on the sail on 1998 craft I examined, eight on the boat from 2002. These slats are usually constructed from *amanau* (Clusiaceae *Garcinia* sp.), the prominent tree in central Muyuw and near an ideal fence stake there. The tree type is selected because, like *kausilay* and *kakam,* its interlocked grains allow it to take much stress without snapping; for this same reason the tree is one of the ideal trees for mid- and small-sized outrigger crossbeam canoes, especially in north-central Muyuw. While the *kunay* and *kaleiy* are expected to vibrate in one dimension, the flapping of the interior of the sail is more complex; hence *pamaloul* are drawn from sources exhibiting qualities that provide a different kind of suppleness. Running the length of the sail, these are about five centimeters in diameter. Two are tied together, usually "top" (*dabwen*) to "top" near the middle of the sail in order to cover the distance. Their ends are notched where they meet the curved ends of the sail; there they are firmly tied to the sail edging, becoming, along with the coconut-wrapped *nit* and the *lawunay* bindings, one of the mechanisms for making the sail ends coherent. Moreover they are also fit underneath a peculiar stitching that runs just inside the defining arch of each sail end. Called *vav takon,* this form should be done with *ayovay* inner bark spun into a circular form. "*Vav*" is the name of a centipede that is known for its painful bite[20]; "takon" in this context references one's chest. "Centipede chest" is the literal translation for which I could get no explanation.

In addition to sail arc ties and the *vav* weave, two different lines hold *pamaloul* in place. These also directly or indirectly help to hold the edges of the sail together as they run the sail's width. One of these is called *latagau*, of which there are three, one positioned in the middle and one toward each end of the sail. This line is "rope" (*skoko yayun*), cordage composed of two or three strands of *ukw* or *ayovay*, if not store bought or found on the beach. Each *latagau* runs from the *kunay* to the *kaleiy* and then back up to the first *pamaloul* above the *kaleiy*. A *sipkwadoy* attaches it to the *kunay*, a *sipbalau* knot to each *pamaloul* and the *kaleiy*. Where it returns and ends at the *pamaloul* above the *kaleiy*, a *sibkwadoy* fastens it. The three *latagau* secure the *kunay*/*kaleiy*/*pamaloul* pieces to make a fluid set, preserving distances between each piece while providing great latitude for movement.

Strands called *taseu* form the second structure across the width. They are made from the inner bark of the *ayovay* tree. About a centimeter wide, they are not twisted into any other form but run the full width of the sail paralleling the *latagau*. While there are only three *latagau*, there are many *taseu* lines. I never counted them on a real sail, but from examining pictures there are at least eighty on the 9.3-meter-long sail noted earlier. These numbers create intervals of eleven to twelve centimeters between each *taseu* line.[21] On the 8.3-meter sail there seem to be more. They are supposed to prevent the sail from shredding. They are cross-stitched and run perpendicular to the sewing that ties individual leaves to one another, lacing that runs the full length of the sail in the same direction as the grains in the *Pandanus* leaves. *Taseu* lines also loosely press the *pamaloul* slats down against the sail matting. At least two pieces of *ayovay* are used for each strand. On the *kunay* side of the sail the material is fastened around the sail edging with a *sibkwadoy* knot encircling the edge of the *nit* so that strands hang down the sail's "inside" and its "backside." To sew the *taseu* lines, two people are stationed, one on each side of the sail, and trade a threaded needle back and forth. The "needle" (*vas*) is made from turtle shell.[22] Usually the "inside" strand is started first and runs the complete width of the sail, encircling the *kaleiy* edge, though not the *kaleiy* (boom) itself, before it starts back up. Then comes the "backside" strand, started first by going all the way around the *kunay* sail edge, though again not the *kunay* (yard). The two strands are tied together using a *siptawau* knot somewhere toward the middle of the sail on the sail's "inside."

The *pamaloul* slats, *latagau* cords, and the splicing ties for the *taseu* lines add to the relative roughness of the sail's "inside" compared to its "backside."

Altogether nine lines attached to the sail position it in the wind. One is the *yawasay*, the halyard, attached to the center of the yard (*kunay*), used to hoist the sail and adjust its height. The original hoisting is often a crowd-drawing event, as two men push the sail up by means of the yard while another holds the halyard as he perches on the craft's *watan* gunwale. He uses the weight of his body to fall backward toward the water, and people call out in unison, "*Vay, vay.*"

The other eight lines consist of four sets, with two ropes in each set. *Sipkwadoy* and *sipbalau* knots attach them to the sail structure. As noted, these boats are designed so that the effect of the wind on the sail makes the boat veer to port, to the side of the outrigger float, itself always facing into the wind, *yiwatin*. All adjustments are made by degrees with this veering in mind.

The eight lines go by four names, two lines to a name, although one of these goes by another name if is left hanging. Starting from the attachments to either end of the *kunay* they are *otan, tabagon, asan,* and *alita/ enay.* While the *alita/enay* couple is uniquely tied to the *kaleiy/*boom that is at the bottom of the sail, the other three sets are part of an integrated unit and emerge organically out of the sail's originating structure. This is complicated; I begin with the way they first appear and stress the lines' individual positions and functions.

Otan are tied to each end of the yard/*kunay*; they would be called "guys" or "braces" in Western sailing terminology. The top *otan* will not be tied to anything; rather, it is left to blow in the wind. The bottom one is, prototypically, tied to one of the stanchions (*watot*) toward the trailing end of the craft; stanchions are the forms attaching crossbeams to the outrigger float. This pulls the bottom end of the sail out over the outrigger float platform while its top end veers toward the opposite side of the craft (*watan*). *Tabagon,* analogous to a "cuningham" or "downhaul" in Western terminology, appear to be tied to an end of a *pamaloul* slat toward the middle of the sail's arcs. Like the *otan,* the top *tabagon* will be left free, though it is often used for pulling the sail down if it is inverted top for bottom when a boat changes its tack. The bottom *tabagon* is said to be fastened to a *seisuiy,* the critically important part of the boat's internal structure, (usually) on the outrigger side of the boat, the part's specification emphasizing its significance.

The *asan,* "tack tackle," appear to be tied to the ends of the *kaleiy,* the boom at the sail's bottom side. As in the case of the *otan* and *tabagon,* the top of the *asan* blows in the wind and may be used for inverting the sail. The bottom one is tied to the mast, low proportionate to the height of the sail.

The *alita,* "sheets," are attached to each end of the middle third of the *kaleiy.* One of these, usually the top *alita,* is tied back to a position along the last crossbeam connecting the hull to the outrigger float. If the other *alita* hangs unattended, it is called *enay*. However, someone posted next to the mast is likely to hold the *enay* if the wind is strong in order to pull the sail opposite the outrigger float. On much of my three-day voyage in 2002, someone held that line because of the strong winds. Sails are partly classed in terms of a contrast between *ilpelop'* and *kanbwatay*, what I translate as "floppy" and "taught." *Alita* govern this, so by tightening or pulling it—"*ibtil alit'*"—one moves the sail from the former to the latter. *Alita* may be—with a *gitimatan* wind—attached to the boat's crossbeam by means of half a *sipbalau* looping the rear *kiyad* while a *lepwason* knot fastens it to a less critical part of the outrigger float platform. Manipulating the tension on this piece is understood to be almost as important for directing the boat as manipulating its "rudder."

These descriptions provide a rough indication of the positioning of these sheets. My informants could easily provide me information in chart-like form specifying kind—for example chapter 4's *gitimatan, duwadul,* and *lilimuiy*—and strength of wind and the relative setting of each line and, related to that, the angles of the mast, boom, and yard. With a strong wind, for instance, the mast is "bent" (toward the outrigger float) and the halyard is "thrown" (loosened so that the sail is lowered). With a *lilimuiy,* wind from the rear, the *otan* ("guy" or "brace") on the lower end of the yard is "pulled," hauled back so that the sail's width tends to be leveled perpendicular to the mast. During a brief voyage I made in 1998 in the face of an oncoming squall, the sail was radically altered from its near upright position to near perpendicular to the mast. So the *asan*, rather than being tied to the mast, was pulled toward the front of the boat and tied to another spring, called *nedin*, arcing over the outrigger platform.

The boat's structure, made especially complex by the outrigger platform, provides a field with many degrees of variation for transforming wind into human intention.

Alita are two separate lines. The other three sets appear to be separate but in fact form a complex set, a fact I did not fully appreciate—though I had the point written in my notes—until I examined one of my sail models then watched Ogis make his second and my third model sail. In describing this I begin with the two *asan*, both of which appear to be attached to the extreme ends of the boom/*kaleiy*; they are for holding down the lowest corner of the sail structure, pulling it in toward, if not right to, the mast. In fact, a *sipbalau* knot is the means by which the line is attached to the *kaleiy* ends. The other ends of the *asan* cords travel along the two curvatures of the sail completely covered by the *nit* that edges the whole sail. The "*asan*" then emerges under the end of the yard/*kunay*. It is held in place because its knot has a diameter larger than the holes drilled in the yard/

kunay ends. For the lower (*wowun*) end of the sail where the *asan* is tied, the structure provides a continuous line of tension across its "base." That facilitates the curvature that becomes the means for "catching" the wind. Wind easily spills out the untensed top

Otan and *tabagon* also appear to be separate but form an ensemble created by one line articulated by a set of *sipbalau* knots. The defining arrangement here encompasses the hole drilled through the end of the *kunay* that facilitates the ultimate *asan* attachment. The *otan/tabagon* intersection encircles the *asan* knot. One end then flows free for the *otan,* the other angles off toward the middle of the curved ends of the sail. It does this by being threaded underneath one or more of the *pamaloul* slats and then by means of a *sipbalau* knot wrapped around one of their ends. This final *sipbalau* knot provides the force and friction that enables the *tabagon* to be manipulated so it can pull the middle portion of the curved sail end. On the model built for me in early 2007 there is a *sipbalau* knot on the first *pamaloul* below the *kunay,* then another on the end of the next *pamaloul.* By contrast, in the pictures I took of a sail from 1998 the *tabagon* seems to run under three or more *pamaloul* before a *sipbalau* knot fixes it toward the bottom third of the sail curve.

Correlations and oppositions define the organization of these lines to structure the sail structure. Although they are relatively autonomous they all enable fine adjustments for the sail in the wind. And while facilitating this operation from different points of the sail, they help turn the sail into a coherent structure by tying it all together. The *asan* enables the positioning of the *kaleiy,* but it is ultimately attached to the *kunay.* That attachment works by means of the cord edging the sail's two curvatures. And that edging is held in place partly by means of the way that *pamaloul* ends are tightly tied to those curves over the coconut mesh (*nit*) edging that encompasses the *asan. Pamaloul* hold down the *asan* structure. The *otan/tabagon sipbalau* knot at the end of the *kunay* holds down the point where the *asan* attaches to the *kunay.* And while the *pamaloul* also hold down the *tabagon,* the latter line remains external to the sail structure, whereas the *asan* is internal, encompassed by the *nit* edging.

Lepwason are not exclusive to sail functioning. They form the essence of another method of tying called *bwabwalita'* that is used to attach a y-shaped structure to the side of the mast, the mast leaning, with reference to it, more or less toward the outrigger float. As in other cases, the point here is a fast knot that can be quickly changed. Thus, this kind of knot is at one end of a massive collective artifact that is the *anageg* boat and *aydinidin* sail.

I could never get clear answers as to how long a sail might last. Three years is the upper limit. Ogis once told me that a sail might only last half a year. Yet sails are patched, so they require continuous attention like the rest of the boat. This is a major reason for the division of labor between producing the boats, having other villages primarily devoted to gardening activities, and other villages still devoted to sailing duties among which boat maintenance is primary. It is no accident that the standard description of a man's new marriage is that he wakes up, realizes he is married, then heads to his in-laws' garden … unless he is in a sailing village, whereupon he goes to their boats and works on them. When Ogis finished assembling the model sail and mast in August 2009, I asked him if a sail's interrelated forms might model the interrelations of a village and its larger associations. He said yes, but the point seemed obvious.

I noted that abandoning these sails generates a wistfulness in some people. Without question sailors appreciate the rush they get from new-fangled *selau* sails. In 2002 when I made the last leg of my journey leaving Nasikwabw for southeastern Muyuw, two other outriggers, both *selau*, departed Nasikwabw with us. The crew and passengers on my boat watched the other two rapidly head out and away from us. Although they were sailing in different directions, closer to the *duwadul* direction and so better positioned for the wind, they were sailing with what appeared to be a more efficient system. But for the intensity of that difference there is a loss of sociability. *Selau* sails are made of cheap plastic that one simply purchases. This tension, the tradeoff of loss of sociability for individual increase in intensity of experience, defines the modern world.

Vabod *and Fishing Nets*

Old sails are not the only devices in which the passing of the social relations of production and use generate ambivalence. Traditional Muyuw fishing nets, *wot*, do the same, and these involve the last "method" reviewed here. These notes supplement material published elsewhere about these concentrated embodiments of labor (see Damon 1990: s.v.; 2000: 56–57; 2005: 87–88). Those publications brought to the fore the regional, composite nature of net forms; to that information should be added the discussion of the tree source for net string, *gudugud*, as noted in chapters 2 and 3.

The 2009 research revealed another net form associated with central Muyuw's Kaulay village. Called *teikw*, this net type is about two meters square and used to gather *tanin*, a sardine-like fish that periodically appears off Muyuw shores in large schools, ostensibly chased there by sharks. Although a single person can operated *teikw*, as with *wot* they are considered more efficient if used with a conjoined group of men coordinating their individually operated nets. Like the *wot* form, weaving, and to a lesser extent assembling them, is considered specialized labor. Kaulay had about twenty *teikw* in 2009, but only one man was considered expert enough to make and repair them.[23] Their details are explored more carefully in the photographic essay accompanying this text.

When speaking of a *wot's* "method," my instructors emphasized two issues. One concerned the order for tying a new net section (*say*), the other the way of tying and counting the floats (*kut*), which are made of *ayovay* wood at the top and shells (*kiyak*) at the bottom of the net (see diagram 5.1, Muyuw Fishing Net Structure).

Net sections for named nets were gradually woven (*kos*), usually by elder men. Each net is named and of a unique size; new sections must fit the "width" (*kuppalan*) of the original with the same sized "gaps" (*mitikan*) composed by the crossing of the structure's cords. If too much time is spent on making regular repairs to an existing net, then responsible elders will insert a new section into the middle of the net. They first cut off the net's older and weaker extremities. When the section is spliced, the left side must be tied first, the right second. And it must be tied from top to bottom with a unique string called *eway*. *Siptawau* knots are used for splicing together the older and newer strands of the sections. The string used to make the net forms diamond-shaped spaces. The intersections are also composed by *siptawau* knots.

The woven structure generates triangle-shaped ends at the top and the bottom of the net. Threaded (*sewak*) through these triangles is a thicker string called an *ovam*. *Ovam* attach floats at the top and shell weights to the bottom of the net.[24] It alternately goes through a float (or shell) and then a *dados*, an intersection to which neither is attached. The threading always goes from left to right, in other words to the right, and the float

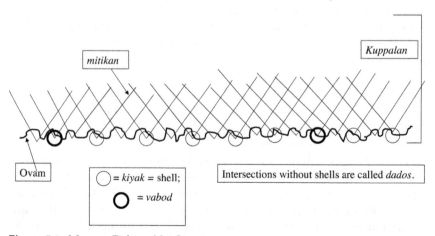

Figure 5.1. Muyuw Fishing Net Structure

(or shell) or *dados* should be held with the left hand while the right hand moves the *ovam*. Two kinds of shells are used, both from the Sulog region.[25] They are threaded by the *ovam* going from the top down through a hole (*tabod*) drilled or knocked into the top of the shell, then over the net above. It then goes all the way around and through the loop. This way the shells can move along the cord. However, every seventh shell is tied with a special knot holding it firmly. The knot is called *vabod*, hence the distinctive "method's" name. The noun classifier for shells is *kwe-*, and there is a unique counting sequence to this system: *kwetanok, kwey, kwetoun, kwevas, kwenim, kwenim kwetanok, vabod. . . .*

If not my description then the pictures accompanying this section should create the impression of a mass of woven material. When this data was gathered in 2002, my interlocutor, Gumiya, the carver and detail person noted in chapter 4, sighed and noted how complex their life had been, and to a lesser extent still was. The complexity was heightened by the apparent ease of going to a trade store and spending money for canned fish; or the seeming effortlessness of zooming out beyond the reefs with a fifty-horsepower Yamaha, trolling for a few hours to catch a boatload of tuna, mackerel, and other large fish. Nobody has to think ahead to spend hours making string, make the *say*, or do the repairs to a net before it can be used. And while an outboard boat with one or two men can obtain a lot of fish (with some luck) trolling by themselves, for a net to be used efficiently, "groups" of men are necessary. Not only is the net a mass of intricately congealed labor, its use congeals more. A net is realized by binding together people already defined by the tensions of specific affinal and generational relations, the obligations of their reciprocal production, i.e. kinship.

The selection of plants used in the magic associated with the net structure illustrates the reciprocal interaction between the production of a net and its use. This magic is only employed when a new net section is added to an existing net, but the magic conditions the net's use until a new section must be added again. The plants consist of one tree and at least four vines. The tree is *ayoyal* (Damon 64, Leeaceae *Leea*), chosen because its sap slightly "burns," "pains," and "irritates" (*ged*) skin and supposedly does the same thing to the net. A modest-sized *ayoyal* tree is staked in or near the center of the village where the new section is held for leaves to be rubbed over it near where the old net is cut (see Damon 1990: 77, 78). The vines are *namonsigeg* (Damon 92, *Passiflora sp.*), *sekita* (Damon 155, Arecaceae),[26] *ulsiyan* (Damon 66, Vitaceae *Cissus*) and *weled* (Damon 72, *Flagellaria sp.*). All have barbs or hooks—*weled*'s are on the leaf ends. And all but *weled* generate sensations like the *ayoyal* tree—slightly agitating the skin. This sensation is conceived to arouse action. Men infuse the net with the plant properties by crushing the leaves together then rubbing them on or into various parts of the net—its center igneous rock weight, the stick/float that marks its top center, and all the woven string. By means of these plants, men infuse ideas concurrent with the intent of the net's purpose into it—catching fish that are rolled up in the net. By their hooks, barbs, and entwining growth patterns, the vine leaves exemplify action that the combination of the (female) net and male labor should effect. The magic models—becomes a mimetic representation for—the conceived net activity, a combination of different forces and relations and things and people for a common purpose.

These net practices allow me to return momentarily to tying in boats. Earlier I noted that methods for tying strakes are named by certain fish. I could get no exegesis on why this was until people pointed out that the fish—small reef fish frequently caught in nets—have barbs on their fins, and one, *kobal*, is notorious for swimming into holes in the reef, making it impossible to get out. With this information in hand, I approached Ogis who recalled that elders requested all ties to be checked when an *anageg* was sailing either north or south carrying one or more pigs (on the outrigger platform). That extra weight put extra pressure on the crossbeams, which would be conveyed to the sides of the boat. The fish names specify the conditions of living.[27]

Just as a sail ties and pulls together diverse materials and persons, so does the production and use of a net. As the binding properties of the *akisi* leaves begin the productive relationships between a man and a woman, so those of the sails and nets transform those people into larger configurations, momentarily or enduring as the case may be. Recall chapter 3's discussion of sago orchards. One of the images I provided there was of orchards or parts of orchards moving between clans or subclans as pigs and people move. It is also said that they move like *vatul*. Men are the active movers in that process, the feminine clans or subclan associations the stable anchoring points. The isomorphism among these conceptions is evident. Tying materials, processes, and products were, and remain, means for and mimetic acts of life.

Perspectives on *Vatul*—*Bitalik Non*

In the classic book that rekindled interest in Pacific sailing, *East Is a Big Bird* (1970), Thomas Gladwin closed his account—chapter 5's "Perspectives on Thinking"—by contrasting patterns of thought made evident in canoe design and navigation with what he presumed was the successful mode necessary in "modern" society. For him the latter were the innovative possibilities that heuristics provided. Arguably these entail the ability to recognize one pattern or relationship and substitute it in another context. This process *is* fundamental to Western thought and practice (see Wallace 1976: 227–39), witnessed in its duller forms with endless commodity substitution, in more important moments the replacement of one form of communication by another. Gladwin argued that Puluwat navigators had "little practice in the art" (225). He may be correct in identifying what they do not do. But for describing what is essential to this region, I suggest that Gladwin's frame of reference is mistaken. The problem in this region is not the logic of substitution but the organization of sequential transformation. This is a major reason why tying, binding, and knots are important. And people practice this a lot.

The terminology of string, knots, and tying takes one not just through body parts but more importantly also to the relations that make them coherent. In this material one sees principles that relate gendered bodies together for the transformative relations of what is customarily call kinship; and in the production of boats, sails, and fishing nets, among the examples reviewed, one sees social effects incorporating kinship's constructed bodies and relations into larger productive units. Foremost in all these associations are transformations. A "string" conceived to attach the heel to the brain facilitates stability; the scent of an erotic flower moves toward "sticking" a loving couple together for their eternity; a knot tied in a hole at the end of a yard holds the "base" of a sail firm by unseen means then appears to become attached to the boom so that it stabilizes the sail, the design converting a power from one direction into a boat heading in another. It is no accident, therefore, that practice for these processes begins early. For many people, the art of tying is not so much second nature as an elaborated form of entertainment, yet one that provides an intellectual point of departure for the culture. String figures are the means for this training. Forming a tactile topology, they are principles conceived and practiced first in the mind that make the proportional world of these places thinkable and doable. And it is with a brief discussion of them that I close this chapter. In doing so I answer that puzzle of the new knot Sipum and Mayal presented to me in 1996: how could they readily recognize something new, think it through, then make it?

Kananik, or "string figures," did not come easily to me. The digital and mental dexterity required to learn and perform their imagery are foreign to me, but I do not think I am alone in this. For these items were often just treated as the stuff of women and children, not serious items people like me have been concerned with. Eventually I realized that "women and children" means everybody, thus raising the question as to what the practice does to cultural agents. And so from my 2002 research stint I began to appreciate how everyone I knew had some capacity and genuine interest in this art. "Interest" might be qualified: in 2009 I came across a variant of the fascination with the practice when I asked my Kaulay host what he thought of string figures—he exploded. He cannot do them, and hates them because when people start them they won't do anything else for weeks … *no* garden work, *nothing!*

Nevertheless, from 2002 I watched my less-than-agile male friends mouth the moves of the expert woman to whom they directed me as she worked her way through an amusing or stunning series of images. Unlike women's tying, that of men is not meant to be beautiful and absorbing; it is meant to hold and release as necessary. Although everybody starts more or less at the same level, women's hands and fingers, through traditional work—sewing, weaving, sorting, and molding[28]—remain "flexible" (*manum*), whereas men's become "hard" (*matuw*). There are as many twists, turns, and inversions in the tying of a sail, mast mount, or hull as there are in many string-figure episodes. However, the boat tying always involves pulling very tightly, and in an *anageg* many strokes are pounded into position. When the mast rigging of the boat I boarded in 2002 was pounded, it drew a fascinated audience no smaller than that gathered for the string-figure displays I witnessed an evening earlier. Yet the contrast with string figures could not be greater. Their whole point entails passing fluidly through a sequence of continuing transformations, ending with its original starting position, a simple closed loop.

The point of string figures is not the isolated moments so often depicted in the literature. Such moments are nevertheless significant, and Muyuw call them "flowers." The flowers are distillations of processes associated with temporal movement, not unlike the shrubs and trees flowering across and marking times of the year. As the passage of time is the experience of temporal cycles, so movement is the point to string figures.

A crescendo of activities in mid-2002 brought me to investigate string figures. I was leery of asking about them when I was waiting in Koyagaugau and Ole for passage to Muyuw, but being delayed there left me with an opportunity too good to pass up. For an entire evening men showed me what they could do. Plenty of men there were interested and competent, yet several women first wowed me with their performances.[29] And indeed they were performing: their demeanors changed; their eyes and smiling faces moved as rapidly and as determinedly as their fingers; they drew audiences. Then there was a memorable night in Nasikwabw with several people showing off their skills. I asked the young Onisimo if he knew the art. Assigned to watch over me when I was in Koyagaugau and then accompany me on our voyage, Onisimo is one of the peripatetic people not infrequently found throughout the Kula Ring: lost souls who wander from here to there picking up necessary work to earn their keep.[30] Some twenty or twenty-five years of age, he then looked like a mid-sized wrestler with a square, well-proportioned body chiseled into form. So, that night in Nasikwabw, I more or less assumed his answer would be no. But I was wrong. He got two strings, gave one to a girl perhaps a decade younger, and the two of them went through an incredible regime of passing the strings back and forth. I was not the only person enthralled by their joyous performance, but I was perhaps the only one who had assumed his muscle-bound shape was incapable of sweeping delicacy; a vision of Astaire and Rogers came to me. The young man and younger girl appeared to demonstrate effortless, elegant complexity—although I cannot believe they ever practiced that dance together, they clearly knew the steps and worked them out with little problem yet evident delight.

It needs to be stated that a single slip will cause the routine to collapse. The game is on the edge of dissolution. It is a concentration of attention, and intention, making it as engrossing as it is encompassing, and, clearly, mimetic. For this is life in this region—the path of a forest-garden-fallow sequence that conditions productive activities and relations; the *consequences* of the arcs and angles of a *kausilay* bole cut for an outrigger canoe; or the engaged intentions that have to be studied in the travels of *this* kula valuable.

In both Koyagaugau and eastern Muyuw there is a prescribed temporal context for *kananik,* after yams have been planted and their vines begin spreading (i.e. during the southern summer, from January to perhaps April). In Muyuw they are followed by the telling of folktales, *kwaneib,* the maturation of the garden vine and minor food source called *boli* (probably *Dioscora bulbifera L.*) operating as the rough transition sign between the two activities. In Koyagaugau people stressed that the women's string-figure play complemented men's boat racing (which is infrequent now); in Muyuw the idea was that the string figures were to be like yam vines climbing their poles, becoming entangled in one another. In Koyagaugau people do not stake their yams, so this association isn't pertinent; but recall that in Muyuw a female (*parawog*) and male (*kuv*) yam should be planted proximate to one another so that their vines climb up the same pole. Informants were explicit about the analogy between the twirling vines, entwined string of the string figures, and folktale storylines. My primary Wabunun illustrator, Aligeuna, associates two of these displays—a string figure episode called *Gumeau,* the Pleiades, and *Mayayan,* "Low Tide"—with the annual order. These two are associated because Gumeau first appears in early June when the low tide occurs at midday. Yam gardens are approaching their full maturity at this time.

The idea here is not just that the convolutions of string figures are like those that should be created in a yam garden. It is also that the former help make the later. The word *kapinanig* is used for the effect that one, *kananik*, is also supposed to induce in the other, the garden. One man explained *kapinani* by twisting and tumbling his hands around and over one another.[31] Both the action of the string and that of the entwined vines are conceived to be beautiful. And in a sense both end in the same way—the simple oval of the string, the ends just easily and simply tied together on the one hand; on the other, grown yams, the ensnarled vines now cut away and gradually drying back to nothing. Yet there is a difference that points to the analytical significance of *kananik* as a mental form and function. In a beautiful garden the growth is so interlocked that its jumbled vines obscure a person inside it; and nobody tries to figure it out. *Kapinanig* is also used with the Kula, but it connotes "confusion" and sooner or later leads to valuables being "knotted up," *kalasik*—bad because nothing moves. By contrast, *kananik* are subject to analysis, well-known, and carefully followed. The form trains the mind to fluidity. In late 2006 I learned a new word related to this form, *atanak,* which is said to be like *sinap,* literally the word for custom. More figuratively and actively employed, *atanak* conveys the idea of being smart. Somebody who schools well is *singaya tasinap,* a very smart person. String figures are said to generate this characteristic, they *bitalik non,* "untie the mind (*non*), they make one smart."[32]

For teaching and knowing this system there are widely known sequential steps. All instructions are given with respect to the names in table 5.3 and the moves in table 5.4. One instruction might be: "*Simveyov bin bipow,*" ("with the index finger" [*simveyov*] "go" [*b-in*] "under a strand and bring it over the top of another" [*b-i-pow*]).

An interesting anomaly in the Muyuw language concerning these terms has a consequence for the practice these exercises impart. Although the generic term for finger, *didi-,* is possessed like other body parts—the noun class a set of suffixes suggesting an intimate, inalienable relation—the terms provided in the table are possessed with the same set of prefixes used for tools, obvious means of production. And indeed for both men and women fingers are tools. In all likelihood, in learning to make string

Table 5.3. Finger Names for String Figure Teaching

Finger	Primary Usage	Alternative Usage, Considered Old Timers' Talk and Rarely Known	Usage
thumb 1	*simvakein*	*Didiyak*	+
index 2	*simveyov*	*Sikatay*	+
middle 3	*simwanay*	*Podoway*	–
ring 4	*moleleis*	*Mwaleleis*	–
little 5	*simkakiw*	*Sikukiw*	+

Table 5.4. String Figure Moves

Step Name	Activity
-suiy	usually index finger (*simveyov*) going under a string between another finger and increasing the connections by 2; usual first step
talkous	thumb (often, not always) hooks from top down and to little finger
-pow	Usually an index finger going under a strand and bringing it over the top of another, often the little finger. With this fingers are spread, with the hands—*nimwotet* = "handwork"—so that the design is multidimensional. It opens up the "flower" so it can be seen.
-nis	to release a strand by lowering a finger or two so that a strand dissolves or disappears; done with teeth (*kwanis/kaunis*) but sometimes finger (*kautinis*)
kwalev/-kilev	This is similar to *-nis*, but is accomplished by spreading out; "*tavag vagan bilal*," "to do so that it flowers."
-kokwavet	pull or take with teeth
katalov/katalev	to change hands, right takes left, left takes right
katuvin	turn upside down or inside out
kilov	release or unfasten, as in "it is finished"
kulagigik/kubtil:	to hold or pull together, to tighten
These expressions are verbs, the last four marked by causative constructions; '-' is person marker, e.g. "i" for him and potentially a real/unreal marker, e.g. "b-."	

figures, people are taking the first steps necessary to turn the body into an object, to make it a tool for the creation of other forms. A formalization of this idea is conceived for both life-cycle courses and kula action. Also in both, the body is conceived to be transferred into the goals of the action, subsequent generations in the case of the life-cycle domain, one's kula partners, and the distillation of one's own name in the Kula: one thing turns into another.

The procedures listed in table 5.4 enable one to move from one "flower" to another in a sequence, a process called *katipel*, "pass it over," "take it across." Processes of doubling, inverting, and doubly inverting are standard fare. These procedures teach sequences of geometrical transformations experienced with tactile figures so that the mind replicates what the body does ... or is it vice versa? "Move" names I recall hearing most often in daily life include *katalov/katilev*, *katuvin*, and *kilov*. *Katuvin*, for example, describes what is done with a sail when inverted 180 degrees so that the boat changes direction to take advantage of the wind from a different angle. *Katilev* may be heard exercising this transformation with the line called *asan* (cuningham controller); it can mean to leave something behind. A variation is used for distributing children in adoption strategies. *Kilov* is used when the *alit'* lines on the sail are released; it is also heard when a voyage's sequences are described, referencing a place just left behind. When I formally inquired of some of these terms in 2012 I was told that this word is also used for firing the cut and dried bush in preparation for planting, that is to say, a stage left and another prepared. Unfortunately it was only long after becoming familiar with these terms that I realized that most people learned them when at a young age, so they experienced them against the backdrop of string-figure images. String-figure language and logic ground everyday life. And that life is not simple. Take the aforementioned *katuvin*. It may be translated as "turn over," "turn upside down," or "inside out." Ogis used this term when unfolding a peculiarity of the structure of kula exchange, in the movement of Kula valuables around the Kula Ring. As noted (Damon 2002), Muyuw conceive valuables moving in a downward direction, say from Muyuw to the southeastern corner and on to the Trobriands. But this model becomes contradictory because, for example, *veigun* come down to Wabunun from points further west and continue going down. Somewhere they have to go up. Ogis used the word "*katuvin*" to explain what happens somewhere about Kitava and Iwa so that valuables coming into Muyuw would be high and could then go low again. The question here concerns the transitivity of kula relationships and movements (see Lévi-Strauss 1963: 309, 323n109).

These tactile forms bear an anticipated relationship to something actual or conceived in the experienced world. And given the often extremely abstracted relationship between the geometrical forms actually achieved and what they are understood to model, the imaginative powers are extraordinary. My list of Koyagaugau and Ole string-figure designs includes the following, in translation: "taro," "old man," "stone axe," "firefly," "mosquito bite," "basket," "joke," "fire," "boat," "a boat leaving (implicitly leaving somebody—often me!—behind)," "coconut," "movement between high and low tide," "a well," "a small bird," "a parrot," and far from last but maybe the least, "snot," the image of the string flowing from the (male) performers nose generating uproarious cackles.

When I started collecting Wabunun's versions, the didactic nature of the forms became increasingly clear, if not as abstract as the kula reference above suggests.[33] One episode called "*bwaloug*" (shellfish) takes one through the counting forms and sequence—in other words it teaches *that* form—used with that set of objects. The numbers of shellfish are depicted with the successive "flowers" of the game's trajectory—hence *sabalitan* (1), *sabaliyu* (2), *sabalitoun* (3), *sabalivas* (4), *sabalinim* (5), and *pwawilen*, the last one. When one person is finished, another puts a finger from each hand in a "shellfish," spreads them, and the whole thing unravels. Another called "fire" (*cov*) generates an image of a fire whereupon another person is told to blow (*yik*) on it to make it bigger. It bursts up, burning him or her on the chin. A series called "women make coconut-leaf skirts" simulates competition between women from Wabunun or Wayavat and Wamwan—women from central Muyuw; the latter always win, thus instilling in model form intervillage hierarchy. Two other series show a man and a woman committing adultery during the day or night (distinct categories of experience). Another makes a paddle (*kalavis*) and ends as the person mimics the motion of paddling. Others illustrate men fishing with a net; the significant tree *bwit* in a dazzling performance that takes over a minute to run through and requires another person to end; the bird *kulawit* (bush hen), whose eggs are avidly sought; and many other objects of everyday life. In considering these images, one recalls Granet, and Mills on Granet (Mills 1941 [1963]) concerning the nature of the Chinese language and writing: the forms are emblematic and pragmatic, collective and moral forces for actions running the gamut from the simple and important to the truly astonishing and elegant.[34] And they illustrate motion.

Although some of the sequences– those requiring more than one person and sometimes more than one string—were breathtaking, the most complex I know is Gumeau. It requires just one string and one performer. I've seen it done with as few as 8 moves, and as many as 27. The complexity may be symptomatic of deep traditions in the history of this region for "Gumeau" refers to the constellation the Pleiades –understood as a single *utun*, "star." For a summary account of the Pleiades in the eastern spread of the Austronesian world see Kirch and Green (2001: 260 *ff.*). For the immediate northern Kula Ring the constellation is one of the more important people follow; some call it the "chief." According to Iwa people its heliacal rising organizes the calendar from Iwa on to the Trobriands. East of Iwa that rising, late May or early June, is observed as a sign for everyone to take a dawn bath. That protects them from the cooler weather soon to arrive. One informant explained the lunar-month name *Kunuwut*, which comes about May/June, as referring to the process of a star having set in the west then swimming under the earth back to the east where it rises, the coincidence here pegged roughly to the rise of this set, the Pleiades/Gumeau. As for that moment, if yams planted the previous planting season are not yet ripe then Gumeau urinates on them and they turn yellow, though this need not affect their size or taste. And a few months later, in August, teamed with the last star in its set, the two determine the intensity and duration of the southeast wind which may have started in April but rarely becomes regular and strong until August and beyond. And although no star nor moon from Muyuw west to Gaw and Kweywata is conceived to formally organize planting sequences, the set in which Gumeau is situated is carefully observed. It marks the passage of time and the horticultural activities that should be organized as the set comes to dominate the evening sky. My most distinguished performer of Gumeau, when asked for the names of the various transformations in the set—they are not all named—switched into the sequences of gardening activities correlated with the movement of the constellation across the sky; you start cutting the small bushes and trees in the forest for a new garden when it is about 45^0 high in the early evening; plant when it is "yanay," zenith at dark; then find "happiness"—*bimasul*—when it rises and it is time to harvest. This period is associated not just with gardens full of new yams, but also fish netted with the low daytime tides of the June solstice. In another couple of months it is *wagabwau*, a combination of the generic term for outrigger canoe, *wag*, and *bwau*, the term for trunks of modest to small-sized trees used as rollers for dragging a boat up on the beach and keeping it on its perch. About this time, August or so, Gumeau and the constellation/ star called *Lakum*, Crab, decide to release their powers, *aniyan*, to effect the hard southeast wind. Boats must be protected from large waves.

The care with which these phenomena are watched, conceived and modeled is directly seen in the states the Gumeau kananik rehearses. Its moves are understood to be so complex, *kapinanigw*, that it needs an extra-long string to control its convolutions. One of its realizations requires the aforementioned katuvin move, many of the sequences have names like "Ulayis" (dawn), "Bwiyam," when it is light and the sequence is over. But not all I could gather necessarily referred to a stage I could easily identify or follow. Some I had to just list as "a few hours later" because the design simulates the continuous passage of time. The most extraordinary sequence goes by the set referenced and often spoken as it is performed:

Ipel kubwan
Tautoul budibud plelidius.
Bwiyam.

Ipel is the verb meaning "it jumps" or rises; *kubwan* is the name of a larger star or planet, the proverbial "morning star." Hence the sequence is about the end of the night. *Tautoul* is very complex physical presence streaming above the horizon, *budibud*, just before dawn's first light. The image was explained to me by saying it was like black clouds (*lov*) that spread out and up from the eastern horizon like the fingers on an outstretched hand. The passage from darkness to light is conceptualized, realized in the physical form of the string figure, by an extraordinary inversion of light and dark. Not really believing the account, I was astonished when I finally saw it on my last night on the island in 2002. Perhaps an iridescent black light streaming above the horizon, the paradox of black streams blocking out the starry night—a darkness negating the darkness—that make visible the stars which then are replaced by *pleledius,* a word and combination of sounds which makes no sense to me. Yet I was told it means something like the clearing effect created by the appearance (*bisunap*) of the sun's (*kalas*) first light (*mitilin*). With this the sequence—*Bwiyam,* daybreak—dissolves into the simple closed loop—it is finished.

At first consideration this stream of motion might just be considered an extraordinary artistic accomplishment, and it is certainly that. Yet it is also the result of two sets of studied relations, one of the progress of stars and lights of the night, the other their replication, their modeling, with the manipulations of the string. The significance of this must be underscored. First, navigation in this region is accomplished by stars and every island in the Kula Ring is associated with a star (course). As often as possible people prefer sailing at night so they may watch them. Courses are set with respect to rising or setting marks of particular stars, some of which are in the same set as Gumeau. Now these stars are at the point used for navigational direction at most once during the night. So after that moment what is calculated is not that star's position, which is always changing, but the relative space it defines where it is sighted with respect to where it rises or sets. In other words, what is used is a configuration of movements. From an early age kananik develop a capacity to think and replicate such configurations. Second, when I talked about these matters with Kaulay people they remarked that they regularly watched the stars rising over the eastern end of their east-to-west aligned village then setting to the west. Major dimensions of their external environment, string figures internalize these forms so that they are intimate aspects of being. It was thus no accident that when I was moved –by Sipum and Mayal—to the woman whose pictures illustrate Gumeau's course they followed her intently and quietly mouthed her sequenced steps. They knew what she demonstrated.

A Closing Note

In this region *vatul* is simultaneously a principle of connection and a vehicle of thought. It is not axiomatic that the world be viewed as if it was composed of string, yet cordage and the intentions derived from it are qualities experienced everywhere...—from vines in forests to veins in the body, between these two poles everything is conceived and made.

Based on slight alterations in sequences, these forms enable systems of transformations, processes no longer obscure for this work. Chapter 1 showed transformation of spaces across the region. In chapter 2 Ogis stumbled upon the tree he should not have seen, and turned into the likeness of a Trobriand chief, albeit one speaking Australian English. More recently my informants have told me one might instead have switched from Muyuw to Dobu, or the Lolomon language, even though in real-time consciousness the speaker might not know those languages. Chapter 3 showed how *tasim* and *sinasop* are inversions of one another. Chapter 4's two similar *Calophyllum*—members of a set of trees whose differentiating criteria include the grains unique to each species—*kausilay* and *apul*, both need open land to develop. Yet one arcs to the sun while the other grows straight. The former leads to *anageg*, the latter to the different form that is the Trobriand *masawa/tadob*. This chapter has culminated with string figures, often not a topic of serious consideration. Yet as a kind of magical geometry, these forms generate the tactile experience of a conceptual core. Literally the beginning of this social system, these practices rehearse its transformations. And as connecting devices they bring us to the ultimate form they enable, the encompassing relation configured in the *anageg* design.

Notes

I present the version first presented of this chapter in May 11, 2004 in André Iteanu's seminar devoted to the heritage of Louis Dumont; thanks to seminar participants for attention and careful questions. Tying and knots are encompassed contraries? Tying invites comparison with Lansing's discussion of the place of Austronesian culture in Bali—underpinning and he thinks feminine (Lansing 2006).

1. Ideas Sipum invoked are probably cognate to the Chinese concept 脈, *mài*, literally "vein" or "pulse."

2. Individual strands are used for tying sails; intertwined strands make the belts for coconut leaf skirts; the same strands are more finely woven into stronger rope. Along with *ukw*, *ayovay* is raw material for heavier-purpose lines on boats. Its wood is used to make canoe paddles and floats for fishing nets, and it is of sufficient quality for small outrigger canoes. Old *ayovay* might be used for making boat ribs. The wood is very "hard" when extracted from the inner part of the bending trunk or branch of an old tree, although it is not considered as "strong" as *kakam*. In 2006 I examined a Wabunun *anageg* in which three such pieces had recently been inserted. The boat's owner said they were fine, but others privately said he should not have used them; weaker than *kakam*, they might only last three to four years rather than six or more.

3. In addition to *ukw*, others are *agwagwam* (*Melochia ordorata*; a poor quality string but a very common early fallow tree regularly used to bundle garden

produce transported to the village); *alvililuv* (*Commersonia bartramiana*; good fastening material); *silaskubay* (Damon 79, *Kleinhovia hospital*); *weylau* (female; Damon 89, *Abroma augusta*); perhaps *malau* (Damon 274, Sterculiaceae or Tiliaceae). At least one person considers these trees a "group." One tree in this set is an exception, *lawoy* (Damon 77, *Heritiera. sp.*), but it is used for making boats.

4. Küchler's essay appears in a volume dedicated to Alfred Gell. Although this chapter draws its inspiration from Kula Ring friends and teachers, Küchler's essay provides an opportunity to acknowledge Gell's stimulation (1992, 1999). Some of his ruminations derive from Roger Green's Pacific archaeology. Since 1998 encounters with Green have enriched my understanding of another corner of the Austronesian world.

5. A man who may be Muyuw's most successful student and practitioner of Western culture, who organized the island's biggest store, and who owned and operated a truck (from approximately 1990 to 2010) and numerous outboard engines belittled my interest in this chapter's topics. His sensibilities are now from a world dominated by mechanical, petroleum power.

6. In southeastern Muyuw, *ayovay* inner bark is used to tie up the much smaller prestations of yams used in analogous exchanges. *Gub* is the name for dividers of *venay* into smaller units of a garden. Those dividers demarcate social/exchange relations, and it may be that the words for the vine and the divider are related notions.

7. *Ipwawun* (*Hernandia ovigera* L.) is not a male tree; however, its bark is considered very "strong," a male quality, and is used in magic for making pigs grow fast and well. And in that context the tree is considered "male."

8. I thank Erik Pearthree for this and other translations of various sailing and boat terms.

9. Probably after *ayovay*, whose bark is used to make it in eastern Muyuw. Central Muyuw use *ukw*.

10. See Horridge (1987: 142–43). Horridge's "A, Madurese Jukung Rig" is the closest example of the *yabuloud* form, which my informants understand to be a simple transformation step from Horridge's "I_1" or "I_2" New Guinea crab claw, making these two sails identical. Horridge suggests the *anageg* form, his P (1987: 142, fig. 83), derives from a Malay sail style. Bugis sailors influence Kula Ring boat forms?

11. Called *kunay*, the yard's center was round and about 7.78 centimeters in diameter. Some 25 centimeters from each end, the *kunay* is carved to be more square or rectangular than round, and it had dimensions of 2.5 x 3 centimeters (to the tree's "top," *dabwen*) and 3 x 4 centimeters (the tree's "base," *wowun*).

12. My sail pictures suggest that the ratios are larger, perhaps closer to 3:1.

13. Through the twentieth century, Muyuw had three or four main sailing communities, from west to east Yemga, Boagis, and Nasikwabw (and Waviay). Their ecologies do not support the production of sail necessities, perhaps excepting *im* sewing line from the *loud Pandanus* tree.

14. Before I first arrived in Muyuw in 1973, that species was moved to the larger island during a moment in the transformation of organic dependencies into an ideal of equivalent units.

15. Iwa uses the sail type Muyuw call *yabuloud*. When on Iwa I was shown the *Pandanus* type for their sails. It grows on the ledges leading from the top of Iwa down to the ocean and is called *kewiba*. This is the same environment in which the Muyuw *loud* grows, but I do not know enough to argue for their identity. Muyuw informants tell me the tree is actually one they call *pwakau*, some instances of which can be found near the Upwason sago orchard. Its size is halfway between the *legis* and *loud*.

16. I thank Erik Pearthree for making this observation. The speeds: with beating, 4.3–7.2 knots with six of the ten recordings under 6 knots; other directions, 4.9–10 knots—of the twenty-one recordings, only five were under 6 knots.

17. Thanks to Emily Wood from the Harvard Herbarium who enlisted Wong Khoon-Meng of the University of Malaya for this tentative identification (received August, 2, 2007).

18. Another tree called *yals* may be used in place of these two. Stevens's tentative identification is Theaceae *Adinandra*.

19. Asymmetrical relations prevail among these forms. Informants have told me *busibuluk* might be used for an *anageg* sail but *ayniyan* would never be sufficiently long and narrow enough for a *yabuloud* sail.

20. From my pictures and description, Allen Allison, herpetologist at the University of Hawaii, suggests it may be *Scolopendra sp.*

21. I measured ten intervals on the 9.3-meter sail in 1998, recording 7, 9.5, 9, 6, 8, 5, 6.5, 7, 6.6, and 6.6 centimeters, averaging 7.1. There are twenty-four *taseu* on the 96.5-centimeter-long model sail made for me in 2007.

22. Needles are about seven centimeters long, tapered to a point, with one side straight and the other arcing to make the point. The needle eye is about 0.5 centimeters wide, a 0.2-centimeter hole drilled through that eye.

23. Because they are smaller than *wot* and thus more easily moved, their ownership is often obscured. Those I examined in Kaulay were held by a senior member of the village to prevent the owner's brothers-in-law from walking away with them.

24. *Teikw* have a cord with the same name encompassing the perimeter of the structure. Kaulay people understand the form to be male because its existence gives form and strength to the structure.

25. One type is called *madin,* and reproduces in the muddy northern end of Sulog harbor contiguous to a significant freshwater stream, making the water less salty than further out. This prized food almost died out during the 1997–98 El Niño drought because the water became too salty. The other type is called *lag,* which grow inside coral formations in the clearer seawaters south of the Sulog region. They have to be pried out of the coral and are only sought for fishing net weights.

26. Most of these plants have other similar uses. *Namonsigeg's* has already been noted. *Sekit'*, a high-forest vine, is coupled with a shoreline bush called *agavagav* (Damon 192, Goodenoaceae *Scaevola sericea* Vahl) to make a powerful concoction that makes the body strong and shiny, both good traits for pursuing women and the kula. This latter plant was also used for aborting unwanted pregnancies and washing and curing tropical ulcers.

27. By intensifying the requirements of everyday life these activities are like the rituals described by Dean and Zheng (2010) for Fujian Province's Putian Plain.
28. Women attending a birth message the infant's face to make it like its father's.
29. Differences exist between Muyuw and Koyagaugau. The woman in Koyagaugau who first impressed upon me her string-figure knowledge told me she learned from her "father," the very famous Mwalubeyay (d. 1995). His wife from 1973 was a Wabunun woman who moved back to Wabunun upon his death. She insisted her husband knew little of these forms. By 2002 one of her daughters had moved back to Wabunun and it was to her, Aligeuna, I was eventually directed.
30. In 2006 he resided in central Muyuw; in 2009 he married into Duau; in 2014 he lived in Wabunun.
31. The expression and motor effect also describes grains in the *Calophyllum kakam* and *kausilay*.
32. A related understanding and concept in Chinese is 了解, based on the sense of 解, jiě which may be translated as "loosen," "unfasten," "untie" or "explain."
33. I started learning these with women I knew best, equivalent to mothers or sisters. Some had no teeth or eyes so they found it frustrating, and I was nudged to people considered experts in the practice. This activity engenders its own rank orders.
34. If, as Mills suggests, Granet overstates the uniqueness of the Chinese language, one must remember the political and cultural significance of its highest form, calligraphy, as landscape beacons cut into stone cliffs and displayed on every temple. Calligraphy is a form that requires practice and generates audiences not unlike the string figures of the Pacific. Damon (2012) continues such speculation.

CHAPTER 6

Geometries of Motion
Trees and the Boats of the Eastern Kula Ring

In truth, the right way to begin to think about the pattern which connects is to think of it as primarily … a dance of interacting parts. (Gregory Bateson, *Mind and Nature* (1979: 13) qtd. in El Guindi 2008)

Fred: When *Lavanay* [a twenty year-old *anageg*/*kemurua*] wears out what will you get next?

Duweyala, the captain and owner: A dinghy and a fifty-horsepower Yamaha.

If string figures are a kind of magical geometry interiorizing a system of concepts, *anageg* are a different kind of mathematical expression. They put into practice a geometry of motion. Both forms are objects of beauty. Although string figures tend to be appreciated only at specified times, *anageg* become a focus of attention whenever they appear. Every time one sails by, those on shore cannot help but watch it—watching the movement moves the watchers. The same may be said every time a new one is drawn ashore. Villagers wait with rollers to place under the keel and outrigger float as they touch the rising shoreline. Then they pull and push the boat up to its resting perch. These beachings—launchings are similar—are full of excitement, laughter and straining bodies. Then everyone becomes quiet as they welcome the crew and examine the craft. Although this moment includes an aesthetic appreciation, it is also a kind of wonder and calculation for how every part is constructed. Each craft is a problem inviting different solutions.

This chapter makes two closing arguments. Concentrating on internal *anageg* structure, it shows that *anageg* are to a regional scale what string figures are to an individual—an exhibition of the art of knowing. The second specifies relationships between their design and other aspects of the social system's local and regional character. The argument combines ideas from Lévi-Strauss and ecological theory concerning keystone species by means of an idea Dumont (1977) draws from Marx. These boats are the epitomizing form of communication in this cultural order, and this chapter shows how they effect an ordering, structure the social system (Hénaff 2009: 191).[1]

Started in 1995, this project's initial purpose became trying to understand assertions concerning the *gwed* (*Rhus taitensis*) tree. Chapter 1 outlined what I learned. Although people built their world up from the ground, that information was fashioned into models at a remove from, if defined by, experience. The well-formed model concerning the sweet and bitter trees delivered a correct directive—plant close to one but not the other—but not a literal description for what is known. Although it first seemed that these practices spanned the northern side of the Kula Ring, small differences emerged. Northern Kula Ring landscapes pattern differentiation. By means of trees, places become articulated parts in a regional system. The *anageg* idea encapsulates these forms, pulls together the region's specificities. Stemming from the Creator, the concept binds the ground from which these boats emerge to the stars that calibrate their making and use. As the craft connect the places that condition their existence, they track how these islands fashion a human order. Following in the footsteps of Pfaffenberger, Lemonnier, and Lansing, I illustrate in this chapter how a technical form crystalizes a cultural order.

A side project that both slowed down and contributed to this work was an exploration of chaos theory's potential contribution to anthropology (Mosko and Damon 2005). That endeavor created a critical dialogue with Jack Morava, a topologist anxious to find mathematical ideas for use in anthropology. But Morava is also a skeptic. The reasons for this were not the usual ones about numeration; rather, he felt that many practices anthropologists consider—mythology, ideological transformations, ritual, spatial patterns, deeply ingrained gender dynamics—entail modes of thought different from those of practicing mathematicians. Using Freudian terminology, he thinks anthropology is largely devoted to describing primary thought processes while mathematical reasoning is secondary, the slow, careful fiddling with terms and relations trying to ferret out uncertain truths.

Ideas of this nature occupy an infrequently articulated place in contemporary anthropology. Consider an important statement from Lemonnier at an early moment in the revival of "technology" and "materiality" as a central ethnographic focus. He writes of the ways by which technical knowledge bridges techniques and society:

> If societies exercise "choices" in a universe of possible techniques—*most often unconscious choices, it goes without saying*—these necessarily leave traces in the systems of representations, and the technical solutions retained must, one way or another, be in harmony with these latter. (Lemonnier 1986: 155, emphasis added)

More recently, rather than an "unconscious" quality of material forms, Lemonnier emphasizes those that are "non-verbal" and "non-propositional" (Lemonnier 2012). This study has repeatedly confronted the relationship between explicit empirical information and collective representations. At first sight the *gwed/atulab* contrast is a matter of simple technical information—an ecological model of good and bad fluid transfers. Yet that became a fiction when I realized that the best gardeners followed the paradigm's invocations but not its reasoning. The same with Wabunun's *teili* tree, which was seemingly a matter of primitive myth and origins until an ENSO drought revealed regional responsibilities fixed in the conscience of its believers—"captured knowledge" to borrow from Joel Gunn. In 2002 I realized acquaintances easily accepted the probability of male and female *kausilay* trees while also presuming the tree is a female person. Consider also the transfer of ritual firewood in life crisis rituals: mere content of a formal process yet the practice marks regional complementarity. Ellen glosses these situations correctly when he writes that such matters "are all cognitive processes (means, agents, instruments) which help us first comprehend the world and then negotiate our way through it" (Ellen 2006: 185). These collective forms are subtle moments in a designed, laminated cultural system. These ideas are more than they seem.

Working up to the 2002 voyaging plans, I wrote Geoffrey Irwin about the advisability of doing a measured drawing of an *anageg*. He encouraged it. I have made modestly detailed drawings and measurements for two *anageg* and bits and pieces for several others. Yet my inexperience aside, I could not manage a measured drawing, and began to doubt its contribution to our knowledge. For I discovered that the boats have a different shape when resting on their shoreline placeholders than they do when in the water—a difference my informants *conceptualized*.[2] The shapes are too complex, for my numbers at least. By means of Morava's invocation of secondary thought processes—the slow, careful fiddling with terms and relations trying to ferret out an uncertain truth—it occurred to me that these boats *are* the formula, they constitute organized reasoning.[3] We might think and act on the proposition that the universe is composed of equations (see e.g. Kolbert 2006: 101), but in the Kula Ring the analogue of that practice is a realized *anageg*.

Consider the crossbeam connecting the hull to the outrigger float. Every *anageg* has ten; *kaybwag*, the middle-sized outrigger class, have six to eight. Without them one has two pieces of driftwood. Although not called *kiyad*, houses too have parts understood to be kiyad-like, significantly half the number of anageg. "*Kiyad*" is also the name for the asterism Orion's Belt in the constellation Orion. A word and concept with deep Austronesian roots (see Ross, Pawley and Osmond. 2003: 164), *kiyad* may hint of invariance transformations across the Austronesian and East Asian divide.[4] So, *kiyad* are not only vital to every boat and house, the idea forms a brilliant component in the night sky. That sky is also likened to a boat. That boat's "crew" are the stars, the triad sky, boat, and garden continued with the Milky Way, which is conceived to be like a garden's *lapuiy*, the smaller east-west lines (north-south in western Muyuw) that create garden *venay* division—an exchange unit and principle of reasoning.

If an *anageg* has passed through western Muyuw, its *kiyad* may be made from the replacement mangrove tree, the slender *igsigis* (Combretaceae *Lumnitzera sp.*). However, when they come directly from Gaw and Kweywata, they should be *alidad* (Damon 278, *Syzygium longipes* [*S. stipulare* (Blume) Craven and Hartley]). An ecotone tree like *kausilay*, *alidad* are understory trees found only in mature *ulakay*-like forests near the water's edge, on Gaw and Kweywata as well as Muyuw. The tree's wood is very hard and "*very* interlocked, like *kausilay*," remarked Dibolel in 2006, having replaced seven of the ten on his *Bwadananakup*. They must be allowed to dry for a few weeks before being formed, otherwise they crack. The hard and interlocked features are important. When worn out they crack rather than snap in half, so they may be tied until they have to be replaced—the tree is its own backup. The closely related but smaller *ayelev* (also a *Syzygium*, Damon 171) as well as *amanau* share the same physical qualities as *alidad*, and they may be so employed on middle- and lower-classed outriggers.

Kiyad are approximately three meters in length, those in the middle longer, those at the end shorter, the difference meshing with the hull's oval shape.[5] Each one is oriented from its *wowun*, base, to its *dabwen* or *matan*, top or tip end. Inside the hull they are tied to *seisuiy*, important pole-like pieces that arc along the sides of the boat; outside the boat the "tips" are tied to the stanchions (*watot*) connecting them to the outrigger float. Tying begins with the base end tied to the *seisuiy* opposite the float, the *watan seisuiy*. The *walam seisuiy* is tied next, on the outrigger float side. Positioning with respect to the two *seisuiy* establishes one of the minute differences between the Gawa and Kweywata *anageg*. In the Gawa model, when the boat is in water, the outrigger side *seisuiy* should be slightly higher than the other side; this differential gives the structure greater strength when loaded. Although they should be evenly spaced along the length of the keel, the gap between the first and second and that between the ninth and tenth should be larger to facilitate placement of the *kavavis*, the boat's steering device. Rather than perpendicular to the keel, the first and tenth *kiyad* angle toward the ends of the float. The distance between the float and where the crossbeam "top" is tied to the stanchions varies, the resulting placement affecting sailing performance. Dibolel shortened that distance on the *Bwadanakup* when he first took it over.

Seisuiy are critically important forms. Some people say they give a boat its strength, its resilience. After strakes have been tied and caulked to the keel, *seisuiy* are threaded through holes cut in the *kakam*-derived *gulu-moms*/ribs that angle each side of the boat. Because they have to arc, and because they must be strong enough to absorb pressures put on them by the crossbeams, *seisuiy* should be made from *igsigis*.[6] The form must maintain its resilience to different pressures when arced. Crossbeams (*kiyad*) transfer the force of the waves on the outrigger float to the *seisuiy*, and through them to the rest of the boat. They also transfer back to the float the hull's reaction to the float that includes the effects of the wave passing to the hull. Simultaneously the arced *seisuiy* give and receive multiple motions spread across their arcs. Crossbeams bear the weight of those multiple motions. If crossbeams crack, they do so where they go through the outrigger side of the hull. Chapter 4 noted that outrigger side strakes should be slightly thicker than the opposite side to withstand wave forces, which reverse the effect of a wave on the float. They should also be slightly thicker because they have to absorb the forces that *kiyad* place on them. *Kiyad* balance contradictory powers and conjoin altercating motions.

There is a loose connection between the *kiyad*, *seisuiy*, and *gulumom* (ribs) that is created by the fact that while crossbeams are tied tightly to *seisuiy*, the *seisuiy* bounce within rib holes. Some people say *seisuiy* are the boat's backbone, an idea deriving from their jointed, flexing nature, as well as their mobility inside the rib holes. Pictures accompanying this section show this.

Recall chapter 4's discussion about how waves suck the float down, pulling the boat along with it. The more obtuse angle on the outrigger float side acts as a lever to force it back up. The float's rising and falling is the most extreme at its ends; therefore, the arcs through which first and last *kiyad* travel are the most extreme. Consequently, while eight crossbeams extend across the interior space of the hull from one *seisuiy* to the other, the first and tenth only go partway into the hull space and so are not tied to the *watan seisuiy*. Unlike the other eight, the strake holes through which these exterior crossbeams pass are not caulked. Although tying assembles these boats, at two critical points there is no tying lest extreme motions rip the craft apart.

There is a further intricacy in the kiyad structure. The relationship among all these parts is bound by a reversal in the ordering of the crossbeams and rib pairs. The mast mount (*kunusop*) centers the boat. There are five crossbeams and rib sets between the bow and the *kunusop* and five between the *kunusop* and the stern. However, a set's relative relation is reversed on either side of the *kunusop*. The reference here is "base" (*wowun*) "top" (*dabwen*/*matan*). Toward the base end, the crossbeam is behind the rib set; toward the top end, it is in front of the rib set. The rib pairs are the same: toward the base end, the outrigger side rib is behind the other one; toward the front, it is in front. Like all knots, a reversal organizes the form.[7]

The forms just described are a matter of calculation—the character of the wood chosen, the shapes into which parts are fashioned, and the tying and positioning define the whole. These concerns articulate one of the two structures expressing the boat's movement in the ocean. If not everyone knew these details, most everyone knew such details had to be known for this social system to work. This chapter first describes structures found in these boats. I show that *anageg*'s internal forms are a mass of calculations. In Plato's theory of knowledge, mathematics and music are at the third level because they define interrelations among the facts and relations they contain. On the eastern half of the Kula Ring, *anageg* play the role of math and music.

The second point moves this argument to the social system at large. I focus on the relationship between the boat's structure and other domains of life and knowledge. This argument combines ideas from Lévi-Strauss and ecological theory concerning keystone species and a discussion Dumont (1977) draws from Marx. Lévi-Strauss makes communication structures—of words, things, and people—anthropology's central inquiry (Lévi-Strauss 1963: 289) because, he thinks, forms of communication organize social systems. Obviously these boats are means for communicating words, things, and people. But they also play a centering, determining role in the social system. They are like the ecological concept of "keystone" species, forms that enable other "species" to take their places in an ecosystem.[8] In a very suggestive synthesis of European culture in North America, Flannery (2001) argued that the successive transportation systems—canals, railroads, interstate highways, and we may add container ships and the Internet—organized the social system (which is now the world), both in their building and in the relations they enabled. The *anageg* form does the same thing. By its necessities, seemingly external realities become a coherent presence.

This brings us to Dumont on Marx. In his critique of the category "economic," Dumont calls attention to what he considered Marx's correct recognition of a hierarchical aspect to society. He writes: "The aesthetic feeling ... in Marx ... marks ... a holistic and hierarchical perception" (Dumont 1977: 162).[9] Following this line of thought, this chapter argues that the *anageg* form constitutes a hierarchizing, aesthetic sensibility. In relationship to it, parts become a moving and transforming whole.

And this shall be my contribution to the expanding discussion of Indo-Pacific sailing craft, a once notable topic (Haddon and Hornell) only recently again gaining the place it deserves (e.g. DiPiazza and Pearthree,

Finney, Gladwin, Horridge, Irwin, Manguin, Southon). That outrigger canoe symbolism is very important is not a new idea in Pacific studies. Although he rarely understood what he was describing, outrigger references suffuse Malinowski's Trobriand corpus; canoes and aspects of their significance are superbly illustrated by Nancy Munn for Gawa, and others elsewhere (Lippsett and Barlow 1997), and may be found or inferred in practically every ethnographic account in the Indo-Pacific region (Manguin 1986). In pointing out that gardens were likened to outriggers and that Muyuw's largest mortuary ritual was configured in their terms, I demonstrated in my earlier work a typical feature of Austronesian life (Damon 1990: 172–78). Muyuw people considered their island and culture formless and infertile until the Creator arrived in an *anageg*. The form brings life (see Needham in Hocart 1970). As they are vehicles for the movement of words, things, and people, these craft synthesize this social system's elements and relations.

Springs as a Boat's Internal Structure: Sequencing the Whole

Anageg move from their original production site to their next home by means of a set of reciprocal exchanges stretching over years. The "end" of these exchanges is the transfer of *kitoum,* individually owned kula valuables, to the producers. In 1974 Wabunun started receiving a new *anageg* from its Gawan friends. From the contribution one elder made toward its acquisition, the new owners named it *One Dollar.* I never learned its original name; it remains known as *One Dollar* to this day. It was meant for the great man Mwalubeyay of Koyagaugau, Wabunun's new connection to the south. In fact, Wabunun used it for many years, so it rotted before it could be conveyed. The *kitoum* Mwalubeyay gave for it became embedded in relationships developed between Wabunun and Koyagaugau by virtue of Mwalubeyay taking a wife from there in 1973. An embarrassing scoundrel when he absconded with the woman in 1973, Mwalubeyay became highly revered and loved by his death in 1995.

In the late 1990s my host into and out of the island, Milel Gisawa, by then a famous radio announcer in Milne Bay Province's Alotau, regaled me with how he sailed *One Dollar* to Mwadau village in western Muyuw in the late 1970s to pick up a woman named Maria and take her back to Wabunun to be Dibolel's wife. For two decades Dibolel's father, Aisi, had been trying to arrange a marriage between Mwadau and Wabunun to gain access to Mwadau's connection to the *anageg*-producing Kweywata. By virtue of slight design differences, Kweywata *anageg* are considered sleeker and faster than Gawan *anageg*. However, elder Wabunun women refused to cede a daughter. Then Aisi learned about Maria, a woman of his own clan and, disputably, subclan. Dibolel went to Mwadau to meet her when he went there for a mortuary ritual after 1975. They liked one another, and an agreement was made. Milel sailed to pick her up.

My affiliations to Aisi and his wife transformed because of witchcraft accusations and the passage of time, so when I to returned for this research I was distributed to Dibolel; Maria—called and known by everyone as "Mwadau"—gradually became my primary host in Wabunun. She went there thinking that she was to populate the village with people from her Sinawiy clan only to be shocked that half the village bore that clan identity. She was stunned when Dibolel first took her to his gardens, now her responsibility. Mwadau village is a serious gardening village, but from slightly different planting regimes between southeastern Muyuw and Mwadau and Wabunun's peculiar ethos (See Damon 1982), she was not prepared for the enormity of this man's work—but she learned.

Beginning to know something about *anageg* by July 1996, I chanced upon a conversation with Mwadau. A defining moment, it remains with me to this day. Women are not supposed to know about boats because they don't make them and do not often sail in them, but by then I was used to overriding minor stereotypes like this. Earlier that year Mwadau helped me on a fence-stake survey. The only other person who responded as confidently as she did about stake tree types was Dibolel, who, although a man of few words—unless the topic is Kula—is analytical and very precise. In any case, with Mwadau I raised the issue of the asymmetry in the sides of *anageg* canoes. Without hesitation she defined hull asymmetry by wave dynamics, using her body to replicate the yaw the forces create. Although no two boats are the same, and it was only evident on the boat I sailed on in 2002 from a certain angle, the two sides are always asymmetrical.

In 2012 a friend made me a model *anageg*, a near-perfect replica just over a meter in length. The strakes were positioned asymmetrically on the keel as they should have been, one side veering toward obtuse. But when the holes for crossbeams were made, which determined the outrigger float would be, they were put on the more acute side. Surprised, I asked if they should be on the other one. He said no. Perplexed, I related this circumstance later that day to Dibolel and Mwadau. Dibolel said nothing, but Mwadau immediately said the model was incorrect. After studying his own boat the next day, Dibolel realized it was constructed like Mwadau said it should be. Later I recounted this episode to Mwadau and Dibolel's son. He was not surprised, saying that the man who raised Mwadau was known to be a boat expert—she probably listened to stories about boat dynamics every day until she moved to Wabunun.

This circumstance puzzles me. Although Wabunun is not a sailing community, two men who have together received five *anageg*, who have disassembled and reassembled all of them at least once, and who are finicky about finer points were nevertheless unaware of a major structural component. I emphasize the finer points. Repeating what my 1999 Yemga informant reported, the man who made the model told me that Gawa and Kweywata boat builders never get the boat's most important part correct. He always redoes it, and prefers the Wabunun-available *aukuwak* to *kaboum*, the tree of choice for most people. Dibolel gave me an analysis of the differences between Gawan and Kweywatan *anageg*, saying how the Kweywata craft look better and are faster ... with no load. But when loaded, the Gawan design sails better.

One conclusion to be drawn from this circumstance is that these are exceedingly complex objects from which people continually learn by means of their experiences with them—though they may never totalize the form.[10]

By the time of my conversation with Mwadau in July 1996, the *anageg* form was pulling me into its intricacies. I had collected most facts synthesized in chapter 4 and wondered about the similarities between *kausilay*'s interlocked grains and cultural aesthetics. For me, then, the boats were first and foremost *kausilay* trees. And beauty in Muyuw—a yam garden as well as a carved canoe prow—follows from how elements blend—*limwelen* is the concept—into indistinguishable wholes, like the grains in *kausilay* trees. Yet two further conditions made me realize that I had to take these craft more seriously. One was the fact that everyone, men and women, seemed to know about them. Their details were not just for experts, though experts there were.[11] The second concerned parts designed to vibrate. All were made from different trees, and I initially focused on three such parts. People would say that without this part pulsing, that part would break. The different forms realize an integrated whole. However inarticulate I was about them, these motions motivated my 1996 discussion with Adrian Horridge in Canberra when we met several weeks after my conversation with Mwadau. By 2002, the form no longer primarily "*kausilay*," I understood more parts in relation to one another. Watching the model constructed in 2012 made evident four springs structured to take and react to articulable forces. The *gulumom/seisuiy/*kiyad*/*watot set just reviewed is a fundamental fifth. Describing these parts sequences the whole in motion. That whole varies from forms so tight they barely move to those not tied but moved at will.

Several words reference qualities of motion built into these forms. One contrast, *mameu/matuw*, soft or pliable/hard, circumscribes the parts I now discuss. Some people used these two terms—and variants modified by causative constructions—to give proportionate statements about all the forms discussed here. If all the pieces are too "hard," too thick or large in diameter, a boat will not just be too slow, it will be unbalanced, going too far in one direction (*buluk*—opposite the wind). Some discussion is couched in the expression "*iyemat wag*," which refers to the oscillation that should occur as the boat goes up and down waves. More important is *igogeu*, "vibrate" or "sway." Mwadau used this word when she talked about the outrigger float countered by a wave as it hit the hull's outrigger side; too much *igogeu* led Dibolel to dismiss *dan Calophyllum* in boat construction. This word is also used to talk about minuscule strake movement, the first part described. With respect the next piece, *kaynikw*, people used the word *ibsipis* for its nearly imperceptible motion. *Gipoiy*, best understood as "springy," refers to *nedin* motion, a part tied fast in its middle, left loose at its ends; it also pertains to the mast, firm at the bottom, rocking at its top. *Bidinidan* covers all these actions. Upon telling one source I didn't understand the word, he said it was like *igogeu*, sway. He illustrated necessary arm and leg movement when a person walks. If a body is too rigid—*mwatamwetal won*—it cannot move; swaying arms and legs facilitate motions like a boat's parts must vibrate for it to move. Coupled with the mast, the last piece discussed in this section is the *kavavis*, the steering mechanism, always made from the tree called *kaymatuw*, "hard tree." It is too rigid to vibrate, so the helmsman completely pulls it out and pushes it into the water as wave forces meet and pass it by.

The Hull: In/Out/In[12]

Held in place by the first things attached to the keel, the parts that move the least are a boat's strakes, *budakay*, a boat's second set of attachments. They will be about eight meters long, fifty to seventy-five centimeters wide, with three staked on one another.

The hull structure begins with the keel, *wag*, the term that gives craft their generic name. Its main dynamics were covered in chapter 4. It is carved from the shape found in the selected tree so that it arcs in two ways: vertically (the keel's center is lower than the two ends) and horizontally (there is a slight turn toward the outrigger float side at each end). People call this arcing *kamwagan*. The sides of the V shape are called *bab*, the shape itself *asinot'*. The slight turns to the outrigger float are called its *sibun*.

Trees used to make strakes—*kausilay* for *anageg*, *antunat* (Sapotaceae *Palaquium sp.*) for *kaybwag*—are chosen because of their interlocked grains. Although *teili* (*Terminalia catappa*) are sometimes used if *kausilay* aren't available,[13] other trees cannot handle the multiple stresses to which the forms are subjected. They are pushed out in their middle, pulled in at their ends and they twist slightly across their width. The shape is daunting.

Keel ends are grooved so that strake ends can be tucked into a slot designed to keep them from falling, while at the same time helping to pull them in to the prow-board slots. These grooves are called *atonen wag*. On the *anageg Bwadanakup*, the distances between the very tips of the keel and the two prow boards are 123 and 125 centimeters.[14] *Atonen* grooves begin just beyond them, running 45–50 centimeters in length. They are stout enough so that they can be holed for tying the strakes.

Strakes are stacked by means of the first parts attached to the keel, the combination *dabuiy* and *kunubwara*, both intricately carved. *Dabuiy*, which go in first, are pounded tightly in place by a tongue (called *ekik*) and groove joint running roughly a meter in from each end of the keel: the keel is grooved, the *dabuiy* tongued. *Kunubwara* are carved prow and stern boards closing each end of the boat. Both an *anageg* and a *kaybwag kunubwara* have vertical grooves carved into their backs into which the strakes fit. Strake ends are referred to as *kudun*, teeth, because they are said to "bite" into *kunubwara* grooves. The pressure from this biting and a tie running from one strake top to the other close to the prow/stern boards pull strakes in at their ends while *kakam*-made *gulumom* (ribs) push them out in the middle. While the prow boards help pull the strakes together, they are held in place by the opposing pressures of the tongue/groove *dabuiy* structure and the strakes that have to be arched in order to fit into the *kunubwara* slots.

Regardless of how much of a piece may be finished before assembly, each new strake requires trimming to get it into the keel structure, or over the strake beneath it, and into groves cut into prow boards helping to pull the ends together. To fit strakes to an *anageg* might require three to four days and a gang of people to press the parts into place. Posts pounded into the ground along the keel help to lever the planks into position.

These keel/hull forms control how the boat moves in the water.

The three *anageg* strakes are named from bottom to top: *budewag'*, *budwanay*, and *silasil*. One source told me the middle one, *budwanay*, is called *in*, which translates to "fish," presumably because fish images should be carved and painted on the side opposite the outrigger float. Part of the stereotyped difference between Gawan and Kweywatan *anageg* is that the strakes of the latter are thinner than the former, thus realizing a different elegance and speed.

Strake positions on the keel are set in relation to holes—called *bab*—drilled into the keel to facilitate tying. That drilling is first organized to set the ribs. There may be five holes between each of those locations for tying the bottom strake to the keel. Tying is called *kowag*, from *koliu*, the presumed tying material described in chapter 5, and *wag*, for boat or keel. The tying goes from bottom to top; from the *watan/katan* side of the boat to the *walam,* outrigger side; from the middle to each of the ends; and from inside to outside. A *sipkwadoy* knot around the keel or rib hole starts each tie so that they pull the strake into the boat. For reasons my sources could not clarify, the ties take on different names. For the bottom one it is *kaltawag,* presumably because the tying is to the *wag,* the keel. The middle one is *kobal*—there are two sets of these, one for the middle strake's attachment to the bottom strake, the other to the top strake. *Gadugwad* is the name for the last set of ties, both to the *gulumom,* the ribs, and the pole that may ridge the top of the boat. Although I could not get an explanation for these names, I could deduce some of their meaning: *kobal* is the name of a fish famous for hiding and staying fast in reefs. This is another instance in which place-marking—a fish hiding in reef holes—inscribes a conceptual order. Often in reference to this named tie, an elder prototypically tells a youth to check a boat in preparation for a serious voyage. Serious voyages usually start or end with a pig tied on the outrigger float platform; the logic is that the pig puts extra pressure on the *kiyad,* which then pressures the strakes. This is metonymic: extra force flowing to a particular boat part commands a thorough examination of the boat's settings. The keel/strake assemblage is a boat's first serious loading. Reference to it stands for the whole.

Once tied, the strakes are caulked, referred to as *kaybas* (*kabilil* in Koyagaugau). For *anageg,* but not the smaller class of craft, the plant used for caulking material is what Muyuw call *akobwow* (Euphormiaceae *Macaranga sp.*), the outer bark of which is extracted from scraped roots and rendered to make the caulk; people to the south use a related tree but take the material from a young tree's bark. Early regrowth after a garden area goes into fallow makes these trees readily available, one of many ways fallows produce materials for maritime requirements.

Fastening strakes to a hull is likened to putting a fence up around a garden, the garden sides called *budakay.* I was initially confused by language employed in public meetings laying out a day's activities: people might talk about tying the strakes to a keel, for example, and then they would then go off to put up a fence. As in all metaphoric usages, a difference exits between these analogous processes. Although fencing a garden is a serious chore, it does not require the finesse that goes into shaping a hull, action that is not unlike that required to arrange a marriage—as Dibolel's marriage to Mwadau testifies, the end result of fifteen to twenty years of his father's scheming.

Kaynikw, Down/Up/Down: The "Heart, or *"Lung," and "Navel"*

Kaynikw is the ambiguous name for a unique spring composed by eight precisely crafted pieces of wood that supports the mast mount, the *kunusop.* In the *anageg*'s most contained position, it is considered the most important structure—"the boat's motor," younger men often told me in English. People say it is the last part made, though that cannot be literally true. Yet the idea fits because after everything else is complete, getting this form correct is the last thing a boat builder does before releasing the boat, and, along with the mast, it is the first part new owners change. If its two central pieces crack or wear out, then the boat's performance sags immediately. Duweyala said that if it cracks while sailing, a noise can be heard moving up the mast; this structure is probably the boat's most complex part.

Selau, the boat style emerging from the south, usually do not employ this form. One Ole/Koyagaugau source said that a reason for this is that many new boats are nailed rather tied together. Most *anageg* fit with a *selau*-type sail do not employ the part. Nevertheless Dibolel used the structure with his *anageg-selau*-sail configuration, insisting that his boat sails better than those without them.

Everyone knows this part by the name *kaynikw.* The same term refers to all internal sections created by a boat's ribs; internal divisions in yam houses go by the same term. "Hold" is a good translation. Some people, especially on Yemga and Koyagaugau, say that "*kaynikw*" is the "big name," something like *kemulua* or *kemululwa,* the "little name," specifying two parts described shortly. This word is close to the southern term for the boats at large, *kemurua,* meaning "wooden thing from Muyuw." That a species name for this part might be a part or whole metonymy is in keeping with the structure's importance.

The name of this part is not its only ambiguous aspect. I have been told by the same people that it is the boat's "heart" or "lung." When I tried to disambiguate the body-part reference, people explained that it was the part that goes faster when you run. From this analogy I also first understood it to vibrate when sailing. Some people denied that. Some say it gives a little when a boat falls into a wave. When I asked another man about its motion, he said it did not vibrate, but noted that when correctly set it makes a high-pitched sound if tapped. It is under great pressure.

Eventually I asked if it is the boat's navel, *pwason,* and so like the navel in a garden, the garden's conceptual center, orienting form, and location for magic if magic is made. Most people said yes. It is not, however, a location for magic. And not everyone follows this idea. In 2009 I anxiously approached a brilliant Nasikwabw man whom I knew to be a boat expert. He could talk at length about the form, but when I asked him if it was the boat's navel, invoking the garden model, he said he didn't know anything about that. The form's details matter, not their correspondence to other domains of knowledge.[15] Nevertheless, just as a garden navel is the centerpiece in a well-formed model, so the boat's *kaynikw*/navel is a concentrated sense of form. The generic word for organizing all the lines and all the tying on a boat is *geneol*. While each form has a name unique to it, this word is most often used as the covering term. As a process it is marked by invoking the *kaynikw* (and *nedin*, the next part considered).

Altogether, the structure consists of three named parts realized by eight carefully shaped pieces of wood. There are two *eyalyal*. These must be made from the mangrove swamp tree *igsigis*—very hard and interlocked wood. There are four *talapal* (likened to a part in a house called *ipobal*), stereotypically produced from the Boagis firewood tree *kaboum*, and two uniquely shaped *kaynikw*, also from *kaboum* and the two pieces whose name goes for the ensemble. The last two are considered male, because they go on top, while the first six are considered female, because they go down first. Figure 6.1 is a side-angle view showing one *eyalyal* and one *kaynikw*; photos contained on the website for this book should be consulted.

Eyalyal poles run inside the carved-out keel under the rib sets that tie strakes to the keel. They go up from near the "base" of the boat to the first rib pair after the boat's center. In a ten-*kiyad* boat, this means that *eyalyal* run under six set of ribs. Approaching five meters, this is roughly half the boat's length. *Eyalyal* are tightly tied to some but not all of the rib pairs fixed across the bottom ledge of the boat. In *Lavanay* they ran under but were not tied to the first two sets of ribs; they were tied to numbers three

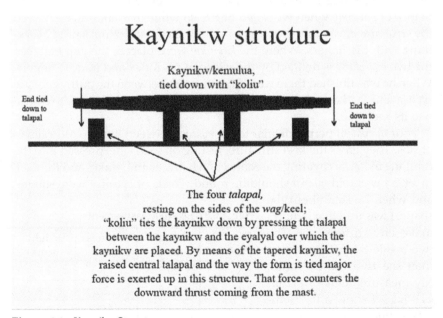

Kaynikw structure

Kaynikw/kemulua,
tied down with "koliu"

End tied
down to
talapal

End tied
down to
talapal

The four *talapal*,
resting on the sides of the *wag*/keel;
"koliu" ties the kaynikw down by pressing the talapal
between the kaynikw and the eyalyal over which the
kaynikw are placed. By means of the tapered kaynikw, the
raised central talapal and the way the form is tied major
force is exerted up in this structure. That force counters the
downward thrust coming from the mast.

Figure 6.1. *Kaynikw* Structure

through six. Their tying is important and should be done with *mamad*, the thick vine prototypically used to tie crossbeams to the posts pounded into the outrigger float.

Eyalyal are meticulously finished and carefully placed in their position. The one opposite the outrigger float should be slightly larger than the one closest to the float. The "hard"(*mamatuw*)/"flexible" (*mameu*) contrast used for describing desired qualities in many boat parts fits here; the former *eyalyal* should be somewhat larger than the outrigger side, which must be more "flexible." I took three circumferential measurements of *Lavanay*'s *eyalyal* and produced the data below:

Table 6.1. *Lavanay Eyalyal* Measurements

Lavanay eyalyal **circumference**	*dabwin* **(cm)**	*kaynik* **area (cm)**	*wowun* **(cm)**
katan side (Opposite float)	13.4	13.1	13
walam side (float)	13.2	13	13

The differences shown here may not be greater than the errors in my measurements. But I doubt this is an issue where it is the thought that counts. Sources say the *watan*/*katan eyalyal* should be larger because when the sail catches the wind, the weight of the mast comes down harder on that side.

Eyalyal sit asymmetrically inside the keel. Like their size, their place inside the keel space is carefully monitored. Duweyal didn't like the performance of *Lavanay* when we sailed north, so while in Wabunun he moved the structure closer to the outrigger side of the boat. He measured everything with his fingers. Where the *kaynikw* were placed, the gap between the two *eyalyal* was defined by the width of his index and middle fingers. When he was finished there were 5 centimeters between the outrigger-side *eyalyal* and the keel edge and 8.5 centimeters between the opposite *eyalyal* and its keel side.

Four *talapal* sit perpendicular to the *eyalyal* between the two central sets of ribs. They cross the keel structure, resting on each side, pressed into the caulking material covering the connection between the strakes and the keel. In 1996 I was told *talapal* should be round. Those of *Lavanay* were square, and when I asked about the shape, I was told that it didn't make any difference.[16] The two inside talapal are higher than the two outside ones. My measurements of *Levanay*'s *talapal* show a 0.7-centimeter difference between the inside and outside ones.

Table 6.2. *Talapal* Measurements

talapal	length (cm)	thickness (cm)	height (cm)
A	29.5	1.6	2.5
B	30.5	2	3.2
C	31	1.7	3.2
D	29.5	1.3	2.5

Talapal are not tied down; instead, they are pressed between the *eyalyal* and the *kaynikw* and forced down by the way the two pieces are tied together, by the *kaynikw*'s tapered shape, and by the difference in their height.

The left, outrigger-side *kaynikw* is tied first after it is placed directly over the outrigger-side *eyalyal*. The right *kaynikw* is off center with respect to the opposite *eyalyal*, veering slightly to the outrigger side. Duweyala separated the two inner *talapal* by his two fingers. Centers are tied first, then their respective ends are *sinev*, or *simok*, "tied down" (with "*koliu*") second. The tying should go back ("base") left, base right, top left, top right. I was told that if the *kaynikw* were not tapered, they would break. *Lavanay*'s *kaynikw* dimensions are found in table 6.3.

Table 6.3. *Kaynik* Measurements

dimension of	length (cm)	middle circumference (cm)	end circumference (cm)	end circumference (cm)
walam kaynik	79.5	8.4	4	3.5
watan kaynik	80.4	8.4	3.8	3

These are similar to all the others I've measured.[17] I do not know if a statistical difference would appear when contrasting Kweywata-made *anageg* with the Gawa type, but I was told that because everything else in the Gawa type is larger and heavier, the *kaynikw* structure should be smaller and more delicate than the Kweywata version. Since this part is adjusted to all of the others on the boat and the proportions must change whenever parts are replaced, I suspect too many variables come into play for the Gawa/Kweywata difference to be anything more than an ideological directive: closely monitor this form. Similarly, people told me that if a set of *kaynikw* works particularly well, they are measured, and replacements are modeled after them. However, since wood textures vary, these measurements are only ideals. What counts is performance in relation to other parts, not previous specifications. This form concentrates coordination with other forms; it is not about previous blueprints on paper or in the mind.

The *kaynikw* is a finely crafted ensemble. With the use of *kaboum*, Boagis's ritual firewood, or its substitute *aukuwakw*, it is composed of the densest and most interlocked trees possible. This is wood that can bend—barely—without breaking. Although now initially shaped by axes and knives, people still employ pumice (*gimut*) as sandpaper to make final, minute adjustments.

The tapering toward each end and differential height of the *talapal* over which they are tied intensifies the upward thrust that the part creates, generating a critical pivot for the boat's structure. Contained and centered by the keel structure designed to cut through the water, the *kaynikw* reverses the pressure that the wind exerts on the mast by means of the sail. This is an extraordinary contained contrary.

* * *

When I returned from Muyuw with my dissertation research complete, I reported to my PhD adviser, Peter Huber, that the Muyuw garden order was "finicky." Eventually he read what I wrote and gave a rare comple- ment for that choice of adjective. I hoped this second research period would find a good reason for that finicky order. Other than inculcating the discipline most cultures require for their existence, I'm not sure there is actually a reason: it doesn't matter which trees are used for the different parts in the garden. Yet a component of a garden's order is the imposed boat design. Not just an ideological container, the *kaynikw*'s structure is an ingenious invention fixing attention to minute, measured differences.

The Nedin: Center Fast, Ends Float

The *nedin* (*nadin* in Nasikwabw) is an approximately eight-meter pole that arcs around the rim of the outrigger side of an *anageg*'s hull, above and outside the edge in its center, tucked inside toward the prow and stern. It is bowed like the strakes and the *kaynikw*; however, it differs from them in that only its center is tied tight. Its ends are pulled or pushed into their position then left loose. It visibly sways when a boat is underway. People say it is the boat's rib (*vilag*), for as ribs must bend for someone to breathe, so the *nedin* must bend for a boat to move. Continuous with the *kaynikw*'s design for modulating the effects of the wind on the mast, the *nedin* is part of a structure created for dynamics derived from the outrigger float in the water. In this it is paired with *kiyad*. By itself just a long pole, the *nedin* represents a moment in a complex structure. Learning about this part and its tree late in 1996 convinced me that I had to take boats and their trees seriously.

Nedin are only made from the tree called *akidus*, identified as a member of the Rubiaceae family. These are *ulakay*, old-fallow trees that may be found in old forests across the island's varying landforms—drier lime- stone soils, north-central Muyuw's wetter zones, the Sulog region; now and again *tasim*. Duweyala needed a new one for *Lavanay* after our trip to Muyuw and found one in Sulog while hunting for a new mast. But he quickly decided not to take it because he was told he could easily find one near Wabunun. He did, and his brother planed it into shape; together they tied it onto their boat just after they finished with the *kaynikw*. *Akidus*, and only *akidus,* are used for this part because they become tall without becoming large in diameter, allowing one to easily trim it to the desired shape. Most important, the wood maintains its tensile strength when bowed.

Discarded *nedins* are sometimes tied to the top edges of the *budakay,* forming a structurally insignificant and unnecessary part of the boat, on the side opposite the outrigger float at least.

Each boat has one *nedin.* Its center is tied on the *duwadul,* the *kakam*-constructed piece whose lower end, the *kunusop,* holds the mast. From the *kunusop,* the *duwadul* runs up the outrigger side of the boat then arcs over the outrigger float platform. The *nedin / duwadul* tie is just beyond the top strake, over the outrigger platform. It is then bent so that its two ends tuck inside the hull near the respective front and back *kunubwara.* I was told that once the part is bent, it will maintain its arc without being tied down. However, in my pictures, admittedly taken while the boat was ashore, some line or stake maintains its bow. When the boat is sailing the *nedin*'s center is stationary, riding up and down from wave dynamics, while the loose ends rock back and forth. What I recorded during my 2002 voyage had more visible rocking a bit further up the line from the ends.

I must step back now and locate this piece in its larger, laminated structure. Please refer to the website pictures accompanying this section.

The *anageg* keel/hull and outrigger float are two forms continuously running through the water. I was always told that they should be the same length, but on every boat I examined the float seemed slightly shorter than the keel structure. Measurements of these dimensions are not simple. I attempted a straight-line measurement from *Lavanay*'s two ends and recorded 10.7 meters; measuring its length by taking into account its arc

Table 6.4. *Lavanay Nedin* Measurements

Lavanay Nedin **Replacement, July 2002**		
	old (cm)	new (cm)
Length	811	795
circumference at end	7.2	8
circumference 20cm from end	8.1	10
circumference 50cm from end	9	10.4
circumference at middle	13.6	14.4
circumference 50cm from end	9.7	11.5
circumference 20cm from end	8.6	11
circumference at end	7.5	8.5
These measurements were made on July 27, 2002. The *nedin* was tied into place on August 2.		

resulted in 11.85 meters. Across the top of the float, which is relatively flat, the number is 11.32 meters. Flat across their tops, floats narrow toward the ends and their bottoms rise. *Lavanay*'s circumferential measurements were: center, 1.27 meters; "base," 0.67 meters; "top," 0.53 meters. The two ends are separated from the keel ends by the same amount. The keel and float ends slightly arc toward each other. I asked Duweyala if this bending was to make the boat turn in a certain direction. He said no; the form's design was to shift some of the weight from the hull/keel to the float. Although this transfer is partly conferred by the rib/*seisuiy*/*kiyad*/*watot* structure outlined earlier, more of it is charged to the dynamics for which the *nedin* is the principal spring. The hull and float are meshed by the structure of the platform between the hull and the float and by one beam inside the hull.

The first connection between the hull and the outrigger float is the *kiyad* form attaching *seisuiy* (a *gulumom* [rib]/strake connection) to *watot*, stations pounded into the outrigger float. The second connection is completed when the *nedin* is tied to the *duwadul*. When the *duwadul* is set, four other pieces (of *kaboum*, like the *watot*) are tied to the *duwadul* at one end while their other ends are pounded into the float. Called *tanuwag*, these pieces are likened to hamstring muscles on a leg (which is what the outrigger float is with respect to the hull). The set is conceived to keep the float from wobbling, and the parts are considered to be more important than the forty pieces of *kaboum* connecting *kiyad* and float. One informant explained "*tanuwag*" by saying the pieces were the *tonewagan* of the boat. Using the idea of a boat—*wag*—to carry its significance, "*tonewagan*" denotes the person or thing in charge of, for example, a process, village, mortuary ritual, garden, or focused kula event. And *tanuwag* are very important. Two go straight down from the top arc of the *duwadul* into the float. Two angle into the float tied from about halfway along the *duwadul*'s length.

In addition to the *tanuwag*, the *duwadul* structure fixes the hull/float connection in two ways. The first is a beam called *auleybwad*. Set after the *kunusop* is tied into place, the *auleybwad* runs across two or more sets of *gulumom* and their associated *kiyad*/crossbeams so that it presses the *duwadul* down firmly toward if not on the inside of the boat.

After the *auleybwad* is set, the second connection begins. Just after the *duwadul* arcs over the edge of the upper *budakay* toward the outrigger float, it is tied through the outrigger float platform to the central piece of a set of beams placed between each crossbeam. These are called *pamanag*. A ten-crossbeam boat has nine of these, each between a *kiyad*. They are tied to and run between two forms covering the length of the outrigger float platform, one of which is the *albeikun*, effectively a board made from *kausilay* and set half a meter or so out from the strakes; the other is called *pilpilitet,* which might be just a pole or a slightly smaller rendition of the *albeikun.* It rests just beyond the *kiyad / watot* limit defining the outer edge of the platform; both of these forms usually have bird heads carved on their ends. Although the sticks—prototypically from the *digadag* fallow tree called *asibwad (Timonius timon)*—used to make the platform flooring are set on the crossbeams, they are only tied to the *pamanag.* But both the *pamanag* and the crossbeams are tied to the *albeikun* and *pilpilitet,* eventually creating a densely interwoven structure. The knot connecting the *duwadul* to the central *pamanag* is called *olau.* It stabilizes the *duwadul* by fixing it to the platform structure. The pictures accompanying this text show the *olau's tageg'eyon,* the wrapping that closes many ties.

The *nedin* attachment to the *duwadul* ends up as a boat's most convoluted knot structure because it entails repeated revolutions between the *nedin* and the *tanuwag* a meter or so up the *duwadul.* It closes with *tageg 'eyon* circular wrapping, which makes the fastening's structure invisible. The lashing is called *analuk* (sometimes *ananuk*). This name is also used for securing a Y-shaped piece called *sowaso* to the mast, which helps position the mast's tilt toward the outrigger float. These two forms contrast: the *nedin* version is very tight, the *soweso analuk* loose. Interestingly, "*analuk*" also names a weather peculiarity: clouds that move in a direction different than the winds experienced at ground level.

After an initial *sipkwadoy* knot around the *nedin,* the whole *nedin-tanuwag analuk* structure begins by putting down "food" for the boat as a kind of magic. First an unnamed weed is placed on top of the *nedin,* then part of a *yagal* leaf (a *Pandanus* tree frequently found in exhausted soils or early fallow garden areas), then a *kakam* leaf (the shoreline *Calophyllum*), and, finally, a wedge-shaped piece of a coconut husk (*cosinuy*). For this last piece, people now substitute a stripe of thick rubber. Then the revolutions up and back to the *tanuwag* set begin, the whole completed by a final wrapping.

Fastening the *nedin* finalizes the connection between the outrigger float and the hull/keel structure. If one end of the *duwadul* form is the way the *tanuwag* are pounded into the float, the other end is the *kunusop.* Thus the *duwadul* structure concentrates some of the forces of the moving float into the *kunusop* form and the *kaynikw* structure beneath it, the mast mount, thus the boat's connection to the wind. When I asked what would happen if there was no *nedin* rocking back and forth with the movement of the boat, I was told the *kaynikw* would break, seriously impairing the integrity of the craft.

What I thought was the opposite end of this spectrum of significance concerns the slender saplings used to provide the flooring for the outrigger platform. These seem to be trivial parts of the boat. However, because they press together connections between the hull and float they make a series of important linkages. One deals with the way *kiyad*/crossbeam link *seisuiy* and *watot*. Another is between the *pamanag* and the *duwadul*, the latter of which is connected to the outrigger float by the *tanuwag*. The outrigger platform flooring bonds these two structures. Finally, the flooring creates an external tie between the boat structure and the environment from which it comes. As I shall show in this chapter's second part, the platform flooring, by coming from a *digadag*, early fallow forest, fills out a set of associations between the idea of a boat and the anthropogenic landscapes it unites.

The Vayiel (Mast): Fixed Base, Swaying Top; and the Kavavis, Which Does Not Bend

The hull is constructed by pushing out the middle of each strake while pulling in their ends. *Kaynik* ends are tied down, their centers are pushed up. The *nedin*'s center is stationary but its ends flutter. The mast is the next variant in this set. Although it pivots inside the *kunusop* that holds it, the form transitions from completely rigid at its base to a two-fold set of motions toward its top. Most bending is achieved by how the mast is fastened to the outrigger platform. That platform, finalized with the tying of the *duwadul* and *nedin*, becomes the means by which lines bend the mast and position the sail in the wind. But once this relatively fixed dynamic is set, a second takes over as the mast must sway toward its very top, visible action that facilitates wind spilling out of the sail. But that is not all. When combined with the sail it supports, the mast works in tandem with but is counterposed to the *kavavis*, the rudder-like structure set to the trailing outside corner of the outrigger platform. The leeboard/rudder's quality differs from this entire set. Because of the wood used to make it—*kaymatuw, meikw*—it cannot bend yet is prone to snap. Incapable of moving itself, the *kavavis* is withdrawn by the helmsman with every new wave and is inserted back into the water when the wave passes. These gyrations provide minute alterations to the craft's movement in the wind and water. The *kavavis* intensifies the whole structure's extreme sensitivity to force and puts that responsibility in the hands of a single person.

Chapter 4 described the trees used to make a mast, the two *Calophyllum ayniyan* and *aynikoy*. They are concentrated in the high, old forests in the Sulog region; their straight grains enable the desired bending. The two types contrast as lighter/heavier (denser) against which mast bending is calculated—if people do not like the dynamics of their first choice, they go for the other. Depending on the wind direction relative to the boat's course, a mast leans more or less toward the leading end of the boat and arcs more or less toward the outrigger float. *Balau* lines manage these adjustments. They are wrapped around the top of the mast on the one hand and around two crossbeams toward the ends of the platform, usually the third and eighth, on the other.

The base end of the mast narrows to a blunt, rounded point that fits into and easily pivots in the *kunusop*. Sometimes leaves that function as a kind of lubrication are put in the *kunusop*.

Carefully and tightly tied to the mast a meter or more above its bottom is the Y-shaped piece of wood called *soweso*. It derives from the mangrove swamp tree, *igsigis*. The base of the Y shape is tied to the "top" (*dabwen*) side of the mast, its other two ends extend over the edge of the outrigger side of the boat bisected by the *duwadul* and usually resting under the *nedin*. A specific knot form called *bwabwalita* fixes the *soweso* to the mast. The *soweso* helps adjust the degree to which the mast veers toward the outrigger float. It pivots in the *kunusop* to lean into position. Two prismatic-shaped pieces of wood called *enam*, the larger the "mother," the smaller the "child," may be inserted in the gaps created by the *soweso*'s Y shape. They rest on the *nedin*. These pieces are made from the early fallow tree that Muyuw call *nilg* (*Mimosa/Acacia*) or the related *vayoun*, the tree used to make the outrigger float. Although bulky, they are extremely light and easily raised, lowered, or withdrawn to adjust the *soweso*/mast angles. If a mast must be angled as far as possible toward the outrigger float, *enam* may be removed.

The very top of the mast is carved in one or more simple decorative forms. But below that are two critical interventions. One is a hole cut completely through the mast into which the short piece of *kaboum* called *amawuw* is inserted—the one I measured was thirty centimeters long. *Balau* lines go around this form, employing the knot form whose name is derived from the line—*sipbalau*. Around the *amawuw* the knot is tight; where it attaches below to a crossbeam on the outrigger platform it is loose, facilitating continuous adjustments for mast angling. Another looping line goes around the *amawuw*, this one charged with pulling up the form right beneath it, the *kuk*. This line is called the *powan balau*, "testicle of the *balau*."

Just below the *amawuw* hole is a notch for a mortise joint for the rooster-shaped *kuk*. The *kuk* bisects the prow/stern direction of the *amawuw*, always pointing away from the outrigger float. The *kuk* is holed so that the *yawasay*, the halyard, can be run through it in order to raise or lower the sail. This pulley-like form may be unique in Austronesian sailing technology (see Horridge 2008). Although the sail is raised from the side opposite the float, it is positioned so that the "back" is against the mast, its "inside" facing the wind's direction, the outrigger float side. As the *balau* lines are used for adjusting the angles of the mast, the *yawasay* is used for adjusting the height of the sail on the mast, its height inversely related to the speed of the wind. As noted in chapter 5, other lines adjust the sail's position.

The wind on the sail and many angles cut into the keel and outrigger float make the boat veer *yiwatin*, toward the outrigger side. The line off the sail configured to effect this the most is called *alita'*. It is fastened toward the trailing end of the outrigger platform, close to where the helmsman will be located. When tightened, *kabtil*, the craft goes more toward the float side; if loosened, it goes less in that direction. This brings us to the form counterposed to the mast, the *kavavis*.

Kavavis are always made from *kaymatuw/meikw* (*Intsia bijuga*) (see Ross et al. 2008: 200). It is a hard wood that may be finely carved into a precise shape (Trobriand carvings employ this tree). Unlike most wood, sun exposure doesn't split it; this is important because *kavavis* lie unprotected on the platform. The wood's weight is finely conceived. *Kaboum* and *gav* (ebony) may be carved as finely as *meikw* but are so heavy they might fall through a helmsman's hands should he doze off. When dried out, *kakam* and *kausilay*, both easily carved, are so light that they are hard to hold down. *Meikw* lies between these poles. Its prime difficulty is that it is brittle, so it is likely to snap beneath the power of a wave as it strikes. This condition emphasizes important characteristics of these craft—they are carefully meshed with, not opposed to, their environment.

Inserting the *kavavis* in the water moves the boat opposite to the force of the wind on the sail, *buluk*. Recall that this is the generic effect if its parts are too *matuw*, hard. The boat's intended course is maintained by modulating the constructed opposites of the wind on the sail and the water on the rudder.

Anageg invariably sail with three *kavavis*, two that are well-formed and referred to as the *ina-*, mothers, and a third less likely to be as precisely finished and referenced as child, *natu-*.

Table 6.5. Sail (*Alit'*)/Rudder (*Kavavis*) Dynamics

		Yiwatin towards the float	*Buluk* away from the float
	kabtil (tighten)	+	−
alit'			
	kasilikw (loosen)	−	+
	kel (lower)	−	+
kavavis			
	gut (raise)	+	−

Kavavis are defined by a twofold set of oppositions. One contrasts their top handle to the bottom, the "base"—*wowun*—to the "tongue"—*mayien*. The second contrasts the "back"—*tapwan*—to the "inside"—*nuwan*. The *kavavis* top may be carved to the shape of a bird, invariably a *dawet*, frigatebird, which will then be called *munudog*. The eyes of the bird always point toward the "top" (*dabwin*) end of the boat even if the *wowun* leads; this fixed relationship follows the *soweso*'s with respect to the mast, always *dabwin*. Although an *anageg* may sail with either its prow or stern leading, the *soweso* and the *kavavis* eyes mark an absolute difference between the two directions.

I measured *Lavanay*'s *tapwan/nuwan* distinction to see if its width measurements would capture its structure, like an airplane wing—they did (see table 6.6). The "back" is like the top of a wing, more rounded; the "inside" is like the bottom, flatter. The *kavavis* "catches" (*kon*) moving wa-

Table 6.6. *Lavanay Kavavis* Dimensions

	Lavanay kavavis	mother1	mother2	child
	length	264	266	242.5
	thickness at handle	2.5	2.6	2.6
	distance from lowest point in the water	Width		
210cm	*nuwan* (flat)	8.4	9.4	8.5
210cm	*tapwan* (rounded)	7.7	9.4	8
200cm	*nuwan* (flat)	11.7	10	9.5
200cm	*tapwan* (rounded)	13	9.8	9
180cm	*nuwan* (flat)	15.5	12.5	13.6
180cm	*tapwan* (rounded)	17	13	14
150 cm	*nuwan* (flat)	18.7	15.9	19.8
150 cm	*tapwan* (rounded)	20	16.5	18
120cm	*nuwan* (flat)	22.9	19	18.3
120cm	*tapwan* (rounded)	23.5	19.2	19
90cm	*nuwan* (flat)	24.7	22	18.9
90cm	*tapwan* (rounded)	25	22	19.5
60cm	*nuwan* (flat)	24.4	23.5	19.1
60cm	*tapwan* (rounded)	24.7	23.4	19.5
30cm	*nuwan* (flat)	21.5	23.2	18.2
30cm	*tapwan* (rounded)	21.7	23.5	19
	bottom	rounded to point		

ter, creating a lift-like effect pulling the trailing end of the boat to the float, *yiwatin,* necessarily moving the front, *buluk,* opposite the float.

The helmsman's location is built into each corner of the outrigger plat-form—a marked place for him to sit and anchor his feet—as is a gap in the platform poles for inserting the *kavavis* into the water. Some boats have grooves in the *watot,* the stanchion connecting *kiyad* to the float, into which the *kavavis* fits as it is lowered and lifted. When in place, the *kavavis* is on a slight angle, the top handle pulled back toward the helmsman. To lift the *kavavis,* that handle is pushed forward. That push releases the pressure thrusting the *kavavis* against the float, making it easier to lift. Since the force of a wave may snap the *kavavis,* it is lifted as a wave approaches. The boat reacts immediately, so helmsmen attempt to guide a boat's angles go-ing up and coming down a wave. *Kel* is to lower a *kavavis, gut* is to raise it. A crew may coordinate by chanting these words these words not so much to give direction as to participate in coordinated action.

These boats rise and fall and twist and turn with every dimension of the wind and water. When people see big freighters that bisect the Kula Ring pass by with only their forward direction visible they often remark that they look like an island in the waves, not moving at all. By contrast and by means of their internal structures *anageg* bend to all the forces they meet. But it is not just that they respond to their internal and external contours—a tensed relation in motion is also part of their structure. For without intended motion the forces of the wind and the waters would soon turn a boat's complementary dynamics into a chaotic, disconnected mess. When at sea, a sail may have to be taken down because the wind is too strong; in that occasion a modest wind break—usually one or more of the woven coconut fronds used to protect the sail when lowered and stored along the outrigger platform—is put up to provide movement. These boats are about motion, and they lack coherence without it. More than anything else, the relationship between the mast and the rudder/lee-board positions the boat on the edge of chaos, a model of resilience across experienced contradictions necessary for their functioning. This is a major reason why these craft constitute such an effective model.[18]

Imagery and the External Structure of the Boat

Yam stakes (*avatam*) are likened to an *anageg*'s mast, the mass of leaves the sail, the vines—"Geliu's *vatul,* the Creator's rope"—the lines attaching the "sail" to the hull-float structure. People claim that a vicious wind turns up every year sometime between February and April (March 3 in 1996) to test how well these forms have been set—if poorly, the garden is flattened just as it should be turning into an enveloping expanse. Because they don't grow to the height of yams, taro are considered the boat's crew; stars in the sky are the crew of the boat there. Along with the idea of a garden moving amid an area's *tasim* like a boat navigating among islands and reefs, these are among the associations people regularly make between the *anageg* form and the garden. *Anageg* intensely model how life should be organized. Following from the constructed complexity of the form itself, they are one of the principal ways by which this social system achieves organization. As chapter 1's description opened up discontinuities in the beliefs and practices across this inquiry's region, this final section shows how those differences are pulled together in the *anageg* design. The consciousness of these interrelated forms suggests they emerge by created, collected knowledge. Outlining these connections brings this study to a close.

The Body, the House, and the Island

Reciprocal images run between the *anageg* form, human bodies, houses, and the island at large. Most of these forms are understood in more than one way, and in some respects are thus inconsistent. Some applications are asymmetrical, and consciously so: a garden's forms are attended to with a care that is similar to that of a boat, but everyone knows that while gardens and boats both have "navels," the trees used for a garden's make little difference while for a boat they are fundamental. Gardens are likened to boats, but boats are not really talked about as if they are gardens. Some of the analogies drawn between houses and boats concern the outrigger platform or *kaynikw* structure and a house's flooring. However, as practically everyone born toward the middle of the twentieth century knew, flooring is a recent phenomenon in Muyuw houses, a product of the colonial era beginning roughly in the 1920s. Body imagery informs aspects of a boat; if the analogies go the other way, I have not heard them (see Austen 1934: 104).

I begin with an association that, by means of tree imagery, ties together houses, bodies, and the island at large. More than one person told me that the "base" of a boat, its stern, is like the legs and feet of a person (*kakein*). This is analogous to the eastern end of the island. Wamwan, Muyuw's central part, is then likened to a person's and a boat's (and a garden's) navel. Some Kaulay friends suggest that the very term Wamwan is derived from *wanuwen*, which means "in the middle." Everyone should pay special attention to their navel, many report, because if it becomes diseased, death follows quickly. This makes the body's navel analogous to the navels in gardens and boats because of the attention paid to them. Central Muyuw (Wamwan), by this reckoning the island's navel, was considered by many to be the highest and most important part of the island. The western end of the island then is the *"dabwen,"* top of the island, like the prow of a boat and head of a person.

Another person said Wamwan was like the sail of an *anageg*, which is in the middle of the craft. The northern middle part of the island furnishes important parts for sails, while the southern middle part, the Sulog region, is the primary source for the *Pandanus* trees used in sail-making. A Wamwan person who conveyed this model also said that Budibud, the set of islands some sixty kilometers to Muyuw's southeast, is the island's *kavavis* ("rudder"); a man from Budibud said the same thing. This is not, however, a rich association, for often the *kavavis* idea is used to talk about things or people that are directing: a good woman is a house's *kavavis*; an elder retired from the Kula yet still directing its activities is called the *kavavis* of the relationships.[19] Yet nobody thinks that Budibud directs the rest of the island.

Some people draw analogies others dispute. It has been suggested that small island villages off the southern coast—Waviay, Nasikwabw, Boagis, and Yemga—are like a boat's outrigger float. But my closest associates deny that.

Many hold that the outrigger float is like the boat's leg, its balancing mechanism. In this regard houses don't have floats, but they have several "keels" (*wag*), the posts called *kakol* that are sunk into the ground with Vs carved into their tops so that they can hold various beams (usually *sol*) from which flooring and roof structures are fixed.

I asked if boats, like bodies, had blood, and I received in response the term *kausilay buyavis*, "the blood of *kausilay*," the reddish liquid that collects in the bottom of an *anageg* as its *kausilay* parts exude their reddish color (noted for different reasons in chapter 4). Houses, by contrast, are not considered to have blood. Chapter 5 covered the point that all the tying materials in a boat, from *"koliu"* to *"mammad"* to the *ukw*-formed ropes are likened to the vessels and veins in a body. These inside/internal phenomena contrast with the most external part of the boat—excepting its various paint-like coverings, which are very temporary—the caulking material, *kaybas*. Some people liken *kaybas* to skin, saying that as skin protects the body from water, so *kaybas* protect the boat. Some people also equate caulking material with *aymalas*, the thin muscles that bind ribcages. The equivalent in a house are its *lokwat*, the outer coverings for roofs, sides, and house fronts and backs formed from sago leaflets; however, some people have told me that *lokwat* is like a boat's sail (*mweg*), and others have said that a boat's *budakay*, its strakes, are like a body's skin.

In the model that makes a boat's outrigger float (*lam*) like the legs of a person—recall the stout beams connecting the *duwadul* to the outrigger float likened to hamstring muscles—the keel is like the *tapwan*, back, of a body or tree. For a tree *tapwan* refers to everything between its base and its branching top, conventionally the tree's bole or trunk. Some people would confuse this with the backbone (*tubwatub*) but then remember that that analogy is usually reserved for *seisuiy*. *Kunubwara*, the name for prow and stern boards, literally means "head/forehead patch" and is likened to the top part of a skull as well as to the slightly angled part of the front and back of a house that runs above the doorway to the roof. Although a *kunubwara* approximates a V shape, the form is inverted when applied to a house. *Kunubwara* are held into position in part by the piece called *dabuiy*, and these are likened to a nose on a face and the tuft of leaves—from sago or meadow grass—that tops the ends of house roofs, especially when they are newly constructed for a mortuary ritual. The ridgeline of a house arcs higher on each (ideally) north and south end than it does in the middle, resembling the arcs always found among the two highest classes of Muyuw outriggers, *kaybwag* and *anageg*.

As noted previously, an *anageg*'s *nedin* is likened to a body's ribs, and people will equate this piece with the *poiy* in a house, as the pieces are constructed from the same tree and held in an arced form crossing the corners in a house's roof. However, it is used to hold down the roof against destabilizing winds—"the four winds"—rather than to absorb wave motion that might destabilize a boat. *Watot*, the stanchions connecting *kiyad* to the outrigger, may be likened to *kakol*, the posts sunk into the ground designed to hold the main flooring structures, as well as the beams running the length of a house and leg bones. But others have told me that *kakol* are like keels (*wag*) because their tops are dug out so that beams may be laid across them. As everything in a boat builds up from its keel, so everything in a house is built up from its posts (*kakol*). Although there are a few substitutes, *kakol* are almost invariably made from *meikw*, the same wood from which *kavavis* ("rudders) are made.

Eyalyal, the understructure for the *kaynikw*, is likened to *livtakon*, a similar set of pieces in a house's flooring. However, nothing lies across *eyalyal* in a boat resembling flooring in a house. It is the same with *dukuduk*, the word for the top layer of a floor structure, the material people kneel or walk on, and *patapat*, the upper flooring on an outrigger platform. Boats and houses have these, human bodies do not.

Although most people say that the *nedin* is a boat's rib, some people have told me that *kiyad* are like a body's ribs (*vilag*). Both crossbeams (*kiyad*) and *vilag* provide form for what is a concavity. However, the relationship between *kiyad* and houses adds a major understanding to the idea of a house. The analogue of a *kiyad* in a house is called *anakan*. Functioning like rafters, *anakan* are often made from a light but stout swamp tree called *amgwalau* (Damon 279, Bignoniaceae). Their bottoms are notched so that they rest over and become tied to what are called *sol* (usually *aku-luiy*) which function like wall plates. *Sol* rest atop posts, *kakol*, the forms likened to a keel because of their V cut. The top of the *anakan* reaches to and overlaps a piece called a *mamwan*, the ridge or hip of the roof, also usually made from *akuluiy*. This beam parallels the *sol* and runs the length of a roofline; it is topped by another beam called *livkayway*. Eventually the ridge conjunction is capped with a layer of individual sago leaflets that cover the sago roofing that will be tied right up to the ridgeline. This is called a *kunumwan*, and the word translates to "head husband" if taken literally. An affine junior to the house's owner should make the *kunum-wan*. In theory, proximate generation male affines—sons-in-law, nieces-in-law, for example—build new houses. In actuality they often just organize roof-making, and even more specifically the final *kunumwan*. If proverbially the person doing this is someone in his twenties, this is not always the case. Sipum made the *kunumwan* for a man who was simultaneously an elder brother-in-law and his wife's mother's brother (*kada-*) until shortly before his death—he was hardly a youth. The task is dangerous because it means edging along the roofline, the task for a nimble junior male affine. The assignment, in any case, speaks to the sociology of a house, because if a house is first for a married couple and their children, duties affixed to its construction make it a node in a social network. Those relationships are part of its internal structure and are carefully attended. An affine may not enter a new house without first participating in a formal *takon*, an exchange entailing a gift of food (cooked in this case), a female item, for a return of something small, now often money or a knife or machete, male item. Originally defined by the early ritual firewood exchanges, these house-building *takon* are context markers used to define relationships (see Damon 1983c; 1990: chap. 4). These details pull the sociology of the house into formal models inscribed in the garden forms outlined in chapter 1.

The *anakan* structure, conceived to be like a *kiyad*, plays a similar role, a part whose understanding inscribes a whole, in this case by means of the understanding of an *anageg*.

When I first learned about the *kiyad/anakan* association, my source carefully explained that every house should have five sets of *anakan*. Each set is composed of two pieces, one going up each side of the roof to the top where they are conjoined. Each pair is likened to one crossbeam. My source then reminded me that an *anageg* has ten crossbeams. And this means, as he diagramed it for me, that every house should be coupled with one on the opposite row of houses. Muyuw villages are supposed to be constructed with two rows of houses, each row going from east to west. Startled and somewhat disbelieving this formulation, I ran it through two other people several days later who confirmed the construct. Presumably this whole would be realized in the old mortuary ritual format that assembled prestations in the shape of a boat so that everyone could see them (see Damon 1990: 117, fig. 5.11).

There is little structural affinity between an *anakan* by itself and the pivoting of forces that crossbeams organize in a boat. *Anakan* are, however, tied into the *poiy* construct that is likened to an *anageg's nedin*, so they are knowingly related to wind forces that affect both houses and boats. Yet the *anakan* idea is used to connect village parts, the rows of houses that, like the relationship between a float and the hull to which it is attached, over time move with considerable tussle in ways that have to be made to mesh by means of small talk, gossip, informal and formal meetings, and the lines of marriage and progeny that manage village life. Invoking the tensed joining of different elements, the analogy instructs.

Recall the fractal organization between rows of houses in villages and rows of villages in eastern and central Muyuw that closed chapter 2. When the *anakan/kiyad* construct was explained to me, I did not inquire how that form might ramify through the form organizing sets of villages in two idealized rows. However, when I asked about this relationship in 2014 my source told me about the taro leaf (*yamwik*)/taro stem (*tatan*) metaphor that governed the relationships between villages along southeastern Muyuw's ridge and those along the shoreline, and how youth would be sent between the two to gather and distribute their complementary resources—in this particular case inner bark from the *gudugud* tree from the "big leaf" area versus betel nut pepper from the "stem" location. Are traveling youth "crossbeams" among such complementarities?

Between the Cock and the Banded Sea Snake

By means of carving, painting, and burning, people adorn boat structures, especially but not exclusively *anageg*, with creatures of the air, land, and sea. By totalizing the experienced environment, these images give their craft an encompassing value. Effecting this description is my purpose, but I presume others will make use of this material to bring the discussion into orbits Campbell (2002), Munn (1986), and others have traced for this region's famous graphic arts.

Carving found on canoes is the region's most ornate. Designs on the lips of some wooden bowls, betel-chewing tools, some house gables, and on bodies are only part of what *anageg* (and middle-sized outriggers) regularly display. My description of this material is necessarily initial and incomplete. There are reasons for this. First, I was not curious about these forms. The ecological orientation to this research made me uninterested in what seemed like symbolism per se. And that focus seemed a sufficient contribution, however limiting it might be; this perspective changed as the symbolism appeared to have ecological content. Yet my original attitude was reinforced by a general disinterest in exegesis common to Muyuw culture. This stance changes as one moves west toward the Trobriands, something Muyuw themselves remark upon. I had realized this circumstance during my 1973–75 research, and I once discussed the matter with Victor Turner, whose dictates (e.g. 1967) about exegetical, operational, and positional/structural aspects of symbolism remain a useful heuristic. Muyuw are not given to exegesis. Once when asking them about the rooster image atop their masts, they told me that people from Gawa must know the answer to that question (and, therefore, I should ask them); only later did I realize that this response was a matter positional significance. For in contrast to the play with the symbolism found from Gawa to the west, Muyuw people consider themselves experts on these craft's "technical" dimensions—a fund of knowledge Muyuw claim people to the west lack. This difference is part of the cultural makeup of the region: more than one person told me Muyuw people know *geneol*, all of the detailed tying and hydro- and aerodynamic carving that goes into these boats. By contrast Gawa and Kweywata people know *lelel*, literally "writing," design work. One of my sources, a token in a type, lets others worry about rooster imagery; he worries about *kaynikw* and liked questions about its technical matters.

Although I am not convinced body imagery on these craft form a coherent message, an example from pursuing the ways boat parts are likened to bodies is telling. When I asked a number of Wabunun people if an *anageg* had a part that resembled a penis, they responded that of course the mast was like a penis, fitting into the *kunusop* that in this context is likened to a vagina. A principle source, however, denied this. But this was the same man who told me that although the tree names *ayniyan/aynikoy* might be different than the word for power, *aniyan*, the entity evident from a star's twinkles and released by a setting star and a corpse, there was a similarity in function inside their slightly different pronunciations: masts facilitate the delivery of the wind's power to the outrigger craft while the tree stands from which they come protect sago trees by absorbing wind gusts that rock the island's igneous core. The issue of *aniyan*, power, partakes in the significance of the trees, *ayniyan/aynikoy*, and the proximity of sounds is part of that ensemble. The same person also told me that sometimes a *kunusop*, the concavity into which an erect mast fits, should be layered with leaves and that water would squirt out of the conjunction if the boat sailed well. The leaves are from *gudugud* and *ayyoval* trees and a vine whose name he could not remember; moreover, they should be placed before sunrise.

He had also added various details concerning the *kuk*, the rooster-shaped mechanism acting like a pulley for raising a sail, and he was the first to tell me that the curled, knotted rope holding up the *kuk* by means of the short dowel running through the mast is called the *balau's* testicle. Details and subtle analysis, yes. Resonating significances? Not interested.

Other than the wood particulars related in chapter 4—*kuk* are made from the *Calophyllum* Muyuw call *kakam*—I learned nothing substantial about this form until I spoke with Gawa and Iwa people in Wabunun in 2006. They were explicit. The rooster shape hovering over a boat is there because roosters keep time, and the most important thing these boats carry are kula valuables, key generators of time in this social system. Not having explored details of the kula with Iwa people, I do not know exactly what they mean by "time." However, the idea is fully consistent with Munn's accounts of the kula in Gawa (Munn 1983, 1986). And it is exactly what Muyuw say about the institution. People exchange kula valuables to make their "names climb." As names rise, bodies age, their withering a sign of accumulated time and successful transformation. In representing time, *kuk* stand over the most important motive in this social system: kula acts and beliefs interconnecting other places and practices.

I could not get Muyuw people to tell me that *kuks* represent roosters because the birds keep time, but once so informed I learned more. Three birds keep diurnal time. Roosters are one, always found in domestic contexts defined by a house no matter how small or isolated, and are the only bird associated with boats that are not classed as "ocean birds" or associated with the shoreline. Among their connotations is the idea that if they crow before midnight, either a person is going to die or the next day a boat will arrive—they mark transitions. The two other birds associated with time are bush hens (*Amaurornis sp.*)—you hear them when they mate—and *payope*, an unidentified bird found in sago swamps. It calls out at evening and dawn, signaling the end and beginning of work in the sago orchards. There may be another reason why bush hens are not used for a *kuk*, but when I asked why, I was told that it is because they do not have the genitals roosters prominently display.

Whether organized or not, whenever two *anageg* sail to the same place they compete to see which one lands first. Once *kuk* genital imagery was settled, I presumed that a main reason for the rooster image there related to rooster competition for hens, an everyday experience of village life. I do not think that scenery is irrelevant. Kula competition is intense. Village roosters' daily life aptly models that dynamic. Yet more important is a rooster's time-keeping function. From dusk to dawn roosters (and hens) perch on available spots outside of house roofs or in trees often found behind houses in every village—*frangipani* and *ayyovay* (*Hibiscus tiliaceus*) in particular. From their height they call out the hours of the night. When the lead rooster calls, others chorus the village. People hear these cries and talk the next morning about whatever happened, or their dreams, according to the enumerated rooster calls (at least three; some people say four to six) during the night. This positioning is relevant to sailing dynamics. Although *anageg* do not always sail at night, night is the favored time for serious departures, especially between Yemga or Nasikwabw south to Panamut or beyond, and from Panamut to the north. Along these routes other islands are often invisible, so star—or sun—courses and time estimates are monitored with extreme care. The rooster up above on the mast provides an imaginary model for the passage of time.

The thick and looped rope called the *balau powan*, "testicle of the balau," keeps the *kuk*/rooster in its place. The top of this cord circles the mast around the short peg (*amawuw*) running through the mast. The bottom goes to the intersection of the rooster's body and neck. *Balau* refers to a knot type discussed in chapter 5 and the rope—two actually—that circles the *amawuw* leading in the opposite direction to the outrigger platform. *Balau* control mast arcs. As the pictures accompanying this section show, the *kuk* and *balau* complement one another by being on opposite sides of the mast. By controlling the mast's arcs, the *balau* facilitates the wind/ sail interaction; by means of the halyard (*yawasay*) that goes through the *kuk*, it keeps the sail in the wind's power. I was told that that if the *balau powan* fails, the *kuk* falls, and since the *kuk* supports the sail, the source of the boat's propulsion, the boat fails. Transforming the wind's power into intended direction follows from the rooster/testicle line structure.

Other images particularize what the *kuk* enables. The ends of an *anageg* keel are called *sibun*, a bird's head cut on each end called *pus*. The design is supposed to be a *bunibwan*, a sea eagle (probably *Aquila gurneyi*). The image is appropriate for two reasons. First, these birds create their nests in high treetops often close to the sea so that they can readily fly over the sea and land proximate to it where they search for their wealth (e.g. "tree fruit" like *Canarium* seeds); second, they fly high looking for islands and so provide a standard for where the keel ends should be, high above the water. According to a story, before boats had these designs, people who sailed looking for wealth never found any. When they started carving these images on their *sibun*, they obtained wealth. The story fixes a hydro-dynamic blueprint in collective thought. Although I do not understand why, one feature concerns the bird's neck. It is "just right," neither too long nor too short.

I shall return to other images running the boat's length but must first note that in contrast to the keel's sea eagle, outrigger float ends model a turtle's throat and head. Neck length is an issue here as well since the turtle head is from one called *tadiyay* (*Chelonya mydas*),[20] which is shorter than another kind called *pau'n*. Unlike keel ends, float ends remain in the water, as does a turtle's head. Moreover, as with a turtle, if the float works correctly, most of it is at or slightly below the waterline while the head just breaks above. The middle- and lowest-class outrigger floats are different, both coming to points resembling the shape of a bullet, their bottoms arcing up slightly. This change in form fits a change in function. Both these smaller craft, and especially the finely hewn middle-sized outriggers, should hydroplane. A float may rise out of the water with fast paddling or a good wind. *Anageg* do not hydroplane.

Moving inside from the sea eagle end, keels have three carved parts: the *tabuiy* (also *dabuiy*), tied atop it a *man*,[21] and the *kunubwara*. The person who carves these pieces is expected to carve their replacements over a boat's lifetime, perhaps twenty years. This is another reason why Muyuw people tend to be dismissive of the interpretive details of these carvings— it is another place's work. The parts are carved from one of several soft *Ficus* trees, *tatoug* or *kwayin*. Straight-grained and easy to carve (*woway*, from "hit"), these trees tend to rot quickly. Only *man* tend to last for an appreciable time because they are only tied to the *tabuiy* top when voyaging; otherwise they are stored inside the boat-owner's house, often tucked inside its roofing.

A grammar governs these pieces. First, whatever the form, all carved surfaces are convex (*katuson*), slanted (*dadan*), or flat (*sipusap*). These are color coded, respectively with *kon* (the color term used for *gunugun*, dark or black), *malak* (a plant name from which a red, *bwabwel*, color is extracted), and *pwakau* (white, originally taken from the same lime used for consuming betelnut). For coloring, or painting, *mwal* is used as a verb. Purchased paints prevail now.

Second, most design forms are understood primarily as "just carving," whether or not they designate something real in the world: tracks that small shoreline crabs make in the sand; coconut husks (C or inverted-C shapes, called *lolau*, which have the grace associated with mango fruit); marks that look like fingernail impressions (also found on pots, old and modern, and bodies[22]). These and other forms are common, although I never met anyone capable of relating a story they might tell. This is in contradistinction to the overall impression they should make, which is that of blended images creating a pleasing impression, exactly like the intermixed growth of a maturing garden. People readily told me whether or not this effect was created, and they did so more eagerly than going into any descriptive analysis of particular designs—the ensemble counts.

There are, however, particulars. I shall discuss the generalities related to birds associated with boats first. These are meaningful. With the excep-

tion of roosters, all birds are considered "ocean birds" (*yol manuwen*) because they eat fish or live proximate to shorelines, either on islets or near
the shore of larger islands. By their imagery they move the land-based,
tree-constructed boat from land to the water, marking boundaries as they
transcend them. The birds include the already mentioned sea eagle; frig-
atebirds (*dawet* and *mwag*)[23]; heron (*boiy*); seagulls (*mwakelikil*), likeness
of which may be fashioned from coconut husks and attached to a sail's
yard (*kunay*) and the top edge of the top strake on the side opposite the
outrigger float; *bayous*; and *kioki*, a kingfisher considered similar to *kelkil*,
the rainbow bird that is conceived to fly between Muyuw and the Trobriands (see Damon 1990: 1). *Mwag* and *bayous* are classed as *abtuvatus* (sign
or marker) birds because they stay close to shore. They indicate an island
that is proximate; navigationally this is only crucial in the middle passage
between Nasikwabw/Yemga and Panamout, between which land may
not be visible. Two birds stand out because their likenesses dominate the
forms associated with them: *boiy* (heron), for the part called *tabuiy*, and
kioki for the part called *man*. Firmly embedded in the keel ends by means
of a mortise and *tenon* form, *tabuiy* press the prow boards into a position
for binding the strakes, yet they lead away along the inclining arc of the
keel toward either of its ends then bend up in what is supposed to be a
graceful form. This is likened to a heron's body shape. Although also considered "ocean birds," *boiy*/herons, of which there are two types ("white"
and "black"), are *aulekel*, "sand" birds, because they frequently walk in
the shallow, sandy water looking for fish. Although the form resembles
a heron, *tabuiy* may also contain other bird images. The photographic
accompaniment to this section displays bird heads that people say are sea
eagles. And like the sea-eagle images on the end of the keel forms, the idea
here is that these parts are to be above the water.

Although for an outside observer the prow and stern *man, tabuiy,* and
kunubwara look the same, they actually differ. I begin with *man*, a word
that literally translates to "bird" or "animal." These pieces are tied to a
shank that extends above the *tabuiy*, always on the side opposite the outrigger float (*watan/katan* rather than *walam*). They may exhibit simple or
ornate carving. Although various "sea birds," especially their heads, may
form their interior designs, the kingfisher, *kioki*, should top the form. *Kioki*
are *wasim manuwen* (on-island birds), understood to fly between smaller
islands. In size and color they resemble, but contrast with, *kelkil*, a "forest
bird" (*wanawoud manuwen*). The prow birds should face one another while
the stern birds should be looking front and back, together representing
the boat's interisland motion and purpose. The stern *man* has a unique
appendage-like form trailing from its middle and dropping to about half
the length of the handle used to tie it onto the *tabuiy*. The message-giving

function of these forms extends to disasters that might befall a crew while at sea. If a boat from Muyuw loses a crew member, the front, *dabwen*, *man* is removed when it returns from the south as a sign of that death: if a southeastern Kula Ring–sector boat loses a crewmember on its return from the north, it removes its *wowun* man.[24]

Whatever the carving on the *tabuiy* and *kunubwara*, cowrie shells are fixed around them. Their work is first to make the whole boat attractive—*katubub wag*. It is understood that as the craft moves through waves, the shells will send spray up and around it. That "smoke," along with the carving, is part of the motion imagery exemplified throughout the craft. Their positioning also embellishes a container/contained theme that invokes the deep structures of the boat itself—recall the all-important *kaynikw* as the most contained part of the boat. Although some stabilizing function may be involved with the cowrie shells, the total image effected by the *tabuiy* and *kunubwara* is likened to a "face," *magin*. The top third of all *kunubwara* are composed of circular-like forms that are said to be eyes, though in some designs the very center is likened to a nipple. Sometimes those centers contain eagle heads understood to be the bird's teeth. Just below them and above the shank whose backside is grooved for the stakes are protrusions said to be "ears," *teigan*; two cowrie shells may be tied to these, configuring *pulakag*, earrings. *Tabuiy* may have a stick tied to them that extends below the end of the bird design on the keel and on which cowrie shells are tied; at the water's edge *bis*, the *Pandanus* leaves that elsewhere function as telltales, are attached. These are called *gilakal*, representing hair on a man's chin.

The prow *tabuiy* and *kunubwara* should be slightly smaller than the stern forms. This difference follows in that the front of the boat is where men might be stationed because they are smaller, lighter, and faster (and can take the waves that crash the front), while the back is where women sit, because they are larger, heavier, slower, and can cover themselves from the elements. This small : male :: large : female set also opposes the "top" to the "base" of the keel, which should conform to the tree from which it was first cut.[25] Other differences, according to some but not all of my informants and pictures, co-vary with these. Various accessories aside, three figures dominate prow and stern *tabuiy* and *kunubwara*. Prominent on both the prow *tabuiy* and *kunubwara* are *suiy*, invariably black and to me—but not my sources—phallic-looking figures that represent squid. This creature is displayed there because it swims so fast. This association is consistent with other aspects of a prow's male-oriented references.

Uvagol and *mwatet*, the nautilus shell (*Nautilus belauensis)* and snake, often dominate stern *tabuiy* and *kunubwara* respectively. Of these the nautilus shell is the most mysterious. Scoditti (1990b) has made much of these in his interpretation of Kitavan art; I could not replicate his ideas in Muyuw. People find the shell baffling—perhaps this is why they have so little to say about it. Since they have never seen the live animal that inhabits the shell in any of their reefs, they assume it is associated with the very deep sea. However, they say that the animal and shell become separated, the animal flying up to mountaintops, the shell eventually washing up on the beach, the midpoint between these extremes. They do not do anything with the shell per se, yet they collect them and break them into pieces to be strung on the ends of *veiguns,* one of the two kula valuables. They function as *dauyoyi,* dangling pieces designed to make a specific jangling noise when moved. This contrasts with those for the other valuable, *mwal,* whose *dauyoyi* are made from hollowed-out seeds (*sasal*) and make a different, muted or resonating sound when they hit one another.

In the, West nautilus shells are renowned for illustrating the Golden Triangle, but nothing leads me to believe that *uvagol* invoke proportionality or boat designs, which *are* illustrations of proportionality. Nevertheless beliefs about the animal pinpoint the limits of their environment, the lowest low and highest high, and the midpoint where vacated shells are found and boats always return.

Muyuw have more to say about snakes. From his study of Milne Bay Province's fauna, Allen Allison (see Krauss et al. 2003) informs me[26] that there are six snakes on the island, including the banded sea snake (*Laticauda colubrine*). In general Muyuw people intensely dislike snakes and beat them to death with a stick—it is forbidden to cut them with a knife[27]—whenever possible. The snake whose image is carved on a prow board is called *mwalek* in central Muyuw or *mwatalalek* in southeastern Muyuw (probably *Boiga irregularis*), one of five snake names I have collected, not counting the banded sea snake. They move at night. During the day they rest amid small leafy plants where they are often encountered when people cut forests for gardens. The snakes find their way into yam and village houses where they steal chicken eggs or chickens. They move across human-defined boundaries. Stories are told about how they swallow eggs whole, climb atop some rafter, then fall off to break the eggs. "Very smart," people say. Other stories emphasize their strength. They capture cuscus (*kwadoy*) or chickens, then wrap around a post or tree until they crush the animal. By virtue of their speed, their extreme strength, and the way they curl up, they are associated with binding materials and the binding that necessarily fastens canoes; this is one of the qualities that the figure brings to its position on prow boards, for they bind, and bound, as well.

Beyond this literalist interpretation is a myth called Mwatitawag, a story that moves the snake from Panaeati to Yemga, Yalab, and eventually Kweywata, a movement understood to trace the *anageg* boat design, from the southeast peripheral to and outside the primary Kula circuit, to Gawa and Kweywata, the middle of its northern side. In this myth, Mwatitawag, the snake form who is the *toniwagan*, captain or controller of the boat, sails north with a crew composed of rats (*simonan*) and kingfishers, *kioki*. Their poles for maneuvering the boat in shallower water are made from sugar cane, obtained from Yemga. These are ineffective, so the boat wrecks at Yalab, which becomes associated with the snake (the snake is as long as the island, which resembles a snake in motion). But they nevertheless end up at Kweywata, where the snake and the kingfisher form become fixed in their respective positions on the boat, the snake on the stern prow board, the *kioki* on the *man*. People understand the snake's relation to *anageg* as an aspect as well as conveyer of a hierarchized understanding befitting the boat: snakes and *anageg* are determiners, definers, moving across boundaries. In a different register this story expresses the same idea as the rooster imagery of the *kuk*. But the story also conforms to a logic suffused through the region's various practices, from a periphery to the center, and often from some kind of whole to a part, in variable contexts metaphors to metonymies and vice versa. Kula employs similar sequential transformations (see Damon 2004), as do ideas associated with the *kausilay* tree.

A remarkable aspect of these boats, from the trees used to fashion them to the figures attached, etched in, or painted on them, is their boundary-marking and transcending qualities. Many of the trees used to make boat parts come from ecotone-like situations, *kausilay* being the prototypical example, and much is the same with the birds used to adorn them—kingfishers fly between islands; *mwag* and *bayous* mark the presence of nearby islands in the midocean; *boiy*, heron, are waders. Fish designs are similar. Although I had frequently noticed them through the years, they are not on every boat—none were on the side of *Lavanay*, the *anageg* on which I sailed in 2002. In the past I had been told they marked the high-water level for a boat, defining its carrying capacity. When I talked to Wabunun people about the designs painted on their *Bwadanakup*, they spoke as if *Bwadanakup* defined a norm. Unfortunately I have no data that would corroborate this or provide a sense of variation that undoubtedly exists.

The fish may only be etched or painted on the middle strake, the one called *budwanay*, on the *katan*, nonoutrigger side of the boat. They should be positioned in the "gap" (*wosas*) defined by the boat's ribs (*gulumoms*) and the crossbeams (*kiyad*) attached to them extending in the other direction to the outrigger float. Some people associate the outside area where the fish are painted with the interior divisions, *kaynikw*. Thus there may be ten altogether. Fish designs on the *dabwen* (top) side of the boat point to its front end by means of the fish's front; those on the other side point to the *wowun* (base) end, the orientations coinciding with the boat's two sailing directions. The fish designs are precise, but when I asked why, I received no answer. However, when I started inquiring about the specific fish and their habitats, a logic appeared. Only three species should be used, and usually all three are on a boat. I cannot distinguish them from my photos, but because their tails vary I suspect Muyuw people could. The three fish are *papis*, *kwakwey*, and *kobal*.[28] Although these fish might be found everywhere, *papis* frequent sandy shoals proximate to islands and are particularly associated with Yemga's sandy lagoon. Wabunun people recall Yemga *anageg* coming to them full of the fish, the load being part of the seafood for horticultural food that typifies the lower-level exchanges between mainland Muyuw and the outer islands devoted to boats and their work. *Papis* have a sharp hook on their tails that causes swelling if it pierces the skin. *Kobal* are associated with reefs near islands but removed from the shoreline. Their tail has two knife-like fins that are also dangerous. Not only is this design restricted to the middle strake, it is also the name of the tying used for that middle strake (and the tying of middle-sized boats that only have two strakes). *Kwakwey* are also associated with reefs but those further away from larger islands like Muyuw. This reference implies the "deeper" ocean, *bwanit*, and the fish are considered shark food and sometimes used as bait for shark.

The spatial implication of these contexts means that a craft's duration is written into the simultaneity of its visual imagery. Other imaginings augment this aspect. The sea eagle that adorns the rising ends of the keel represents a boat's purpose—looking for wealth on other islands; *kioki* represent that interisland motion; heron, the waders that the *tabuiy* represent, harken to the position a boat finds itself just before it is pulled ashore, or when it is pushed into the water. The turtle shapes found on the ends of the outrigger float mark where it should be.

In many respects a form that should be burned along the length of the outrigger float finalizes this patterning. Although today it is rare to see boats with burned floats I have seen several, one in 2006. When I actively asked about the practice in 2009, almost everyone told me that every boat's float should be burned and "was in the past."[29] I first inquired of this practice after seeing a banded outrigger float as a pig was being scorched on the nearby beach prior to being butchered. Since everything in Muyuw is smoked or singed—mothers and babies, land in preparation for planting, sago trees, pigs—I first tried to figure out how an outrigger float was like a body, sago tree, pig, or plot of land. And eventually people told me that the treatment helps preserve the wood, similar to how fumes infuse a baby's body with a tree's qualities or how lightly torching instills strength into sago trees. However, that is not the primary purpose.

Before burning, a float is wrapped with coconut leaves so that some of the wood is covered and some is exposed. The bare parts become charred, thus creating white and black bands. The appearance is classed as *limwelan*, mixed or blended. And the banding should make the float look like a banded sea snake, *mwatabwalay* (*Laticauda colubrina*). As one of my Muyuw advisers, Leban Gisaw (see note 28), informed me, the systematic literature considers them highly venomous (neurotoxic) marine reptiles. This point seemed consistent with other ideas about the animal. It is considered the "chief" (*guyau*) of the sea, and all other sea creatures, including shark, defer to it. One person said that below the surface of the sea fish will look up, see a passing canoe but mistake it for a banded sea snake, and, I presumed, shudder. Today people recall instructions from their elders that they must never hurt the animal, because, like other *guyau*, it has special powers, *aniyan*, that, when released by death, cause disturbances in nature. Muyuw only come across *mwatabwalay* when they are fishing on reefs, and should they harm one, the subsequent *aniyan*-caused storm might imperil them.

Mwatabwalay might be seen swimming anywhere, but it is widely known that their "place" (*kaban*) is an island in southeastern Muyuw called Eyon, a few kilometers south and west across the lagoon from Wabunun. I arranged to be taken there in 2009. Within a few moments of arriving, one young man, who had quickly jumped out of the boat and headed down the beach, sauntered back with a live snake wrapped around his neck. This moment inverted the vicious image I had deduced from Muyuw attitudes toward other snakes and what I had learned about this one—that it was a deadly animal positioned atop animal hierarchies in the ocean. Although people should not kill them and do not eat them—no woman would sleep with a man who had eaten one—occasionally they dissect them. Stories circulate about the surprisingly large fish found in their gut. They are significant ocean predators; their chiefly attributes are not just a cultural imposition. But the modeling implications of the animal gradually became apparent when the man holding the snake lay down on the sand. The snake slowly crawled over and away from him, and even more laboriously headed for the tree and rock line that separated the sandy beach from the forest. There the animal found a crevice into which it curled up to hide from the sun—that is what they do when they come ashore, drink fresh water and hide from the sun. *Kakam* (*C. Inophyllum*) root structures often create crevices in which the animals hide, according to my informants, and this is part of the received knowledge associated both with the tree and the snake. This is not, however, a fixed relationship; the photos I provide of a hiding *mwatabwalay* are from another tree. But the hiding place above the wave reach is not the only relevant factor here.

Other snakes are only glimpsed as they beat a path away from humans who they know will likely beat them to death. By contrast the banded sea snake's tortuous movement approaches the comical. Although they slither like other snakes, they struggle to gain leverage on the sand, moving so inefficiently that they almost come to a full stop as they gather to make another surge. Yet when in the water they swim rapidly across its surface—their tails look and function like paddles.

Their motion and immobility makes this "chief of the sea" a perfect mimetic for outrigger floats and the way that the float stands for the whole boat structure. Like the snake, a boat should move rapidly across the surface of the sea. But when moving from the sea to its resting place out of the sun's drying forces and waves' wrecking capacity, a boat's motion is a strain. The smallest outrigger craft may be dragged up the sand or rocks by a single person who picks it up by the leading crossbeam then laboriously pulls it up, its other end dragging behind. Middle-sized outriggers require available sticks to be used as rollers, and at least two people, usually more, move it up and out of or down into the water. *Anageg* are different. A dozen or more, sometimes many more, people gather to move them up or push them down. Larger-trunked trees are collected for rollers, and as many as ten younger men get under the outrigger platform so they can grab a crossbeam to simultaneously push up as well as toward the landing or the ocean; and when pulling a boat in, usually one rope or more is tied carefully around some sturdy part then made available so that many people can join the task. If stronger and younger men tend to focus on the sides and undercarriages of the craft, children as well as women and older men often find a place along the rope. If the momentum achieved from the first motion isn't sufficient, then people stop, collect themselves for a few moments, and start over. These events are ritualized and often turned into hilarious experiences.[30] But they are also part of competitive relationships between villages, for one of the ways primary gardening villages compete for the privileges and responsibilities due to, and from, the sailing villages is by being able to gather the human power necessary for handling the craft. Sailing communities know and watch for which villages can support them and which cannot. Being able to pull a boat out of the water and protect it from the sun, as a banded sea snake must protect itself, becomes internal to the structure of intervillage dynamics (see Manguin 1986).

These qualities and conditions make boat imagery more than just symbolism. The capacity to handle a boat indicates political efficacy. As a *guyau*, big man, is so by his ability to gather persons and things into his fold, so an *anageg* incorporates the environmental analogue of that action. Not coincidentally, a constituted *anageg* redounds back to the fact that so many tree names, and socioecological knowledge sets about trees, configure the socioecology of this region's organized places. The imagination that assembles these boats organizes places into a moving synthesis.

From the Land, By the Stars

Iwa is the pivot across the northern side of the Kula Ring. *Anageg* ply the waters east of it, yet its inhabitants believe they start the New Year sequence that winds its way west through the Trobriands. It is the furthest west of the islands that participate in the ritual firewood practices exhibiting complementary differentiation across the ring's eastern half. Wabunun people attribute to Iwa their founding in a practice—their *teili* tree and "don't forget us"—that orients the whole area to the El Niño Southern Oscillation (ENSO) troughs, "big suns." Dependent upon the places to its east for the reproduction of its root crops, Iwa starts yam cultivation practices (pruning) that reach their epitome in the Trobriands. Its people eat the birds that are harbingers of plentitude to the islands on either side. Iwa represents the discontinuities through which this social system achieves coherence.

Chapter 1 opened this book by introducing the problem of modeling, the sweet and bitter trees, and the problem of discontinuous forms—landscapes—composing the northern half of the Kula Ring. I close the discussion by showing how trees turn those discontinuities into a viable whole by means of the *anageg* form, a product of the contrast between the two primary outrigger forms derived from the two different *Calophyllum* species, *apul* and *tadob* to the west, *kausilay* and *anageg* to the east.[31]

The form is not a simple construct. As chapter 4 noted, in addition to the *apul* : *kausilay* :: *tadob* : *anageg* contrast, another change in the positioning of the *tadob's* sail turns into another outrigger form, *lakatoi.* That model dominated the waters just outside the Kula Ring's southern reach yet made an occasional appearance inside it. *Anageg* are made on Gawa and Kweywata, though occasionally elsewhere; in 1974 a central Muyuw man told me he had made one. Similar craft were made in Budibud, though they rarely become items of interregional exchange like the Gawa and Kweywata craft. But Gawa and Kweywata are not just production sites. They constituted different versions of the form, and if Yalab people make one, it is assimilated to one of the two ideal forms.

Although the boats leave these two islands conforming to well-wrought rules, they come neither complete nor perfected by the standards of their next stops. Some parts come directly from these islands. The craft's first principle, the *kausilay*-generated keel, derives from Gawa and Kweywata, whether its ecotonal conditions are generated by forest/garden practices or the overhang effects of each island's steep cliffs. *Kakam,* the shoreline

Calophyllum, is also an ecotonal tree. From it people make *geil* or *gulumom,* boat ribs; the *duwadul* (mast mount and its extension to the float); *kuk. Kiyad,* crossbeams, also come from these islands and they should be *alidad* (now *Syzygium stipulare* [Blume] Craven and Hartley), an understory trees found in mature forests near the water's edge, thus also an ecotonal position. But the same cannot be said of other parts.

Consider first *kaboum.* It is the source of the stakes that secure crossbeams to the outrigger float and the spring underlying the mast mount. Although I was shown one planted in Iwa's interior, they are prototypically found on sandy beach locations (*umon*) common along western Muyuw's shoreline. It is the ritual firewood tree for Boagis, the sailing village at western Muyuw's southern tip. Two giant *kaboum,* along with a third, *meikw* (*Intsia bijuga*), the source for *anageg* rudders, tower over Boagis. Since both Gawa and Kweywata shoot up out of the water and do not have sandy shorelines common to western Muyuw, I asked people from those places that I met in Kaulay village in 2009 where they got *kaboum.* According to them, both islands have *kaboum* and use them when they can. But they said they regularly "begged" (*nitoug*) the wood from western Muyuw. Wabunun people knew these facts when I repeated them, but they added that while Kweywata had more *kaboum* than Gawa, in both places it grew on the *dibwadeb,* limestone outcroppings that line shores rather than the preferable sandy *umon.* These details matter. Boagis's ritual firewood marks a regional relationship, with other places dependent upon conditions it exemplifies.[32]

Kaboum's social and material significance raise the question of *eyalyal,* the undercarriage to the *kaynikw*/mast-mount spring, and *seisuiy,* the two important rod-like parts that fit through the ribs and onto which crossbeams are tied. Both are made from *igsigis* (*Lumnitzera sp.*), the tree found in mangrove swamps. Although Kweywata has a small bay in which the tree can be found, I was told that boat builders obtain them from western Muyuw's vast mangrove region. Wabunun has access to these trees in the mangrove swamps in south-central Muyuw, yet they associate the trees, and the boat parts, with western Muyuw. This is not the association of externally related ideas; it is the reflex of a structure's vital relations. The part carries the regional significance of those places.

A wood substitute for *watot* and *seisuiy anageg* is *amanau* (*Garcinia sp.*), what one north-central source called a "*tasim* tree." Uncut areas of more mature forests from Kaulay to Dikwayas, their *tasim,* are full of *amanau.* Frequently used as fence stakes around Kaulay gardens, they also serve as a *kiyad* in smaller outriggers because of their interlocked grains. And they constitute the tree of choice for the seven or eight slats (*pamaloul*) that run the length of *aydinidin* sails, the original *anageg* sail form. As saplings they are supple (*woweu*) and will bend without breaking. One Wabunun friend told me that new boats with new sails coming from Gawa or Kweywata sometimes arrived with *akidus* trees for these parts; he substituted *amanau* as soon as he could. Moreover, although a few *amanau* may be found in the lower forests near Wabunun, he said that he would get them from central Muyuw.[33] Kaulay people told me they furnished Gawa and Kweywata with the *pamaloul/amanau* parts they needed for their sails, an assertion confirmed by the people from there. In one conversation about analogies between the island's shape and boat forms, some Kaulay people said the middle part of the island was the sail. As noted previously, between the Sulog area in south-central Muyuw and the Dikwayas-Kaulay area to the north, people find all necessary parts for sails.

Generated by the environment in north-central Muyuw, this sail part and tree bring us to the more interesting and complicated question of *lam,* outrigger floats. The Kaulay-Dikwayas region is the source for outrigger floats.

Outrigger floats come from a tree called *vayoun* (*vayoul*), in the Fabaceae family, probably a *Mimosa* or *Acacia*. The tree flowers and fruits in the oldest forest class, but the seeds need sun to germinate, so they usually remain dormant until big trees fall or are cut and burned for a garden. It appears in abandoned fields very quickly but only begins to reach the appropriate size when a forest reaches the *oleybikw* stage, the second fallow class and the ideal garden fallow in the Kaulay-Dikwayas region. I was shown three *vayoun* in 2009 in an area kept in a permanent early fallow stage, and the trees were dying. Gawa and Kweywata, however, do have these trees. Wabunun's Dibolel told me they grow like their *nilga,* another *Mimosa/Acacia* tree common in early fallows throughout this region and appropriate for an outrigger float for a middle- or lowest-size outrigger.[34] Kaulay friends agreed that Gawa and Kweywata have them, but they insisted theirs were better. This locational specificity is marked. Although Wabunun people know about the tree, they did not recognize a leaf specimen I collected in 1996 and showed them; they said their region doesn't have the tree, so they don't pay it attention. To an elder in Unmatan, a village some six kilometers west of Wabunun, I repeated the Wabunun idea that the tree did not grow in southeastern Muyuw. However, he contradicted this thought by saying that Unmatan had them, yet he added that theirs didn't grow straight and large enough to be turned into appropriate *anageg* floats. In addition to the favored fallow class, the soils in north-central Muyuw produce the tree's preferential growing circumstances—but it is not just the soils in general. Kaulay people understand their region by contrasting *tawan* with *ulakay*. Although large trees typical of the oldest fallow class may define both regions, "*tawan*" refers to the land sloping down to the ocean, whereas *"ulakay"* references forest growing toward the unoccupied center of the island. Tawan soils tend to be dryer, like the soils generated by the shortest fallow length; *ulakay* soils are wetter. The demarcation between these two zones is referred to as a *kalivas:* a meeting point of two differences, a mixture of *ulakay* and *digadag* characteristics and the location of both Kaulay and Dikwayas villages. This zone, another ecotone, is where the best *vayoun* grow, and it is where I was taken to see them in the Kaulay vicinity. *Yayoun's* ecological context thus complement's *kausilay's.* And together this contrast emerges: *vayoun* are white and light, *kausilay* red and heavy; the latter qualities are part of the female-person identity given to *kausilay,* while the former attributes are masculine—although nobody considers *vayoun* to be men.

Although the symbolic implications of this contrast are rich,[35] different associations stood out for my instructors. Although their carving is never as precise as that invariably found with keels or strakes, like *kausilay, vayoun* are relatively easy to shape. More important is the wood's quality. In addition to being very light and buoyant, it is very *awoyan,* soft or spongy. This makes it difficult to chop, and it becomes more so as it gets deader/dryer. However, it rarely splits, and this is the critical issue for the two main purposes that the wood fulfills. One is for making warfare shields, *kaulawala.* Likenesses of these are known from Trobriand dancing and other regalia; I've never seen one in Muyuw. But people told me that if a spear hit a shield it might get stuck, but the shield wouldn't split, nor would the spear travel through it. This is precisely the quality needed for the float. An *anageg* float should have forty-four pieces of *kaboum* pounded deeply and snugly into it. Forty derive from the four for each of the ten *kiyad.* This makes two sets of twenty on the same line running the length of the float (the four others coming off the *duwadul* are on different angles and lines)—enough to split most wood. Yet *vayoun* absorb insertions without splitting.[36]

The tree's sociology shows its absorption into the social fabric. Land around Kaulay and Dikwayas is nominally owned by a specifiable sub-clan. But independent of that, these trees seem to be possessed—called *kavayaw,* a noun—by whoever first sees one growing and takes title to it. In 2009 I was shown several claimed by a man who said he first saw them between 1979 and 1982.

When a tree is selected for cutting, Kaulay people usually help cut it down, assist in the initial trimming, and pulling it down to the landing where visitors beach their boats. Although much of the trail drops a hundred meters in elevation, the distance between where a tree is cut and Kaulay's usual landing area could be five kilometers. *Anageg* owners are responsible for the final finishing.

Although none of the Gawa and Kweywata men I met in 2009 had plans to obtain new floats, they recalled four boats fit with new floats from that region in 2004. Later in July 2009 Kweywata people coming for a United Church event were planning on "changing" (*katilevau*) their floats. Trees I saw in 2009 had been turned into floats by 2014, as had one I had not seen. My primary Kaulay host from 1995 easily recalled three boats from Yemga coming to Kaulay to cut three new floats. Bubibud outriggers were usually fit with floats made from *nelau* (*Burkella obovata* [Forst.] Pierre) until they sailed to Kaulay where Budibud's products—pigs, coconut leaf skirts, sleeping mats, and coconuts—were exchanged for both a *vayoun* float and sago.[37] One Kaulay man, however, recalled getting two personally owned kula valuables, *kitoum*, from Budibud. *Kitoum* center the Kula system. In another case an elder exchanged a Kaulay *vayoun*/float for a Budibud pig. He gave his pig to a younger brother who used it for a mortuary obligation to his wife, a Wabunun woman. In ceding the pig in a mortuary ritual he created a debt (which the recipients have repaid). *Vayoun* transferals capitalize other relationships. Boats built on Gawa and Kweywata also received new *vayoun* as floats in exchange for *kitoum*, thus capitalizing kula relationships in Kaulay (or Dikwayas).[38] One of my elder female sources in Kaulay married into the sailing village of Boagis when she was young, she said, to facilitate the acquisition of floats.

By 1974, when familiar with minor dialect differences, I was surprised that the "Muyuw" spoken in the Nasikwabw sailing community—where the first language remains the language spoken on Misima—more closely approximated that spoken in Kaulay than in Wabunun. (My Wabunun hosts agreed with my dialect recognition.) I had thought that Nasikwabw was spatially and socially closer to Wabunun. Yet their talk followed from the necessary intercourse between Nasikwabw's sailing responsibilities and Kaulay's function in the regional system. After World War II Nasikwabw people created the village Waviay on the little island across the lagoon from Wabunun. One elder who had had a wife from Kaulay spoke Muyuw with a Kaulay accent, as did his children in the 1970s; one of them was living in Kaulay in 2009. Such relationships continue. A son of my primary Kaulay host from the 1970s until 1995, Vekwaya, was "divided out" (*katilev*), adopted, to Nasikwabw and was living there in 2002. Vekwaya's sister's son married a woman from Nasikwabw, now living in Kaulay.

These relationships work in other directions as well. Vekway married a woman from Moniveyov, the northernmost western Muyuw village with very close ties to Kweywata. This relationship facilitated a 1970s *anageg* transfer. Named *Umaye*, the boat was made by a man named Baledi of the Kwasis clan. He gave it to Mesiau, an important Kubay clansman from Moniveyov, who was Vekwya's wife's mother's brother (*kadan*). The mother's brother gave the boat to his niece in Kaulay; she gave it to Vekway; he gave it to his father; his father gave it to a Memiau from Kavatan, the eastern village closely related to Kaulay; he gave it to David, the elder (Kubay clan) in Nasikwabw at that time; and David gave it to Dagoyau (Dawet clan) in Ole, the southeastern corner of the Kula Ring. Following a Kula route, the transfer through Kavatan was technical and not real. The boat was in Kaulay for about a year during which a new float was cut and a new sail "burned." No *kitoum* were transferred for the float because at that time Kaulay already "owned" the boat. In 2014 Vekwaya claimed that five *kitoum*, a basket (*kaynad*) full, were transferred to the boat owners.

This information shows how words, things, and people become bundled together in the structures these boats entail and realize, in this case beginning with *vayoun* trees coming out of the ground.

Although I learned incidental facts concerning outrigger floats in the 1970s, I started to gather the information intensively in 2009. At one point my principal Kaulay informant extended the analogies between boat parts, places, and fallow regimes to southeastern Muyuw. I feared he was working off my searching for (structuralist) patterns. What he added to my knowledge was southeastern Muyuw's role in this production. It was to furnish *patapat*, the small tree trunks used to make the flooring for the outrigger platform. These derive from *digadag* garden fallows, southeastern Muyuw's ideal fallow regime. My source was emphatic about this paradigm: in addition to *seisuiy*, Nayem, western Muyuw, furnished *watot* (and *kaynikw*) wood from mature trees whose age, though not context, is like the ideal western Muyuw fallow structure; Wamwan, in central Muyuw, produced *lam*, mature in the forest class that produces Kaulay's ritual firewood; and Muyuw, generated *patapat* wood from a tree in the fallow class that produces its firewood. He also brought up Sulog's role for making masts to which I will return. But for him boat parts mapped complementary regional resources.

In Nasikwabw in 2002, the *Lavanay* crew planned to drop me off in Wabunun then return to Nasikwabw to refit their boat; then they planned to sail home. The wind didn't cooperate, and after several weeks they decided to refit in Wabunun. They were directed to a fallow area to cut new *patapat*, of which they stripped the bark on Wabunun's beaches and then tied them down. The photographic accompaniment to this volume shows some of the work. I was told what tree was used and where they were from—one of Dibolel's gardens used seven years earlier; by 2009 the appropriate trees were still there but they were too big for that usage. But in 2002 I thought little of this: Nasikwabw garden fallows and many of their trees approximated Wabunun's, so this seemed incidental. Hence I tried to discount my Kaulay informant by reviewing his model with Wabunun people.

Dibolel, my principal source for sorting out potentially conflicting information, said my Kaulay source was correct. The ideal southeastern Muyuw fallow regimes, from Guasopa to Wabunun, systematically produced the right-sized tree (*asibwad*, Rubiaceae *Timonius timon*) for the outrigger float's platform (as well as another tree used for caulking). Immediately after an area is abandoned and the gardener moves to next year's field, it appears by the score (in early fallow gardens). Although it does not participate in the sweet/bitter contrast invoked by *gwed* and *atulab*, it has characteristics similar to the latter. It is hard to kill and has a dense root mass that makes it bad for crops growing next to it. Yet once it is tied into place on a platform, the saplings form light (*atulab* is heavy), strong, and supple flooring relatively resistant to wind and water. But Dibolel went on to explain a crucial fact about relationships of this type: they are loosely determined. Nothing would prevent any crew from refitting its craft wherever appropriate trees were available. Everybody in Wabunun knows that the best outrigger floats are found in the Kaulay-Dikwayas region. But that didn't prevent Wabunun's *Bwadanakup's* crew from cutting a new *lam* on Normanby Island (Du'au) when they visited there between 2009 and 2012; nor did it stop the *Lavanay* crew from taking advantage of a resource designed for their needs while in Wabunun.

Nobody along the southeastern shoreline can guarantee the availability of the right-sized tree every time a needy friend might have to refit his craft. Recall that sailing villages—Waviay, Nasikwabw, Boagis, Yemga—lining Muyuw's southern arc are the principal intermediates between Muyuw in the northeast and the Lolomon region in the southeast Kula Ring. All the important men along the southern Muyuw shoreline compete with one another for boat owners and their services. Yet they are also engaged with one another in small-scale kula relationships designed to offset other relationships, affinal or kula, which might pit them against one another (see Damon 1983a: 331–33). Consequently, if one person did not have a fallowing garden as a good resource, he almost certainly would have another person, in his or a nearby village, to whom he could send his boat friends. (Dibolel was not host for the *Lavanay* crew, yet he let them use his resources.) Social and temporal relationships overlap just as the fallows across the landscapes of the northeastern Kula Ring became cross-cut to produce different resources for the vehicles that bind them together. As motion is built into the boats to facilitate contradictory dynamics, so are social relationships coordinated with respect to complementary divergences.

The interaction between my Kaulay and Wabunun instructors transformed what seemed like incidental, accidental events into a structured and reciprocating model between these islands' qualitatively different spaces and the intricacies of the *anageg* design. And the issue is not just the replacement of parts, which the southeastern Muyuw *patapat*/float platform might suggest. This goes back to the difficulty in learning about carving and painting on the boats: my Muyuw informants thought that was really the carvers' work—their responsibility concerned wood details and how the pieces fit together. Similarly, Gawan and Kweywatan boat builders release boats that only appear to be their own finished products. Very likely they had to get materials from western Muyuw. Then when the boats move to their next position systemic refitting and replacing begins resulting in a superior construct. *Anageg* compose sequenced places and skills. They are similar to rice practices Lansing describes for Bali. There minute inputs of skilled labor create a whole that is a massive engineering project only in its final result (Lansing 2006: 41–42).

Anageg inscribe relations that cross these socially physical spaces. Although it was not conceptualized as such, when I first arrived in Muyuw in 1973, the area was coming out of an ENSO drought. I watched Wabunun people double and quadruple their yam and taro plantings over the next two years, but I did not then connect that fact, along with all the food that southeastern people exported to Yemga and Nasikwabw, to a routinized understanding. During that time, however, I learned that people understood the placement of villages across the island in relation to lower/ wetter and higher/dryer ground, and that there were regular exchanges between these places that balanced the differential consequences of extremely dry or extremely wet weather for yams and taro. Wabunun, a higher and dryer place, exchanged yam seeds for taro seeds with Kauwu-way, a lower and wetter place along Muyuw's southwestern shoreline. Two Iwa *tadob* paddled into Wabunun at some point and, along with hearing disparaging comments about Iwa people, I learned about Wabunun's *teili*. But like a single figure carved or painted on a canoe, these facts were only points; I lacked a well-designed mosaic that assembled the relationships, like the geometrical garden forms reviewed in chapter 1 that synthesize "kinship" dynamics. The systemic forms that appear in chapters 2 and 3 started becoming apparent before my attentions turned to boats. Although they were made from wood, and my topic was trees and what was done with them, outriggers and their complexity were beyond my competence (a circumstance that never changed). Yet new material that I slowly gathered transformed the situation. Assembling facts about *Calophyllum*, chapter 4, became pivotal. Munn (1986) brought to my attention the symbolism inherent in the tree called *kausilay* (eastern Muyuw pronunciation); as my ecological studies proceeded during 1996, I learned that this tree came out of ecotonal spaces that were systematically created.

Only gradually did the whole construct appear. That the northeast corner of the Kula Ring is an anthropogenic environment is not surprising now. Landscape activities that non-Western societies effected are increasingly part of our achieved wisdom (e.g. Gammage 2011), but in 1996 such possibilities were just gaining disputed currency. And the synthetic, conscious models by which these relations are realized unfortunately have tended to remain beyond our grasp.

Among the trees used to build these boats, there is a fundamental contrast between the *Calophyllum* species by which the keel is made, *kausilay*, and *ayniyan/aynikoy*, the trees turned into masts. *Kausilay* are designed with respect to the movement of water, *ayniyan/aynikoy* with respect to air. Like other trees, *kakam*, *kaboum*, *igsigis*, and *vayoun* for other principle parts, and maybe the *Ficus* species used for the ornate carved parts, these are ecotone trees. *Kausilay* is so paradigmatic because its ecotonal structure is a human artifact. By contrast, *ayniyan* and *aynikoy* are products of distinct biomes. The same could be said of the plants used for tying (from meadows), the caulk source (*akobwow*), and the tree used for outrigger platforms (from early fallows). These plants come from unique places. Although it can be found in practically any environment on the island as long as it has not been recently cut, *akidus*, the tree used to make the spring attached by its center to the *duwadul* and also used as a spring in roofs, might also be said to be a product of a biome. But what marks *ayniyan* and *aynikoy* is that humans do not transform their environment. Sulog's ritual firewood emphasizes the point: for that tree is from a very old forest, and it is so hard to cut that the original residents preferred not to deal with it. Sulog trees are close to out of touch.

Ayniyan and *aynikoy* are associated with the higher part of the island, a distinction marked by the differences between the two: *ayniyan* are lower than *aynikoy* (*koy* = mountain). Both are used to help pull the wind's power into the boat's human intentions, according to how their differential properties are deemed to mesh with the rest of the boat. And as one of my informants noted, their names associate them with the idea of power, *aniyan*, and manifold associations to which it attends. One of these concerns is the "power" emanating from setting stars, the discontinuous patterns of which model periodicities used to turn forests into gardens and back into forests whose trees go into the boats. This returns us to the ways by which gardens and fields are imaginatively turned into an image of boats, coursing around reefs and islands like gardens move through the years around the patches of high forests, *tasim*, thus defining horticultural practices.

Entailed in these relationships are parallels and ties among the land— in other words, the trees in the boats, knowledgeable men and the stars.

Muyuw call the Milky Way *lapuiy,* a word that also refers to the east-west (north-south in western Muyuw) lines of sticks and stones that define the garden unit called a *venay.* Over the course of the year the sky's *lapuiy* swings from parallel to the island (*kadumwal ven*) during the calm period—the Yavat or Uwal, centered around January—to perpendicular (*kanbilabal ven*) during the windy period, the Bobweilim, configured around June. These spatial and temporal referents are garden images, and the similarity between the heavens and gardens finds a parallel in an understanding of the sea. *Kwab* is a common term throughout this region used to refer to plots of land that are or will be turned into gardens; "field" is an approximate translation, though some *kwab* are forested. In north-central Muyuw, named plots of land are individuated as *kwab.* In southeastern Muyuw the term, *bidiwakwab,* is more generalized and refers to flat land appropriate for gardens, not necessarily individuated named plots. This same term is used for the open sea, especially the broad stretch that runs between Nasikwabw and Panamut.

Tree shapes give these forms order, hence the tree "base" (*wowun*)/"top" (*dabwen*) contrast applied to fields and villages, as well as stars. The apparent east-to-west movement of a star is likened to the shape of a tree, its eastern horizon point its "base" (*wowun*), the western setting point its "top" (*dabwen*). As the land is demarcated by tree types and forms, so the is sky.

The named "stars," *utun,* are those that stand out, and they are understood to be like "big men" in that they are organizers. Their movement across the night sky loosely defines lunar month names and gardening periodicities. In the 1970s a Wabunun elder walked into the center of the village one evening to make a meeting pointing to the asterism Kib in the night's sky. He said its position meant that they had to get their next year's gardening activities underway. Later when I realized people insist they use Gumeau (the Pleiades) to garden, not Kib, I was told that one could easily refer to one star to indicate the position of another. Another man told me he made turtle-hunting calculations according to the faint constellation/star called Lakum. However, since he could not see it, he followed another star more visible for its placement.

Recall the pattern. Muyuw liken the dominant stars to big men (*guyau*). When *guyau* die, a "power" is released, realized in stormy, windy, and rainy weather. So too when "stars" die, when they are no longer seen at dusk, a period of wind and rain follows. This pattern conceptualizes an oscillation between sunnier and rainier weather. A good gardener uses this oscillation to organize cutting, burning, and planting crops. This means the whole pattern of forests and fields, and *tasim,* their temporal directionality, follows star periodicities.[39] Of course, like wind eddies that disrupt sailing courses, so too do "Big Suns" introduce a different magnitude of discontinuities and new strategies for action. Cognizant of what should happen, good gardeners, like helmsman watching *bis* on the sail in front of them, remain attentive to what is happening.

Stars, like big men, now and again dazzle in their appearance. For stars, this is primarily experienced at their "base" and "top," when they rise in the east and set in the west. This, it is said, is when their *bis, Pandanus* streamers, their twinkling, are most evident. This redounds to big men in that when they are dressed up for Kula or other official occasions, they should attach such streamers to their armbands and, most importantly, to the end of their combs (*sinat,* which is also the name for the constellation, *utun,* Scorpio). These combs were often twenty to thirty centimeters long or more, often stuck in the tuft of hair by the combs tines, the handle angling up and in front of the person's head. A streamer should be tied to that end. As the person walked or the wind blew, the streamer bounced, drawing attention to him.

Sailors use stars to set courses. A course set by sight, using, for example, the sun, wind, or another landmark such as a lighthouse or a mountain viewed over a tree line, is called *mumumov.* A course set by a "star" is called *kut.* Once a course is set, waves may be used for maintaining courses because stars are not always visible; constancy of the feel for the boat in waves is taken as an approximate guide. A change in the size of waves—which are larger toward islands—is used for estimating a position with respect to land forms. *Anawat,* "its belt," is the expression used to describe the increased size of waves found proximate to islands.

Table 6.7 lists the star "courses," *kut,* collected over the years. As implied by this list, people associate every island with a star and can chart this representation in the sand (see photos for this section). As for the chart, sailing from Panamout to Nasikwabw (or Muyuw), for example, would entail setting a course by the constellation Kib, more or less what we mean by Delphinus; when I asked if the Pleiades might be used, I was told no, that would take one to Budibud. Instead the Pleiades should be kept *walam,* to the outrigger side of the boat, presuming the wind is from the east (southeast) so that the outrigger float is into the wind. As implied, the discontinuous shape of the canoe structure facilitates organizing courses. The information I have on the courses is not sensitive to that so I do not know what leading part of the boat is used to maintain all these courses. Some people have told me that they could use the movement of the sun over the oval shape of a sail for maintaining a course during the day. Duweyal used it during our 2002 trip from Panamout to Nasikwabw even though it was an overcast day. The geometry for this usage is daunting, and I do not understand it. For with the sun, what must be known includes the particular trajectory of the sun on specific days of the year, meaning when the sun is north or south at the zenith. I believe the mental training required here returns us to the cerebral gymnastics *everyone* is

Table 6.7. Star Courses (*Kut*)

Course	Indigenous Name	Western Name (Approximate)	Source (Person)
Koyagaugau (Gaboyin/ Dawson Island) to Nasikwabw	Kib	Delphinus	Muyuw
Muyuw to Lolomon	Tanaboub	Southern Cross	Muyuw
Lolomon to Budibud	Gumeau	Pleiades	Muyuw
Muyuw-Budibud-Muyuw	Kiyad	Orion's Belt	Muyuw
Lolomon to Yemga & Nasikwabw to Boagis	Kuluwit	The Big Dipper	Muyuw
Koyagaugau (Gaboyin/ Dawson Island) to Panamut	Kib	Delphinus	Koyagaugau
Koyagaugau (Gaboyin/ Dawson Island) to Muyuw	Kuluwit	The Big Dipper	Koyagaugau
Nasikwabw to Misima	Tanaboub (at its base) or Sinat	Southern Cross or Scorpius	Nasikwabw
Nasikwabw to Lolomon (Koyagaugau, etc.)	Tanaboub (at "noon")	Southern Cross	Nasikwabw
Kitava to Yemga	Tanaboub	Southern Cross	Muyuw

subjected to in their youth when they learn, or watch, string-figure manipulations: people are trained to follow complex geometric transformations.

Some people differentiate calculations based on the relative shapes and positions of a star over the course of a night. For example, when the Southern Cross rises at its base, it tilts so its top is to the left, east; when it is at "noon," its zenith, it stands straight up pointing south; at its "top" it sets tilting to the west. The initial condition taken for a course (*kut*) is the determining factor in setting one, even if the targeted star is never seen rising from that point during a journey. Other stars are used to estimate where the position actually is—this is "prototype" thinking that suffuses the culture's knowledge system. This is why the number of officially designated stars is deceptively small. Since the named star for a particular course may never be visible, another, named or not, is used in its place. Stars were used to navigate a voyage between Panamut and Nasikwabw in 1998. The boat's captain turned the steering wheel over to Wabunun's Ogis. The night was cloudy, so the full sky was rarely visible. Yet Ogis obsessively looked up at whatever star peeked through an opening in the clouds. He used them to check the boat's compass and gauge where his *kut*-star was, that particular star being Kib, which was never visible.

Parallels among the several phenomena should be noted here. If at one level the thinking that dominates might be characterized as "prototype," it is also predicated on a continuous series of "initial conditions." These are original backsights of the place of departure and then the foresight when the target island comes into view. For constructing gardens it is the same. If star appearances and star deaths provide initial time referents and models for patterned sun/rain alternations, once a garden is underway its current conditions determine what comes next. The same is true with making a boat: the initial *kausilay* arc provides the immediate dimensions, but new and more refined determinations follow from and become increasingly exact as (tree) parts and their possibilities govern.

Out of the darkness of night comes one of the more profound memories these boats create, not by their actual sight but by the sound of shells announcing their appearance. When a boat arrives for a ritual, a trumpet shell should be blown for each pig that the crew carries to its landing. This captivating association is not one just for me. Men tell me women weep at the thought of that sound, now rarely heard. But also rarely seen for different reasons is the final preparation for every *anageg* when sailing for something important, usually a kula or mortuary ritual. For such an event the boat's hull should be covered with white lime then dotted black from burned coconut husks. The white is the blackness of night, the black dots the vibrant stars. This thought, usually erased by the waves, wind, and rain the craft meets during its journey, is the inverse of the starry night, which, as noted before, is also conceived to be a boat with its crew composed by stars. The boat's launching is an initial moment not unlike a star's rise or setting—both fluttering sights as they start their coursing. As in the case of many other models, it is a thought that measures, that reverberates across the domains of this social system. Men make these craft connecting the land, which has been made to produce their very materials and which they bind by their travels, by means of the heavens whose motions control their time and space. *Anageg* assemble into images of the whole the patterned sequences that enable their life.

Exploring the distant implications, historical and typological, of this mediation of the sky and the land is another work.

* * *

I left Muyuw in August 2009 fearing I would never again see Sipum's wife, Bwadibwad, for her loss seemed certain to follow his. But time brings new stories. By my return in 2012 her fate remained unsettled. Yet because of my close association with Bwadibwad's deceased husband, as a courtesy not a requirement, Sipum's remaining mother asked me to approve Bwadibwad's remarriage. In principle I did, though no new husband had been found. A strong, determined person, Bwadibwad was not ready for that next move, and she became angry with the permission I granted, and together we broke down in tears. Along with Ogis, Gumiya, and others, whenever I leave we hug for the last time.

Notes

1. Iteanu (2009) is very important for thinking through this order.
2. When *anageg* are beached, the hull tilts (*kantageg*) toward the outrigger float; when launched it straightens up, balances, *kanbwatay*.
3. An important encounter with civil engineer Chris Potulski managing Auridium Gold explorations in 1996 solidified this conclusion. After explaining the reasoning for the boat's tree selections, Potulski said, "Fred, the sea doesn't lie."
4. Connoting "participation" or "join," the Chinese character 参, *cān*, derives from 曑, the three "suns" on the top referencing Orion's Belt. On these matters I am indebted to Anne DiPiazzi, Erik Pearthree, Andrew Pawley, Zhang Bin, Luo Yang, and Wang Mingming.
5. The 1998 Nasikwabw *anageg* I examined was 23.5 centimeters wide at its bow prow board (*kunubwara*) and 29 centimeters wide at its stern, and 162.5 centimeters wide at its center. *Lavanay* dimensions were 34/30 centimeters, the distance across its width for the two central *kiyad* 183 and 182 centimeters.
6. All sources prefer *igsigis,* but the tree called *akidus* is satisfactory, and my Wabunun sources said *amanau* is a useful substitute. All have the necessary granular qualities, but *akidus* of sufficient length are thicker and require more trimming to go through the rib holes.
7. Lower-class craft do not show this reversal.
8. I partly follow Ellen (2006), as do Cristancho and Vining (2004), written in association with Ellen. Stéphane Breton's reaction to a Paris presentation started this line of thought.
9. Dumont draws from McLellan's version of the *Grundrisse* (1971). It became better known through the Martin Nicolaus translation and Foreword (Marx 1973).
10. An end of this technology came about 1973 when people stopped racing middle-sized outriggers.
11. Men younger than thirty always provided fast answers to questions; more experienced people offered slower, conditional answers, because they knew more variables.
12. I saw a mid-sized outrigger assembled in the open space in Wabunun's center, *takoven*, in the 1970s. This description derives from watching the model in 2012 and much discussion. A mid-sized outrigger is much easier to assemble than an *anageg*. It only has one prow board, *kunubwara,* into which the strakes must be slotted; a *kaybwag*'s "base" is a *matsibod,* a kind of cup making a different closure.
13. Budibud strakes are made from a tree called *nelau* (*Burkella obovate* [Forst.]

Pierre) and replaced with *kausilay* when a Budibud boat sails to Kavatan or Kaulay. The straighter-growing *kausilay* near the beach at Kaulay's Ulgulag are the replacements.

14. On *Lavanay,* 112 centimeters separated the "tip" of the keel and the *kunubwara;* 121 centimeters came between the "base" end and its *kunubwara.*

15. Nasikwabw people consider themselves from Misima, not Muyuw, and are sailors (this man was also a pastor), not gardeners. The garden model is not important for many Nasikwabw people, this man among them.

16. A Boagis craft had round *talapal,* one of which was thirty-one centimeters long, nine centimeters in circumference. I did not measure other three, because I didn't realize the two central ones should be higher.

17. A Boagis *kaynikw* was 74 centimeters long with a 4.5-inch circumference near an end tied to the *talapal,* a 7.6-inch circumference near its center.

18. "In general, it pays to design with nature's pulses rather than to confront them" (Odom, Odom and Odom 1995: 554).

19. Since 2002, this is said of Tamdak, the reigning elder in Yemga (see Damon 1983b).

20. I thank Leban Gisawa for providing *tadaya*'s identification (February 15, 2015). Wabunun-named turtle types are *tadiyay, pwanowan, pau'n,* and *wedal,* characterized by their colors, sizes, thickness of shell, egg-laying time and place, taste (smell) of meat and eggs, how long the smell might stay in one's body and house, and whether or not they are eaten—*wedal* usually not. Although *tadiayay*'s head is smaller than some, especially *pwanowan*'s, the animal is usually large, the meat tastes good, and its smell is gone by the following day.

21. Kaulay, central-Muyuw people, call these parts *sikusak,* after their official Ugwawag term, the name Muyuw people give the islands to their west.

22. Young women should gouge their lovers' thighs when they make love, and men proudly display the impressions the next day.

23. A picture displayed in the photographic section for chapter 1, taken in 2002 as our boat approached Nasikwabw, was identified by a Wabunun elder as a *mwag;* Bruce Beehler identified it as "a frigatebird, one of two possible species—Greater Frigatebird Fregata minor or Lesser Frigatebird Fregata ariel." These birds vary significantly by gender and maturity.

24. When someone dies, villagers destroy a tree planted in or immediately proximate to a village. Arriving boat crews scout a village's flora looking for such signs to deduce how to comport themselves when they step ashore. If a "friend" is not at the shoreline for the customary ritual arrival, that means that person has died or has suffered death calling for a specific funeral practice to reopen a path blocked by a fallen tree—the death.

25. Once when I asked why betel nuts are considered female while betel pepper are classed as male, an informant responded, "Because females go down first," alluding to intercourse positions, which chewing betel nut models.

26. Allen Allison, personal communication, August 25, 2009.

27. When a snake called *mutuiyou* is caught and killed, its tail should be snipped off and put in the bottom of a *kunusop/mastmount,* which is said to help the boat go faster.

28. I identified these according to names Muyuw people gave me during my 1973–75 research using Munro (1967). The Wabunun marine biologist and Inshore Fisheries manager in Port Moresby's Papua New Guinea National Fisheries Authority, Leban Gisawa, identified them this way (personal communication, January 26, 2014): *papis*, streamlined spinefoot (*Siganus argenteus*); *kwakwey*, blue sea chub (*Kyphosus cinerascens*); and *kobal*, striated surgeonfish (*Ctenochaetus striatus*).

29. This is a question I had to explore because the practice is so infrequent now. One person told me *anageg* floats were never burned; Dibolel denied that and said they were regularly burned.

30. The esprit also makes them dangerous: I witnessed a small child trip in the excitement and have its leg crushed by a roller.

31. Although he has not focused on the details of Trobriand canoes, Mosko's work (e.g. 2013) shows that they clearly play a vital role in the area's paradigmatic forms.

32. Gawan and Kweywatan men report that *kaboum* is difficult to work with because it is so hard and cross-grained. Knot-free branch sections or boles—perhaps the size of a large coconut tree—are cut into blocks approximating the length of the needed part, then split from the top down and finally adzed and sanded to their final shapes.

33. Since the advent of Muyuw's mining epoch (ca. 1895), the primary route between southeastern Muyuw and Kaulay/Dikwyas is by boat along the southeast coast up to the northern end of Sulog harbor, then by creek and swamp trail to Kulumadau, and finally by path, now bulldozed to a road, to these old villages. Still in the minds of many Kaulay people, a shorter more direct path goes east from Kaulay to the Ocelio River. It quickly puts people in sight of Kweyakowya, specifically and recently Sinamat on the northern hill overlooking southeastern Muyuw. Sinamat is no more than an hour's walk from villages that have for 1,500 years lined the southeast coast, Gusoapa in the east to Unamatan in the west. Twentieth-century communication routes make southeastern and north-central Muyuw farther apart than they once were.

34. Vayoun burn like *nilg'* too, very fast, but not hot.

35. Suggesting this structuring set: float : keel :: mast : mast mount :: prow : stern :: male : female.

36. In north-central Muyuw the tree is also used for *kuwu*, the smallest class of outrigger boat. *Nilg* used as floats elsewhere are sometimes turned into *kuwu* in eastern Muyuw.

37. Budibud *anageg* were often refit by means of their relatively regular visits to Kavatan, the only place where most people were capable speakers of the Budibud language.

38. And moving *kitoum* from Gawa and Kweywata, where there is a concentration of them because of *anageg* production, to central Muyuw, which rarely produces new *mwal*.

39. Fearing that too much of my calendrical and gardening information came from Wabunun, in 2012 I explicitly tested its model in Kaulay only to be told, "If you follow that pattern you find food" (*kan*).

References

Abers, Geoffrey. 2001. "Evidence of Seismogenic Normal Faults at Shallow Dips in Continental Rifts." In *Non-Volcanic Rifting of Continental Margins: A Comparison of Evidence from Land and Sea*, ed. R. C. I. Wilson, R. B. Whitmarsh, B. Taylor, and N. Froitzheim, 305–18. London: Geological Society, Special Publications 187.

Abers, Geoffrey, Aron Ferris, Mitchell Craig, Hugh Davies, Arthur L. Lerner-Lam, John C. Mutter, and Brian Taylor. 2002. "Mantle Compensation of Active Metamorphic Core Complexes at Woodlark Rift in Papua New Guinea," letters to *Nature, Nature* 418: 862–65.

Almeida, Maurow, B. De. 1990. "Symmetry and Entropy: Mathematical Metaphors in the Work of Lévi-Strauss." *Current Anthropology* 31(4): 367–385.

Anderson, J. A. R. 1980. *A Checklist of the Trees of Sarawak*. Forest Department, Kuching, Sarawak: Vanguard Press Sdn. Bhd.

APG III. 2009. "An Update of the Angiosperm Phylogeny Group Classification for the Orders and Families of Flowering Plants." *Botanical Journal of the Linnaean Society* 161: 105–21.

Atran, Scott. 1994. "Core Domains versus Scientific Theories: Evidence from Systematics and Itza-Maya Folkbiology." In *Mapping the Mind: Domain Specificity in Cognition and Culture*, ed. Lawrence A. Hirchfeld and Susan A. Gelman, 316–40. Cambridge: Cambridge University Press.

———. 1998. "Folk Biology and the Anthropology of Science: Cognitive Universals and Cultural Particulars." *Behavioral and Brain Sciences* 21(4): 547–69.

———. 1999a. "Folkbiology." In *The MIT Encyclopedia of the Cognitive Sciences*, ed. Robert Wilson and Frank Keil, 316–17. Cambridge: MIT Press.

———. 1999b. "Itzaj Maya Folkbiological Taxonomy: Cognitive Universals and Cultural Partiuclars." In *Folkbiology*, ed. Douglas L. Medin and Scott Atran, 119–203. Cambridge: MIT Press.

Austen, Leo. 1934. "Procreation among the Trobriand Islanders." *Oceania* 5: 102–13.

Baldwin, Suzanne L., Paul G. Fitzgerald, and Laura E. Webb. 2012. "Tectonics of the New Guinea Region." *Annual Review of Earth and Planetary Sciences* 40: 495–520.

Balée, William. 1998. *Advances in Historical Ecology*. New York: Columbia University Press.

Barrau, Jacques. 1959. "The Sago Palms and Other Food Plants of Marsh Dwellers in the South Pacific Islands." *Economic Botany* 13: 151–64.

Bateson, Gregory. 1979. *Mind and Nature: a Necessary Unity.* New York: Dutton.

Bayliss-Smith, Tim, Edvard Hviding, and Tim Whitmore. 2003. "Rainforest Composition and Histories of Human Disturbance in Solomon Islands." *AMBIO: A Journal of the Human Environment* 32(5): 346–52.

Bellwood, Peter, G. Nitihaminoto, G. Irwin, Gunadi, A. Waluyo, D. Tanudirjo. 1998. "35,000 Years of Prehistory in the Northern Moluccas." In *Bird's Head Approaches, Modern Quaternary Research in Southeast Asia* 15, ed. G. J. Bartstra: 233–75. Boca Raton: CRC Press

Berlin, Brent. 1992. *Ethnobiological Classification: Principles of Categorization of Plants and Animals in Traditional Societies.* Princeton: Princeton University Press.

Bickler, S. H. 1997. "Early Pottery Exchange along the South Coast of Papua New Guinea." *Archaeology in Oceania* 32: 11–22.

———. 1998. "Eating Stone and Dying: Archaeological Survey on Woodlark Island, Milne Bay Province, Papua New Guinea." Unpublished PhD dissertation, University of Virginia: Charlottesville. (Also published by UMI Microfilms Ltd.).

Bickler, S. H., and B. Ivuyo. 2002. "Megaliths of Muyuw (Woodlark Island), Milne Bay Province, PNG." *Archaeology in Oceania* 37: 22–36.

Bickler, S.H., and M. Turner. 2002. "Food to Stone: Investigations at the Suloga Stone Tool Manufacturing Site, Woodlark Island, Papua New Guinea." *Journal of the Polynesian Society* 111: 11–43.

Boyer, Pascal. 1996. "What Makes Anthropomorphism Natural: Intuitive Ontology and Cultural Representations." *Journal of the Royal Anthropological Institute* (N.S.) 2: 83–97.

Burgess, P. F. 1966. *Timbers of Sabh (Sabah Forest Records No. 6).* Forest Department, Sabah, Malaysia.

Burke, Kenneth. 1969. *The Grammar of Motives.* Berkeley: University of California Press.

Burkill, I. H. 1935. *A Dictionary of the Economic Products of the Malay Peninsula.* London: Governments of the Straits Settlements and Federated Malay States by the Crown Agents for the Colonies. 2 volumes.

———. 1944. "The Early Economic History of the Tree Mesua Ferrea (Guttiferae)." *Proceedings of the Linnaean Society of London* 156(2).

Campbell, Shirley F. 2002. *The Art of Kula.* New York: Berg Press.

Chomsky, Noam, and Morris Halle. 1968. *Sound Patterns of English.* New York: Harper and Row.

Coe, F. G., and G. J. Anderson. 1996. "Ethnobotany of the Garifuna of Eastern Nicaragua." *Economic Botany* 50: 71–107.

Coley, John D., Douglas L. Medin, Julia Beth Proffitt, Elizabeth Lynch, and Scott Atran. 1999. "Inductive Reasoning in Folkbiological Thought." In *Folkbiology,* ed. Douglas L. Medin and Scott Atran, 205–32. Cambridge: MIT Press.

Craven, L. A. 1990. "One New Species Each in *Acmena* and *Eucalyptopsis* and a New Name in *Lindsayomyrtus* (all Myrtaceae)." *Australian Systematic Botany* 3: 727–32.

Cristancho, S., and J. Vining. 2004. "Culturally Defined Keystone Species." *Human Ecology Review* 11(2): 153–64.

Crumley, Carole L. (ed.). 1994. *Historical Ecology*. Sante Fe: School of American Research Press.

Cunningham, Clark. 1964. "Order in the Atoni House." *Bijdragen tot de taal-land-en volkenkunde* 120: 34–68.

Damon, Frederick H. (MS) "Fifteen Years Among the Scientists: A Social Anthropologist Reflecting on his Encounter with the Natural Sciences and Mathematics." Conference paper for Conference 2006 "Challenges to Interdisciplinary Collaborative Research," December 14–16, Taipei, Taiwan Institute of Ethnology, Academia Sinica

———. 1979. "Woodlark Island Megalithic Structures and Trenches: Towards an Interpretation." *Archaeology and Physical Anthropology in Oceania* 14: 195–226.

———. 1980b. "The Problem of the Kula on Woodlark Island: Expansion, Accumulation, and Overproduction." *Ethnos* 45: 176–201.

———. 1983a. "What Moves the Kula: Opening and Closing Gifts on Woodlark Island." In *The Kula: New Perspectives on Massim Exchange*, ed. J. W. Leach and E. R. Leach, 309–42. Cambridge: Cambridge University Press.

———. 1983b. "The Transformation of Muyuw into Woodlark Island: Two Minutes in December, 1974." *Journal of Pacific History* 18(1): 35–56.

———. 1983c. "Muyuw Kinship and the Metamorphosis of Gender Labour." *Man* (N.S.) 18(2): 305–26.

———. 1989. "The Muyuw Lo'un and the End of Marriage." In *Death Rituals and Life in the Societies of the Kula*, ed. Frederick H. Damon and Roy Wagner, 73–94. DeKalb: Northern Illinois University Press.

———. 1990. *From Muyuw to the Trobriands: Transformations along the Northern Side of the Kula Ring*. Tucson: University of Arizona Press.

———. 1997. "Cutting the Wood of Woodlark: Retrospects and Prospects for Logging on Muyuw, Milne Bay Province, Papua New Guinea." In *The Political Economy of Forest Management in Papua New Guinea*, ed. Colin Filer, 180–203. NRI Monograph 32. Hong Kong: National Research Institute and International Institute for Environment and Development.

———. 1998. "Selective Anthropomorphization: Trees in the Northeast Kula Ring." *Social Analysis* 42(3): 67–99.

———. 2000a. "From Regional Relations to Ethnic Groups? The Transformation of Value Relations to Property Claims in the Kula Ring of Papua New Guinea." *Asia Pacific Journal of Anthropology* (formerly *Canberra Anthropology*) 1(2): 49–72.

———. 2000b. "To Restore the Events? On the Ethnography of Malinowski's Photography." Review of *Malinowski's Kiriwina: Fieldwork Photography 1915–1918*, by Michael W. Young. *Visual Anthropology Review* 16(1): 71–77.

———. 2005. "'Pity' and 'Ecstasy': The Problem of Order and Differentiated Difference across Kula Societies" In *On the Order of "Chaos": Social Anthropology and the Science of Chaos*, ed. Mark Mosko and Fred Damon, 79–107. New York: Berghahn Books.

———. 2007. "A Stranger's View of Bihar: Rethinking Religion and Production." In *Speaking of Peasants: Essays on Indian History and Politics in Honor of Walter Hauser*, ed. William Pinch, 249–76. New Deli: Manohar Publishers.

———. 2009. "Afterword: On Dumont's Relentless Comparativism." In *Hierarchy: Persistence and Transformation in Social Formations*, ed. Knut Rio and Olaf H. Smedal, 349–59. New York: Berghahn Books.

———. 2012. "'Labour Processes' across the Indo-Pacific: Towards a Comparative Analysis of Civilisational Necessities." *Asia Pacific Journal of Anthropology* 13(2): 163–91.

Davis, Mike. 2002. *Late Victorian Holocausts: El Niño Famines and the Making of the Third World*. London: Verso.

Dean, Kenneth, and Zhenman Zheng. 2010. *Ritual Alliances of the Putian Plain*. Vol. 1, *Historical Introduction to the Return of the Gods*. Leiden: Brill.

Di Piazza, Anne, and Erik Pearthree (eds.). 2008. *Canoes of the Grand Ocean*. BAR International Series 1802. Oxford: Hadrian Books Ltd.

Douglas, Mary. 1966. *Purity and Danger*. New York: Frederick A. Prager.

Dumont, Louis. 1977. *From Mandeville to Marx*. Chicago: University of Chicago Press.

Ehara, Hiroshi Slamet Susanto, Chitoshi Mizota, Shohei Hirose, and Tadashi Matsuno. 2000. "Sago Palm (*Metroxylon Sagu*, Arecaceae) Production in the Eastern Archipelago of Indonesia: Variation in Morphological Characteristics and Pith Dry-Matter Yield." *Economic Botany* 54(2): 197–206.

El Guindi, Fadwa. 2008. *By Noon Prayer: The Rhythm of Islam*. Oxford; New York: Berg, 2008.

Ellen, Roy. 2004. "The Distribution of *Metroxylon sagu* and the Historical Diffusion of a Complex Traditional Technology." In *Smallholders and Stockbreeders: History of Foodcrop and Livestock Farming in Southeast Asia*, edited by Peter Boomgaard and David Henley. 69–105. Leiden: KITLV Press.

———. 2006a. "Fetishism: A Cognitive Approach." In *The Categorical Impulse: Collected Essays in the Anthropology of Classifying Behavior*, 166–89. Oxford: Berghahn.

———. 2006b. "Local Knowledge and Management of Sago Palm (*Metroxylon sagu Rottbeoll*) Diversity in South Central Seram, Maluku, Eastern Indonesia." *Journal of Ethnobiology* 26(2): 258–98.

Fernandez, James. 1974. "The Mission of Metaphor in Expressive Culture." *Current Anthropology* 15(2): 119–45.

———. 1998. "Tress of Knowledge of Self and Other in Culture: On Models of the Moral Imagination." In *The Social Life of Trees: Anthropological Perspectives on Tree Symbolism*, ed. Laura Riva, 81–110. New York: Berg.

Flannery, Timothy F. 1994. *The Future Eaters: An Ecology History of Australasia Land and Peoples*. Chatswood, NSW: Reed.

———. 2001. *The Eternal Frontier: An Ecological History of North America and Its Peoples*. New York: Grove Press.

Fox, James J. 1971. "Sister's Child as Plant: Metaphors in an Idiom of Consanguinity." In *Rethinking Kinship and Marriage*, ed. Rodney Needham, 219–52. New York: Tavistock Publications.

Fox, James J. (ed.). 1993. *Inside Austronesian Houses: Perspectives on Domestic Designs for Living*. Canberra: Anthropology, Research School of Pacific Studies.

Fox, James J., and Clifford Sather (eds.). 1996. *Origins, Ancestry and Alliance: Explorations in Austronesian Ethnography*. Canberra: Australian National University.

Freeman, J. R. 1999. "Gods, Groves and the Culture of Nature in Kerala." *Modern Asian Studies* 33(2): 257–302.

French, Bruce R. 1986. *Food Plants of Papua New Guinea: A Compendium*. Australia Pacific Science Foundation. Published by author.

Gamble, J. S. 1972 [1902]. *A Manual of Indian Timbers*. 2nd edition. Dehra Oun: Bishen Singh Mahendra Pal Singh.

Gammage, Bill. 2011. *The Biggest Estate on Earth*. London: Allen and Unwin.

Gell. Alfred. 1992. *The Anthropology of Time: Cultural Constructions of Temporal Maps and Images*. New York: Berg.

———. 1999. *The Art of Anthropology: Essays and Diagrams*. Edited by Eric Hirsch. London: Athlone Press.

Gladwin, Thomas. 1970. *East Is a Big Bird*. Cambridge: Harvard University Press.

Gregory, Christopher A. 1982. *Gifts and Commodities*. New York: Academic Press.

———. 1997. *Savage Money: The Anthropology and Politics of Commodity Exchange*. Amsterdam: Harwood Academic.

Gunn, Joel. 1994. "Global Climate and Regional Biocultural Diversity." In *Historical Ecology*, ed. Carole L. Crumley, 67–99. Santa Fe: School of American Research Press.

Haddon, A. C., and J. Hornell. 1936–38. *Canoes of Oceania*. Honolulu: Bishop Museum Press, Special Publication Nos. 27, 28, and 29.

Hecht, Susanna B. 2004. "Indigenous Soil Management and the Creation of Dark Earths: Implications of Kayapo Practices." In *Amazonian Dark Earths: Origin, Properties and Management of Fertile Soils*, ed. Johannes Lehmann, Dirse C. Kern, Bruno Glaser, William I. Woods, 355–72. Springer Netherlands: Kluwer Academic Publishers.

Hecht, Susanna B., and Darrell A. Posey. 1989. "Preliminary Results on Soil Management Techniques of the Kayapó Indians." *Advances in Economic Botany* 7: 174–88.

Heinsohn, Tom. 1998. "Captive Ecology." *Nature Australia* 26(2): 36–53.

Hénaff, Marcel. 2009. "Lévi-Strauss and the Question of Symbolism." In *The Cambridge Companion to Lévi-Strauss*, ed. Boris Wiseman, trans. Yves Gilonne and Jean Louis Morhange, 177–95. Cambridge: Cambridge University Press.

Hide, R. L., R. M. Bourke, B. J. Allen, T. Betitis, D. Fritsch, R. Grau, L. Kurika, E. Lowes, D. K. Mitchell. 2002 [1994]. "Milne Bay Province: Text Summaries, Maps, Code Lists and Village Identification." Agricultural Systems of Papua New Guinea Working Paper No. 6, Canberra, Department of Human Geography, Research School of Pacific Studies, Australian National University.

Hocart, A. M. 1970. *Kings and Councilors: An Essay in the Comparative Anatomy of Human Society*, ed. and intro. Rodney Needham, Chicago: University of Chicago Press.

Horridge, Adrian. 1987. *Outrigger Canoes of Bali and Madura, Indonesia*. Honolulu: Bishop Museum Press.

———. 2008. "Origins and Relationships in Pacific Canoes and Rigs." In *Canoes of the Grand Ocean,* ed. Anne Di Piazza and Erik Pearthree, 85–105. BAR International Series 1802. Oxford: Hadrian Books Ltd.

Howard, Alan. 1995. "Rotuman Seafaring in Historical Perspective." In *Seafaring in the Contemporary Pacific Islands: Studies in Continuity and Change,* ed. Feinberg Richard, 114–43. DeKalb: Northern Illinois University Press.

Irwin, G., S. Bickler, and P. Quirke. 1990. "Voyaging by Canoe and Computer: Experiments in the Settlement of the Pacific Ocean." *Antiquity* 64: 34–50.

Irwin, Geoffrey. 1994. *The Prehistoric Exploration and Colonisation of the Pacific.* Cambridge: Cambridge University Press.

Iteanu, André. 2009. "Hierarchy and Power: A Comparative Attempt under Asymmetrical Lines." In *Hierarchy: Persistence and Transformation in Social Formations,* ed. Knut M. Rio and Olaf H. Smedal, 331–48. New York: Berghahn Books.

Kennedy, Jean, François Wadra, Urgei Akon, Roselyne Busasa, Josephine Papah, and Matthew Piamok. 1991. "Site Survey of Southwest Manus: A Preliminary Report." *Archaeology of Oceania* 26: 114–18.

Kirch, Patrick V. 1989. "Second Millennium B.C. Arboriculture in Melanesia: Archaeological Evidence from Mussau Islands." *Economic Botany* 43(2): 225–40.

———. 1995. *The Wet and the Dry: Irrigation and Agricultural Intensification in Polynesia.* Chicago: University of Chicago Press.

Kirch, Patrick, and Roger Green. 2001. *Hawaiki, ancestral Polynesia: an essay in historical anthropology.* New York: Cambridge University Press.

Kirch, Patrick, and Terry Hunt. 1997. *Historical Ecology in the Pacific Islands: Prehistoric Environmental and Landscape Change.* New Haven: Yale University Press.

Kolbert, Elizabeth. 2006. *Field Notes from a Catastrophe: Man, Nature, and Climate Change.* New York: Bloomsbury USA.

Kraus, Fred, Allen Allison, Thane Pratt, Dan A. Polhemus, John Slapcinsky, Kristofer Helgen. 2003. "Wildlife Resources Survey and Analysis of Milne Bay Province." Final Report to the Department of Environment and Conservation, Papua New Guinea.

Küchler, Susanne. 2001 "Why Knot: Towards a Theory of Art and Mathematics." In *Beyond Aesthetics: Art and the Technologies of Enchantment,* ed. Christopher Pinney and Nicholas Thomas, 57–79. New York: Berg.

Kurin, Richard. 1983. "Indigenous Agronomics and Agricultural Development in the Indus Basin." *Human Organization* 42(4): 283–94.

Lansing, John Stephen. 1991. *Priests and Programmers: Technologies of Power in the Engineered Landscape of Bali.* Princeton: Princeton University Press.

———. 1995. *The Balinese.* Fort Worth, TX : Harcourt Brace College Publishers.

———. 2006. *Perfect Order: Recognizing Complexity in Bali.* Princeton: Princeton University Press.

Leach, Edmund R. 1957. "The Epistemological Foundations to Malinowski's Empiricism." In *Man and Culture: An Evaluation of the Work of Bronislaw Malinowski,* ed. Raymond Firth, 119–37. London: Routledge and K. Paul.

Lemonnier, Pierre. 1986. "The Study of Material Culture Today: Toward an Anthropology of Technical Systems." *Journal of Anthropological Archaeology* 5: 147–86.

————. 2012. *Mundane Objects: Materiality and Non-verbal Communication*. Walnut Creek, CA: Left Coast Press.

Lévi-Strauss, Claude. 1963. "Social Structure." In *Structural Anthropology,* trans. Claire Jacobson and Brooke Grundfest Schoepf, 277–323. New York: Basic Books.

————. 1966. *The Savage Mind*. Chicago: University of Chicago Press.

Lipset, David M., and Kathleen Barlow. 1997. "Dialogics of Material Culture: Male and Female in Murik Outrigger Canoes." *American Ethnologist* 24(1): 4–36.

Mahias, Marie-Claude. 1993. "Pottery Techniques in India: Technical Variants and Social Choice." In *Technological Choices: Transformation in Material Cultures since the Neolithic,* ed. Pierre Lemonnier, 157–80. New York: Routledge.

Manguin, Pierre-Yves. 1986. "Shipshape Societies: Boat Symbolism and Political Systems in Insular Southeast Asia." In *Southeast Asia in the 9th to 14th Centuries,* ed. David G. Marr and A. C. Milner, 187–215. Singapore: Institute of Southeast Asian Studies; and Canberra: Research School of Pacific Studies, Australian National University.

————. 2012. "Asian Ship-Building Traditions in the Indian Ocean at the Dawn of European Expansion." In *The Trading World of the Indian Ocean, 1500–1800,* ed. Om Prakash, 597–629. History of Science, Philosophy and Culture in Indian Civilization 3, part 7. Ed. D. P. Chattopadhyaya. Delhi: Pearson Education and Centre for Studies in Civilizations.

Mann, M. E., Z. Zhang, S. Rutherford, R. S. Bradley, M. K. Hughes, D. Shindell, C. Ammann, G. Falugevi, F. Ni. 2009. "Global Signatures and Dynamical Origins of the 'Little Ice Age' and 'Medieval Climate Anomaly,'" *Science* 326: 1256–60.

Marx, Karl. 1973. *Grundrisse: Foundations of the Critique of Political Economy,* trans. with a foreword by Martin Nicolaus. New York: Vintage Books.

McClatchey, Will, Harley I. Manner, and Craig R. Elevitch. 2006. "Metroxulong amicarum, M. paulcoxii, M. sago, M. salomonense, M. vitiense, and M. warburgii (Sago Palm)" in *Species Profiles for Pacific Island Agroforestry*. www.traditionaltree.org.

McLellan, David (ed. and trans.). 1971. *Marx's Grundrisse*. London: Macmillan.

Mehrotra, S., R. Mitra, and H. P. Sharma. 1986. "Pharmacognostic Studies on Punnaga, Calophyllum inophyllum L., Leaf and Stem Bark." *Herba Hungaria* 25(1): 45–71.

Mills, C. Wright. 1963. "The Language and Ideas of Ancient China." In *Power, Politics and People: The Collected Essays of C. Wright Mills,* ed. and intro. Irving Louis Horowitz, 469–520. New York: Ballantine.

Mintz, Sydney. 1985. *Sweetness and Power: The Place of Sugar in Modern History*. New York: Viking.

Mohd, Wahid Samsudin, M. Nor Ibrahim, and Ikram M. Said. 1998. "The ASEAN Review of Biodivdersity and Environmental Conservation." Article I, "Composition of the Steam Volatile Oil from Calophyllum inophyllum." http://www.metla.fi/archive/forest/1998/05/msg00161.html.

Mosko, Mark. 2013. "Omarakana Revisited, or 'Do Dual Organizations Exist?' in the Trobriands." *Journal of the Royal Anthropological Institute* (N.S.) 19: 482–509.

Mosko, Mark, and Fred Damon (eds.). 2005. *On the Order of "Chaos": Social Anthropology and the Science of Chaos*. New York: Berghahn Books.

Munn, Nancy D. 1977. "The Spatiotemporal Transformation of Gawa Canoes." *Journal de la Société des Océanistes* 33: 39–54.

———. 1983. "Gawan Kula: Spatiotemporal Control and the Symbolism of Influence." In *The Kula: New Perspectives on Massim Exchange*, ed. J. W. Leach and E. R. Leach, 277–308. Cambridge: Cambridge University Press.

———. 1986. *The Fame of Gawa*. Cambridge: Cambridge University Press.

———. 1992. "The Cultural Anthropology of Time: A Critical Essay." *Annual Review of Anthropology* 21: 93–123.

Munro, Ian S. R. 1967. *The Fishes of New Guinea*. Sydney, NSW: Victor C. N. Blight, Government Printer.

Nelson, Hank. 1976. *Black, White and Gold: Gold Mining in Papua New Guinea, 1878–1930*. Canberra: Australian National University Press.

Nordholt, H. Schulte. 1996. *The Spell of Power: A History of Balinese Politics, 1650–1940*. Leiden: KITLV Press.

Norris, Christopher A. 1999. "Phalanger lullulae." *Mammalian Species* (620): 1–4.

Notis, Christine. 2004. *Phylogeny and Character Evolution of Kielmeyeroideae (clusiaceae) Based on Molecular and Morphological Data*. Gainesville: University of Florida. Internet resource.

Nunn, P. D. 2000. "Environmental Catastrophe in the Pacific Islands about AD 1300." *Geoarchaeology* 15: 715–40.

Nunn, P. D., and Britton, J. M. R. 2001. "Human-Environment Relationships in the Pacific Islands around AD 1300." *Environment and History* 7: 3–22.

Odum, William E., Eugene P. Odum, and Howard T. Odum. 1995. "Nature's Pulsing Paradigm." *Estuaries* 18(4): 547–55.

Osmund, Meredith. 1998. "Fishing and Hunting Implements." In *The Lexicon of Proto Oceanic: The Culture and Environment of Ancestral Oceanic Society. 1 Material Culture*, ed. Malcom Ross, Andrew Pawley, and Meredith Osmond, Series C.152: 211–32. Canberra: Pacific Linguistics, Research School of Pacific and Asian Studies, Australian National University.

Oxford University. 1987. "Woodlark '87: The 1987 Oxford University Expedition to Papua New Guinea." Expedition members: Chris Norris, Sally Haiselden, Rachel Lambert, Fraser Smith, Ilaiah Bigilale.

———. 1988. "Woodlark '88: The 1988 Oxford University Expedition to Papua New Guinea." Expedition members: Sally Haiselden, Rachel Lambert, Katherine van der Lee, Paul Carson, Roger Price, Ilaiah Bigilale.

Pawley, Andrew, and Medina Pawley. 1998. "Canoes and Seafaring." In *The Lexicon of Proto Oceanic: The Culture and Environment of Ancestral Oceanic Society. 1 Material Culture*, ed. Malcom Ross, Andrew Pawley, and Meredith Osmond, Series C. 152: 173–209. Canberra: Pacific Linguistics, Research School of Pacific and Asian Studies, Australian National University.

Perry, Lily M., with Judith Metzger. 1980. *Medicinal Plants of East and Southeast Asia: Attributed Properties and Uses*. Cambridge: MIT Press.

Person, R. S., and H. P. Brown. 1931. *Commercial Timbers of India*. Calcutta: Central Publication Branch.

Petitot, Jean. 2009. "Morphology and Structural Aesthetics: From Goethe to Lévi-Strauss." In *The Cambridge Companion to Lévi-Strauss,* ed. Boris Wiseman, 275–95. Cambridge: Cambridge University Press.

Pfaffenberger, Brian. 1992. "Social Anthropology of Technology." *Annual Review of Anthropology* 21: 491–516.

Posey, Darrell A. 1985. "Indigenous Management of Tropical Forest Ecosystems: The Case of Kayapó Indians of the Brazilian Amazon." *Agroforestry Systems* 3(2): 139–58.

Posey, Darrell A., and Kristina Plenderleith (eds.). 2002. *Kayapó Ethnoecology and Culture.* Studies in Environmental Anthropology. New York: Routledge.

Randhawa, M. S. 1969. *Flowering Trees.* New Delhi: National Book Trust.

Reyes, Luis. 1938. *Philippine Woods.* Manila: Bureau of Printing.

Rival, Laura. 1993. "The Growth of Family Trees: Understanding Huaorani Perceptions of the Forest." *Man* (N.S.) 28(4): 635–52.

———. 1998. "Trees, from Symbols of Life and Regeneration to Political Artefacts." In *The Social Life of Trees: Anthropological Perspectives on Tree Symbolism,* ed. Laura Rival, 1–36. New York: Berg.

Roads, James W. 1982. "Sagopalm Management in Melanesia: An Alternative Perspective." *Archaeology in Oceania* 17: 20–27.

Robbins, Joel. 2004. *Becoming Sinners: Christianity and Moral Torment in a Papua New Guinea Society.* Berkeley: University of California Press.

———. 2009. "Conversion, Hierarchy, and Cultural Change: Value and Syncretism in the Globalization of Pentecostal and Charismatic Christianity." In *Hierarchy: Persistence and Transformation in Social Formations,* ed. Knut Rio and Olaf H. Smedal, 65–88. New York: Berghahn Books.

Ross, Malcom, Andrew Pawley, and Meredith Osmond (eds.). 1998. *The Lexicon of Proto Oceanic: The Culture and Environment of Ancestral Oceanic Society.* Series C. 152 1 *Material Culture.* Canberra: Pacific Linguistics, Research School of Pacific and Asian Studies, Australian National University.

———. 2003. *The Lexicon of Proto Oceanic: The Culture and Environment of Ancestral Oceanic Society.* Series C. 152 2 *The Physical Environment.* Canberra: Pacific Linguistics, Research School of Pacific and Asian Studies, Australian National University.

———. 2008. *The Lexicon of Proto Oceanic: The Culture and Environment of Ancestral Oceanic Society.* Series C. 152 3 *Plants.* Canberra: Pacific Linguistics, Research School of Pacific and Asian Studies, Australian National University.

Ruddiman, William F. 2003. "The Anthropogenic Greenhouse Era Began Thousands of Years Ago." *Climatic Change* 61, 261–93.

———. 2010. *Plows, Plagues, and Petroleum: How Humans Took Control of Climate.* Princeton Science Library. Princeton: Princeton University Press.

Sahlins, Marshal. 1992. "The Economics of Develop-man in the Pacific." *RES:* 13–25.

Sather, Clifford. 1993. "Posts, Hearths, and Thresholds: The Iban Longhouse as a Ritual Structure." In *Inside Austronesian Houses: Perspectives on Domestic Designs for Living,* ed. James J. Fox, 65–115. Canberra: Department of Anthropology in association with Comparative Austronesian Project, Research School of Pacific Studies, Australian National University.

Scoditti, Giancarlo M. G. 1990a. *Kitawa: A Linguistic and Aesthetic Analysis of Visual Art in Melanesia.* New York: Mouton de Gruyter.

———. 1990b. "The 'Golden Section' on Kitawa Island." *Culture and History in the Pacific.* Helsinki: The Finnish Anthropological Society: 233–66.

Scott, William Henry. 1982. "Boat-Building and Seamanship in Classic Philippine Society." *Philippines Studies* 30(3): 335–76.

Seligman, Charles. 1910. *The Melanesians of British New Guinea,* with a chapter by F. R. Barton and an appendix by E. L. Giblin. Cambridge: Cambridge University Press.

Shugart, H. H. 2014. *Foundation of the Earth: Global Ecological Change and the Book of Job.* New York: Columbia University Press.

Soerianegara, I. and RHMJ. Lemmens, RHMJ (eds.). 1993. Timber Trees - Major Commercial Timbers Plant Resources of South-East Asia, No. 5(1). Wageningen, Netherelands: PUDOC Scientific Publishers.

Southon, Michael. 1995. *The Navel of the Perahu.* Canberra: Australian National University Department of Anthropology.

Stevens, Peter F. 1974. "A Review of Calophyllum L. (Guttiferae) in Papuasia." *Australian Journal of Botany* 22: 349–411.

———. 1980. "Revision of the Old World Species of Calophyllum (Guttiferae)." *Journal of the Arnold Arboretum* 61.

———. 1994. *The Development of Biological Systematics: Antoine-Laurent de Jussieu, Nature and the Natural System.* New York: Columbia University Press.

———. (2001 onwards). Angiosperm Phylogeny Website. Version 12, http://www.mobot.org/MOBOT/research/APweb/.

Swap, Robert J., M. Garstang, S. A. Macko, P. D. Tyson, W. Maenhaut, P. Artaxo, P. Kallberg, and R. Talbot. 1996. "The Long Range Transport of Southern African Aerosols to the Tropical South Atlantic." *Journal of Geophysical Research* 101: 23777–91.

Taylor, Christopher. 1992. *Milk, honey, and money: changing concepts in Rwandan healing.* Washington: Smithsonian Institution Press.

———. 2001. *Sacrifice as Terror: The Rwandan Genocide of 1994.* Bloomsbury Academic.

Teeler, Sharon. 2001 "Feelings and Forces behind the Flow of Trees in Japan," PhD qualifying paper. University of Virginia.

Tenzer, Michael. 1991. *Balinese Music.* Singapore: Periplus Editions Inc.

Tochilin, Clare, William R. Dickinson, Matthew W. Felgate, Mark Pecha, Peter Sheppard, Frederick H. Damon, Simon Bickler, George E. Gehrels. 2012. "Sourcing Temper Sands in Ancient Ceramics with U–Pb Ages of Detrital Zircons: A Southwest Pacific Test Case." *Journal of Archaeological Science* 39(7): 2583–91.

Tone, Nat J. 1903 "Canoe Making in Olden Times." *Journal of the Polynesian Society* 11: 124.

Traube, Elizabeth. 1986. *Cosmology and Social Life: Ritual Exchange among the Mambai of East Timor.* Chicago: University of Chicago Press.

Turner, Terence. 1990. "On Structure and Entropy: Theoretical Pastiche and the Contradictions of 'Structuralism.'" *Current Anthropology* 31(5): 563–68.

Turner, Victor. 1967. *The Forest of Symbols: Aspects of Ndembu Ritual.* Ithaca, NY: Cornell University Press.

Uchiyamada, Yasushi. 1998. "'The Grove Is Our Temple:' Contested Representations of *Kaavu* in Kerala, South India." In *The Social Life of Trees: Anthropological Perspectives on Tree Symbolism*, ed. Laura Rival, 177–96. New York: Berg.

Vitousek, P. M., T. N. Ladefoged, P. V. Kirch, A. S. Hartshorn, M. W. Graves, S. C. Hotchkiss, S. Tuljapurkar, and O.A. Chadwick. 2004. "Soils, Agriculture, and Society in Precontact Hawai'i." *Science* 304: 1665–69.

Wallace, Anthony F. C. 1978. *Rockdale: The Growth of an American Village in the Early Industrial Revolution.* New York: Knopf.

———. 1987. *St. Clair: A Nineteenth-Century Coal Town's Experience with a Disaster-Prone Industry.* New York: Knopf.

Waterson, Roxana. 1990. *The Living House: An Anthropology of Architecture in Southeast-East Asia.* New York: Whitney Library of Design.

Wheatley, Paul. 1983. *Nagara and Commandery: Origins of the Southeast Asian Urban Traditions.* Chicago: University of Chicago, Deptartment of Geography.

Whistler, Arthur. 1992. *Polynesian Herbal Medicine.* Hong Kong: Everbest Printing Co. Ltd.

Wills, Garry. 2002. *Mr. Jefferson's University.* Washington, DC: National Geographic Directions.

Wilson, Edward O. 2002. *The Future of Life.* New York: Alfred A. Knopf.

Winterhalder, Bruce. 1994. "Concepts in Historical Ecology: The View from Evolutionary Theory." In *Historical Ecology*, ed. Carole L. Crumley, 17–42. Sante Fe: School of American Research Press.

Woodford, C. M. 1890. *A Naturalist among the Headhunters.* London: Philip George and Son.

Yen, Douglas. 1990. "Environment, Agriculture and the Colonization of the Pacific." In *Pacific Production Systems: Approaches to Economic Prehistory*, ed. Douglas Yen and J. M. J. Mummery, 258–77. Canberra: Department of Prehistory, Australian National University.

Young, Michael. 1998. *Malinowski's Kiriwina; Fieldwork Photography 1915–1918.* Chicago: University of Chicago Press.

Index

Aesthetic, 45, 112, 197, 228, 232–33, 266, 296, 301, 304, 325

Aisi, xiv, 4, 93, 141, 218

Akobwow (*Macaranga tanarius [l.]Muell.-Arg*), 126–7, 307, 345

Akewal (*See also* D. Arnold), 55, 67, 99–100, 108, 112, 128

Akidus (*Rubiaceae*), 157, 160, 312–13, 339, 345, 350n6

Akuluiy (Mastixiodendron *smithii M.& P.*), 55, 110, 126–27, 141, 156, 159–60, 183, 222, 324

Alidad (*See also* crossbeam. S. *stipulare* (Blume) Craven & Hartley), 98–99, 118n16, 299, 338

Aligeuna (String Figure expert), 285, 294n30

Allison, Allen, 243n36, 294n21, 332, 351n26

Alocasia sp. (Vesop), 55, 66–67, 81

Alotau (Milne Bay Province Capital), xiii, 6–7, 15, 27, 141, 248, 254, 256, 302

Amanau (*yamanau, Garcinia sp.*), 106–07, 117n5, 155, 160, 274, 299, 338–39, 350

Anacardiaceae, 38, 91

Anageg, 5, 6, 7, 9, 15, 44, 98, 133, 137, 145, 151, 193, 193, 200, 215, 219, 223, 234, Chapter 6
  Bwadanakup, 299, 306, 333, 343
  Design, 9, 193–94, 229, Chapter 6
  Gawa versus Kweywata, 234–5, 299, 302–04, 306, 311, 337
  Kaiyelu, 241n18
  Kemuyuw/Kemurua, 7, 138, 172, 219, 229, 230, 297, 308

Lavanay, 8, 193, 230, 296, 309–14, 319, 333, 342–43 350n5, 351n14
  Shape, 298, 350n2, 299
  Tying method (*see also kumis*), 264, 265, 267
  Umaye, 341–42

Aniyan as power, 218, 290, 326, 335, 345

Antunat (*Palaquium sp.*), 126–7, 239, 305

Apul (*See also* C. *apul* P.F. Stevens), 198–205, 220, 222, 226, 227, 243n31, 244n50, 244n52, 291

Arboriculture, 61

Asibwad (*Timonius timon*), 55, 66, 315, 343

Atran, Scott, 20, 81, 87–88 91, 116

Atulab (*Alstonia brassii*), 54–55, 58, 66, 71–74, 128, 167, 298, 343

Aukuwak (*Ixora cf.* asme Guill.), 111–2, 119n33, 126, 304

Aunutau (*See also* tree distribution), 128, 176n9, 238–29

Australian National University, xii, 5

Austronesian, Austronesian Studies, xii-xiii, 4, 83, 90, 112–13, 141, 174, 186, 248, 260–61, 289, 292n5, 298, 302, 317

Aynikoy (*See also* C. *soulattri*), 191, 208–219, 220, 231, 232, 326, 345

Ayniyan (*See also* C. *goniocarpum* P.F Stevens), 191, 196, 208–219, 220, 231, 232, 273, 326, 345

Ayovay (*Hibiscus tiliaceus*), 249, 250, 262, 272, 273, 274, 275, 292n3, 293n7, 293n10, 280

Balau (back- and forestays, *sipbalau knot*), 210, 216, 262, 266–67, 272, 275–78, 317

    Balau powan (testicle), 327–28

Balée, William, 20–21

Bali, Balinese, xi, 18, 33, 112–4, 179, 292n1, 344

Banded sea snake, (*Laticauda colubrine, mwatabwalay*), 332 335–36

Barrau, Jacques, 142

Base/tip, 43, 49, 113–6, 125, 141, 161, 165, 168, 173, 196, 210, 228, 237, 250, 261, 264, 265, 269, 271, 274, 299, 300–01, 309, 311, 319, 333–34, 346

Base/tongue, 269, 319

Bayliss-Smith, Tim, 205

Beauty, beautiful, 44, 104, 135, 217, 247–48, 283, 285, 296, 304

Bêche-de-mer, 26–7

Beehler, Bruce, 351n23

Bellows, Laura, 33n9

Berlin, Brent, 20, 87–88, 108, 111, 116

Betel nut, 55, 102, 122, 166, 178n33, 259

Betel pepper, 55, 66–67, 115, 251

Bickler, Simon, xiii–xiv, 5, 6, 22, 34n19, 77n24, 177n21, 221, 244n46

Big men (*See also guyau*), 218, 346

Biome, 21, 128, 345

Bird, 51, 88–89, 123–24, 133, 192, 209, 241n17, 257, 263, 288, 315, 319, 328

    Bird life, 13, 14

    Birds on outrigger forms, 328, 351n23

    Birds and sago, 143

    Birds and time, 327

    Birds and tree distribution, 127, 238

    Chickens, roosters (*See also* kuk), 195

Bis (*See also* streamers, telltales), 12, 14, 217–18, 331, 347

Boagis, 129, 151, 170, 172, 195, 215, 293, 309, 311, 322, 3, 338, 341, 343, 348, 351

Bod (*See also* group), 90, 92, 93–107

Body imagery, 210–11, 229, 248, 259–60, 300, 305, 308, 312, 314, 317, 321–325, 326–27, 331

Boom (*See* Kaley/kunay (boom/yard))

Bourke, Mike, 38, 76n19, 77n31

Breadfruit trees, 61, 129

Breton, Stéphan, 350n8

Bryson, Reid, 21, 34n16

Budakay (*See also* strake), 45, 152, 307

Budibud Is., 22, 31, 50, 77n31, 129, 144, 150, 169, 193, 197, 225, 235, 268, 270, 341, 350–351n13, 352n37

Bungalau (*See also* meadow), 69, 139, 144–49, 203, 206, 208

Bwaboun (Imperial pigeon), 123–24

Bwanabwana, 7, 33n5, 264

Calendar, 202, 289, 352n9

Calophyllum, 9, 14, 20, 42, 87, 127, 133–4, 142, 177n26, 178n35, 180–244, 250, 256, 273, 345

    C. apul P.F. stevens (*See also apul*), 177n27, 190, 198–205, 242n25

    C. euryphyllum, 205

    C. inophyllum (Beach C. *See also kakam*), 9, 180–81, 188, 190–98, 242n19, 244n48, 295n32

    C. leleanii P.F. Stevens (*See also kausilay*), 37, 107, 190, 219–240

    C. obscurum, 189

    C. goniocarpum P.F Stevens (*See also ayniyan*), 190, 208–219

    C. peekelii L., 190, 199, 242n25, n26, 243n28

    C. soulattri (*See also aynikoy*), 186, 187, 190, 241n9, 208–219, 243n42

    C. vexans P.F. Stevens (*See also* Dan), 142, 145, 147, 190, 205–208

    Comparison with sugar, 183–4

    Medicinal uses, 185–86, 188, 241n8

    And *Mesua ferrea*, 186–88

Campbell, Shirley, xiii, 85, 179n42, 325

Canarium sp., 123, 147, 175n4, 177n20

Canoe (*See* outrigger)

Carbon, Carbon Isotopes, 63–64, 78n35

Casuarina littorale (*See also* yay tree), 83, 176n11, 166

Central Muyuw (South or North...) (*See also* Wamwan), 12, 14, 26, 29, 32, 36, 37, 40–42, 49, 51, 56, 58, 60, 64, 67, 71–72, 77n27, 92, 95, 100–01, 103–06, 110, 115–6, 117n2, 118n24, 119n34, 129–30, 132, 139, 141, 146–47, 155, 158, 161, 172, 176n7, 177n23, 177n24, 178n28, 178n31, 179n38, 198, 203, 208, 221, 225, 229, 256, 259, 274, 279, 288, 312, 321, 325, 332, 337–39, 342, 346, 352n33, 352n38
    Tasim in C. Muyuw, 83, 133–34

Chaos, Chaos theory, 120n38, 219, 229, 297, 316, 320, 349

China, Chinese, xiii, 4, 17, 19, 292n2, 295n33, 295n35, 299, 350n4
    Fujian Province, xiii, 294n28
    Granet, M. (C. W. Mills), 289
    Quanzhou, xiii
    Yunnan Province, xiii, 17

Christensen, Rolly, 26, 28, 198, 200, 204

Christianity, Christian Missions, 23, 27–30, 35n26, 36n27, 59, 236
    Catholic Church, 29
    Christian Revival Church, 29

Clan/subclan (*See also* Kum/dal), 12, 43–44, 49, 76n17, 90–91, 94, 113, 118n13, 121, 128, 141, 144, 151–53, 175n4, 238, 251, 259, 303, 340–41

Classification, 20, 87–113

Classifiers (noun), 49, 89–90, 107, 115, 141, 244n53, 250, 280, 286, 288

Clay pots, pots, potshards, 9, 22–23, 30, 34n20, 40, 77n30, 118, 122, 151, 154, 173, 175, 178, 222–23, 270, 329

Climate epochs, 21, 23
    Little Ice Age, 21, 23
    Medieval Climatic Optimum, 21, 23

Coconuts, leaves, plantations, trees, 59, 82–83, 104, 129, 150 169
    Coconut husks, 315
    Coconut leaf skirts, 252, 262

Container/contained, 43, 308, 311–312, 331
    Contained/contrary, 312

Copra, 3

Coral outcrop (*See also* sasek), 133, 222

Cornell Laboratory soil tests, 62–63, 68–73

Cowry shells, 267

Craven, Lyn, 118n16, 241n4, 243n32

Creation mythology, 45, 59, 64, 117n8, 174, 302

Creator (*See also* Geliu), 17, 18, 42, 125, 133–35, 139, 143, 250, 259, 265

Crossbeam (*See also* kiyad), 106, 262, 264, 265, 267, 274, 277

Crumley, Carole, 20, 85, 108, 276

Cunningham, Clark, 156

Curses, 25–26

Cuscus (*See also* Phalanger lulluae), 41, 94, 134, 177n18, 223, 231–32, 263

Dal (*See also* kum, clan/subclan), 44, 152

Dabuiy/Tabuiy (*See also* prow board), 220, 233, 306, 323, 329–31, 331

Dan (*See also* C. vexans), 142, 145, 147, 190, 199, 203, 205–208, 220, 231 243n31, 305

Davis, Mike, 179n45

Dean, Kenneth, 294n28

Demiavek, dum (*See also* Swamps, sago), 103, 126, 139–153, 166

Dibolel, xiv, 54, 105, 118n23, 141, 154, 196, 212, 226, 244n51, 302–03, 307, 308, 352n29, 339, 342

Digidag (*See also* garden fallow), 42, 60, 92, 97, 109, 111, 126–29, 135–38, 147, 155, 161, 177n27, 238–39, 315–316, 342

Digim'Rina, Linus, 36, 40, 76n16, 136, 179n42, 244n52

Dikwayas village, 51, 106, 115, 157

Dioscorea esculenta, 40, 52

DiPiazza, A, 301, 350n4

Direction (*See also* wind), 43, 117n11, 261

Dob (*See also* skirt), 104, 252, 262

Dumont, Louis, xii, 296n1, 296, 301–02, 350n9

Duwadul (Boat part), 195, 242n20, 242n21, 265, 277, 313–17, 322, 338, 340, 345
    (*See also* wind direction) 214–15, 277, 279

Duweyala, 8, 12–15, 210–215, 308, 311–12, 314

Dysoxylum, sp. 55–56, 66, 99, 118n21, 128, 130, 178n36, 183, 256

Eastern Muyuw (southeastern), 5, 29, 32, 36, 40, 42, 45, 45, 50, 54, 56, 57, 58, 64, 67, 77n29, 84, 85, 92, 95–98, 101, 104–07, 111, 115, 117, 119n26, 125, 143–44, 147, 155, 157–61, 170, 173, 175, 176n7, 177n16, 178n28, 179n38, 191, 203, 223–24, 229, 293n7, 303, 332, 339, 344, 346, 352n33
    Tasim in eastern Muyuw, 127, 133

Ebony (Ebenaceae), 26, 166–67, 200, 183, 226, 227, 242n24, 274, 318

Ecotone (*See also matantalas*), 21, 123, 299, 333, 337–40, 344–45

Ehara, Hiroshi, 142

Ellen, Roy, 140, 177n20, 180, 219, 298, 350n8

El Niño Southern Oscillation, 3, 19

El Niño, 3, 64, 143, 174, 182, 200, 204, 208, 218, 294n26
    ENSO, 3, 15, 130–31, 137, 146–48, 173, 179n45, 298, 344

Encompassing/encompassed, 34n21, n23, 43, 265

Endospermm mdeullosum (*akilim*), 183

Eucalyptopsis, 183, 241n4

Female/male contrasts, 104, 122, 149–50, 175n4, 236, 239, 247, 256, 257–58, 261–62, 263, 281, 293n8, 294n25, 309, 340, 352n35
    "Flexible"/"hard", 283
    With prowboard symbolism, 331

Fence stakes, posts, 86, 106, 134, 155–156
    Fences as strakes, 307

Fern, 53, 76n22, 144, 146, 177n26, 252, 265

Fernandez, J., 244n55

Finey, Ben, 14

Fire, firewood, 19, 39, 40–41, 58, 60, 97, 102, 138, 144, 146–149, 152, 165–174, 177n24, 179n39, 270, 288, 334–35, 342
    Fired land, (*sigob*), 58

Fish, fishing, 4, 22, 26–27, 41, 49–51, 88–91, 109, 112, 117n10, 121, 134, 141, 150, 169, 188, 242n21, 252–54, 274, 352n28

Fish imagery, 306–07, 330, 333–35

Fishing line sources, 103–04, 115, 173–74

Fishing net (wot & *See also* tying), 103, 115, 118–9n24 149, 173, 250–58, 260–62, 274, 279–282, 289, 292n3, 294n26

Flannery, Tim, 24–25, 34n16, 34n22, 301

Flora, floral, xi–xii, 3, 5, 16, 20, 25, 31–32, 57, 60–61, 83, 86–86, 108, 113, 116, 121, 126, 129, 153, 232, 258, 351n24

Fox, James J., xiii, 23, 75n8, 87, 174

Fractal, 28, 116, 247, 267, 325

Freeman, J. R., 133, 177n17

Frogs, 178n31

Gammage, Bill, xiii, 345

Gap, 39, 97, 123, 130, 138, 204, 222, 270, 280, 320, 333

Garden (*kwab*), 36, 346
    And boat resources, 307
    Digadag, 37, 39, 42, 59
    Forms & transformations, 43, 46–47, 58, 133, 139, 140, 259, 299
    Garden fallows, 37, 57, 128, 155, 170
    Idal, 58
    Kadidulel, 58, 66
    Oleybikw, 37, 39, 56, 60, 66
    Stages, 52, 58
    Ulakay, 37, 39, 42, 66–

Gawa, 31, 36, 41, 46, 51–53, 56, 61, 70, 94, 100, 114–5, 127, 131, 132, 136–38, 151, 165, 171–73, 189, 192, 204, 223, 225, 234–35, 237, 268, 273, 302, 327, 339–41

Geil, gulumom (*See also* boat ribs), 193, 227, 304

Geliu (*See also* Creator), 42, 236

Gell, Alfred, 34n23, 292n5

Gender, gendered, 75n7, 85–86, 153, 178n28, 236, 259, 261–62, 282, 297, 351n23

Geology, geological, geologist, 17–18, 33n8 226

Gideon, 8, 14

Gisawa, Leban, 27, 117n10, 351n20, 353n28

Gisawa, Milel, xiv, 7, 302

Gitamatan (*See also* wind direction), 214–15, 228, 277

Gladwin, Thomas, 63, 282

Global warming, 33–34n16

Godelier, Maurice, x

Gold, gold mining, gold exploration, 3, 18, 24

Grains, wood, interlocked/straight (*See also* kasiliu), 6, 98–99, 107, 111, 178n36, 188–91, 194–6, 198, 209–10, 214, 219, 232, 247, 249, 267, 274, 275, 291, 295n32, 304–05, 316, 329, 338, 352n32

Granet, Marcel, 289

Green, Roger, xiv22, 292n5

Gregory, Chris, xii, 5, 24, 155

Group, grouping (*See also* bod), 90, 189, 256, 292n4

Guasopa, 4, 45, 115, 125, 127, 169, 171, 174, 343

　　Guasopa airstrip, 4, 125

Gulumom/geil (*See also* rib), 304, 307

Gumiya, 159, 212, 221–23, 243–44n45, 280, 350

Gunn, Joel, 20–2

　　Captured knowledge, 173, 298

Guyau (Big Man), 105, 218, 335–36, 346

Gwed, gweda (*See also* Rhus taitensis), 5, 19, 30, 36–42, 52–56, 60, 62, 64–66, 71–74, 75n7, 86, 92, 94, 115, 128, 130, 134, 138, 155, 160–61, 163, 166–67, 173, 179n38, 254, 297–98, 343

Haddon, A. C. and J. Hornell, 6

Halyard, 195, 273, 276–77, 317, 328

Hard (Materials quality; *See mameu/ matuw*), 87, 92, 94, 97–98, 102, 110–11, 118n23, 137, 155, 158, 170–71, 175n4, 191, 254, 258, 259, 273, 283, 292n3, 299, 304–05, 309–10, 318, 345, 352n32

Harvard Herbarium, 20, 31, 38, 117n5, 119n33, 176n8, 181, 185, 243n35, 293n18

Heinsohn, T. and Captive ecology, 177n18

Hénaff, Marcel, 297

Heterarchical/hierarchical, 85, 108

Hide, Robin, xii, 35n30, 75n6, 76n20

Historical Ecology, 20

Ho, Ts'ui-p'ing, xiii

Horridge, Adrian, xxii, 5, 196, 229, 293n11, 317

House, 49, 84, 106, 110, 134, 137, 147, 150, 155–158, 259, 262, 298, 304, 321–325

Howard, A., 241n12

Hull, 303, 305–307, 350n12

Hydrodynamics, 102, 193–4, 197, 201, 212, 216–18, 231, 233–5, 266, 303, 305, 312–14, 316, 318, 320, 328, 343

Igsigis (*Lumnitzera sp.*), 162, 299–300, 309, 350n6, 317, 338, 345, 350n6

Incest (tanagau), 109, 121–22 16

Initial conditions (*See* chaos)

Intsia bijuga (*See meikw & kaymatuw*)

Irwin, Geoffrey, xii, 22, 34n19, n20, 243n43, 298

Iteanu, André, xii, 34n21, 292n1, 350n1

Iwa, 31, 36, 41, 46, 51–53, 57, 60–61, 64–73, 76n15, 77n31, 78n35, 81, 86, 94, 114–15, 118n12, 127, 131, 132, 136–37, 154, 155, 158, 163–165, 171, 173, 180, 189, 192, 199, 201, 204, 237, 288, 293n16, 327, 337, 344

Kabat (Mt. Kabat), 176n6, 176n11, 142, 146

Kaboum (Manilkara fasciculate), 127, 162, 170, 172, 195, 216, 304, 309, 311, 317–18, 338, 340, 345, 352n32

Kadidulel (exhausted soils), 58, 66, 145, 174

Kakam (See also C. inophyllum), 181, 190–98, 215, 226, 244n48, 268, 274, 313, 315, 335, 337–38

Kalabwadog/Kuban, 116, 170

Kaley/kunay (boom/yard), 196–97, 214, 242n24, 272–78, 293n12, 330

Kalopwan, 116

Kalow/man ankayau (See also tree distribution), 127–8

Kasiliu (See also grains), 190, 198, 299

Kaulay village, 3 50, 51, 56, 58, 77n31, 83, 84, 105, 106, 115, 126, 130, 134–35, 144, 157, 170, 177n6, 178n31, 198, 203, 225, 291, 339–42

    Kaulay village net *(teikw*), 279–80

Kausilay (See also C. leleanii P.F. Stevens), 14, 67, 134, 190, 193, 200, 200, 209, 219–240, 274, 291, 298, 304, 305, 337, 344, 351n13

    Kausilay arcs, 227–28, 233

Kavatan village, 45, 49, 77n31, 84, 91, 97, 118n15, 118n, 118n20, 105, 124–25, 142, 144, 154, 169, 172, 208, 223, 237, 342, 350–351n13

Kavavis (See also rudder), 11, 14, 49, 110, 158, 178n36, 215, 235–36, 299, 305, 316–320, 351n19

Kauwuway, 116, 344

Kaybwag, canoe type, 196, 219, 234, 235, 265, 298

Kaymatuw (See meikw)

Kaynen (trees of…), 109, 124–26, 134, 176n7, 203

Kaynikw, 160–61, 305, 307–312

Kaypwadau, 110–11

Keel, 14, 147, 152, 220, 230, 232–33, 238, 252, 257, 269, 305

    Carved imagery, 328

Kelkil (See also vakiya), 52, 132, 330

Kemurua (See anageg), 7

Kennedy, Jean, 184–5, 205, 241n5, 243n33

Keystone species, 296, 301–02, 350n8

Kinship, k. relations *(See also* Clan/ subclan), 94, 119n30, 198, 281–83, 344

Kioki (See also vakiya), 52

Kirch, Patrick V., 18, 33n13

Kirch, P. V. and R. Green, 289

Kitava, 31, 41, 46, 52, 75–76n13, 76n16, 131, 136, 153–54, 164, 180, 189, 201, 288

Kitoum, 175n4, 302, 341–42

Kiyad (See also outrigger crossbeam), 117n6, 155, 262, 298–301, 304, 307, 314

Knot (*sip*), xiii, 119n35, 243n41, 247, 248, 249, 261–283, 285, 301, 307, 315, 317, 327

Kokoyit (fern), 146–7, 252, 263

Koliu (fern), 146–47, 177, 252, 265–66, 306, 311, 322

    kibkweway method, 264–65

Koyagaugau *(See also* Ole), 7, 8, 31, 75n3, 132, 154, 171, 209, 236–37, 307

    And string figures, 284, 294n30

Kubub (See also wind reversals), 217

Kubwag (Ficus), 56, 67, 71–73, 101, 130

Küchler, Susanne, xiii, 247, 249, 292n5

Kuduk, Max, 38, 181, 189

Kuk, 195, 210, 266, 317, 330, 326–28, 333, 338

Kula Gold Ltd., 24–5, 34n24

Kula Ring, xi, xiii, 3, 4, 7, 181, 187, 188, 196, 201–02, 248, 288, 291, 293n11, 333

    Regions of, 7, 167, 202, 229, 231, 297, 343

Kula, 5, 29, 35n27, 132, 150–1, 153–54, 175n4, 165, 218, 237, 253, 286, 303, 341, 343, 347, 349

    Kula Open, 8

    Kula valuables, 7, 122, 177n25, 175, 237, 240, 257, 261, 327, 332, 341

Kum (See also clan/subclan, dal), 43–44

Kumis/kibkeway, 264–266

Kunai grass (*Imperata cylindrica*), 60, 65, 76n16, 78n35, 136, 146–47

Kunubwara (*See also* prowboard), 269, 306, 313

Kunusop (*See also* mastmount), 195, 241n21, 308, 351n27

Kuv (*See also* yam), 46

Kuwu, 219, 234

Kwadoy (*See also* cuscus and knot), 41, 94, 134, 231–32, 255, 263, 332

Kweyakwoya, 127, 173, 204

Kweybok, 121

Kweywata, 114, 127, 132, 151, 173, 225, 241n18, 234–35, 339–41

Lae Herbarium, 38

Lam, 214, 239, 339–42
    Katan/watan, (*buluk*) away from float, 117n11, 214–55, 233, 305, 318, 320
    Walam, to the float (*yiwatin*), 214–5, 233, 320

Lae Herbarium, 31, 93, 103, 119n33, 176n8, 181, 184, 189, 199, 242n25, 242n26

Lakatoi, 196, 202

Lansing, Steve, x–xi, 18, 179n44, 292n1, 297, 344

Lapita pottery, 34n20

Lapuiy (garden line; *See also* Milky Way), 44, 127, 159–60, 299, 346

Lawrence, Debora, 77n32

Leach, E. R., 79

Lee board (*See also* kavavis, rudder), 11, 12, 215

Legume, 19

Lemonnier, Pierre, xii, 297–98

Lévi-Strauss, C, 108, 180, 244n55, 249, 288, 296, 301–02

Likilok (*See also* Trochus shell), 129

Lidau village, 73, 115, 170

Life forms, 88–91

Lilimuiy (*See also* Wind direction), 214–15, 228, 231, 277

Luo, Yang, 350n4

Macaranga sp., 85, 126, 162

Macko, Steve, x, 19

Madagascar, 186, 187

Magic, 5, 247, 253–54, 256–57, 293n8, 294n27, 315
    And fishing nets, 281

Macintyre, Martha, 8

Malinowski, Bronislaw, Malinowskian, 21, 53, 79, 196, 201, 302
    Coral Gardens, 141

Mamad (*kumis* method), 252, 264–65, 310

Mameu (*manum*)/*Matuw,* 283–84, 304–05, 310

Mamin/manina (*Syzygium sp.*), 97–98, 128, 135, 168, 170, 203

Man (*See also* prowboard), 329

Mann, Michael, 34n16, n17

Manguin, Pierre-Yives, 242n19 243n43, 302, 336

Marriage, 44, 302–03, 307, 341

Marx, Karl, 96, 301–02

Masawa (*Tadob*), 190, 196, 201–02, 204, 219, 229, 231, 268, 291, 337, 344

Mast (*See also* vayiel), 12–15, 106, 159–56, 195–96, 200, 208–19, 266, 278, 316–320
    Mast-mount (*See also* kaynikw), 235, 265, 273, 277, 300, 307–08, 310

Matan- (eye/end of…, tip), 102, 113–16, 122–4, 141, 165, 261, 271, 299–300
    Matantalas, 224, 227, 240

Math, mathematical expressions, xiv, 230, 296–301

Matuw (*See mameu/matuw*)

Maylu, 196, 202, 242n23

McClatchey, Will, 140, 178n32

Meadow (*See also* sinasop), 39, 52, 68–73 109, 110–11, 121, 129, 139–153, 203–04, 208, 231

Meikw (*Intsia bijuga, kaymatuw*), 110–11, 126, 155, 158, 160–4, 171–72, 183, 221, 305, 316, 318, 323, 338

Meliaceae (Mahogany family), 38, 56, 77n25, 90, 99, 108, 118n18, 169, 185

Mesua ferrea (*See also* Calophyllum & *M.f.*), 186–87

Miliyout, hull-outrigger float gap, 201, 243n29

Milky Way, 299, 346

Milne Bay Logging, logging activities, 25, 118n18, 145, 182, 1983, 189, 221

Milne Bay Province, 131, 248, 302

Mines, mining, 24, 34n21, 139, 145–46, 177n19, 204, 350n3

Mintz, Sydney, 183–4

Misima Is., 12, 24, 154, 166, 172, 178n33, 207

Modeling, models, xi, 37, 45, 54, 73–75, 83, 167–174, 196–97, 200, 210, 231, 237, 248, 259, 270, 288–91, 297–98, 335, 337, 345

    Fish as modeling, 307, 334

    Kaynikw as modeling, 312

    Outrigger canoe as model, 320–349

    Roosters as models, 328

    Tying as modeling, 282–292

Moniveyova village, 59

Moon, 109, 163–164, 165n10, 244n47, 289

Morava, J., 297–301

Mortuary rituals, practices, 29, 75n11, 46, 105, 109–110, 141, 144, 150–51, 153, 167–68, 198, 218, 220, 247, 262, 302–03, 323, 331, 349, 351n24

    Lo'un, 17

Mosko, Mark, xii, 114, 244n52

Munn, Nancy, 5, 153, 172, 228, 237, 302, 325, 327, 344

Muyuw (Woodlark Island), 41, 52

    As Eastern/Southeastern region, 342

    Gardens (fields) of, 36, 64

    Understandings of, 18, 114

Mwadau Is (*See also* Western Muyuw), 31, 129–31, 173

Mwadau (Maria), 302–05

Mwadau village, 46, 77n31, 135, 156, 158, 161–63, 170, 239, 302–03

Mwalubweyay, 7, 33n3, 294n30, 302

Mwatabwalay (*See* banded sea snake)

Mweg (*See also* sail), 196, 267, 268–70, 322

Namonsigeg (*Passiflora sp.*), 111, 257–58, 281

Nasikwabw, 7, 8, 31, 77n31, 84, 112, 114, 123, 129, 151, 154, 178n33, 166, 171, 225–28, 235, 236, 270, 279, 308, 341–42, 351n15

Nautilus shell (*N. belauensis, uvagol*), 331–32

Navel (*See also* Pwason), 43–44, 113, 133, 308, 321

Navigation, 201, 291, 347–48

Neate (family members and history), 25–26, 118n18, 183

Nedin (a boat spring), 158, 277, 305, 312–316

New Year ceremony, practices, 49, 57, 114

Nitrogen (fixing n.), 19, 37, 62–68, 76n22

Normanby Is., 12, 15, 343

Notis, Christine, 214, 240n2, 243n27

Nunn, P., 178n34

Obsidian, 173

Odom, Odom & Odom, 351n18

Ogis, xiv, 6, 81–83, 108, 154, 212, 218, 287, 348

Ole (*See also* Koyagaugau), 7, 31, 75n3284

Oleybikw (*See also* Garden fallows), 56, 66 97, 101, 134, 161, 339

Onosimo, 7, 8, 13, 284

Origin mythology, 59, 85, 121

Origin Place, 113, 174, 175n4

Orion's Belt, 298

Outrigger (boat, canoe), x-xii, 5–7, 12, 15, 30, 39, 41, 44, 45, 52, 102, 104, 105, 111, 132, 134, 157

    Outrigger Animal symbolism, 325–337

    Caulk, caulking, 126, 251, 307, 343

Outrigger (boat, canoe), continued
  Outrigger crossbeam (See also
    kiyad), 98, 106, 117n6, 155, 262,
    298–301
  Float carving and imagery, 328,
    334–35, 340, 352n29
  Hull and float dynamics, 313–16
  Outrigger Design, (Construction),
    198, 202, 205, 233–34, 262, 303,
    332, 343–44
  Outrigger float (See also lam), 8,
    11, 58, 104, 105, 130, 145, 195,
    214–15, 235–36, 239, 265, 267,
    313–14, 339–43
  Outrigger imagery, 45, 320–349
  Outrigger types, 9, 221, 234,
    260–261, 264, 308

Pacific Chestnut (vitivit), 125, 127176n6
Paddle, 49, 81, 99, 130, 158, 171, 178,
  201, 235, 265, 288, 292, 336, 344
Panamut, 8, 12, 22–23, 346, 348
Pandanus, 90, 118–9n24, 123, 144, 150,
  163, 196, 214, 217, 243n37, 250–256,
  259, 268, 270, 273, 293n16, 315
Panaeati, 154, 172, 196, 268
Parawog (See also yam, Dioscorea
  esculenta), 40, 46
Patch, patches, patchiness, 21, 79, 83,
  85, 91, 93, 124–28, 138–39, 153, 167,
  214, 223–25
Pawley, Andrew, xiii, 112–3, 119n35,
  n36, 350n4
Pearthree, Erik, 14, 293n9, 293n17, 301
  350n4
Pfaffenberger, B, 297
Phalanger lulluae (See also Cuscus), 41
Phosphorous (P), 70–73
Photographic accompaniment, xii
Pig, 4, 9, 44–5, 51, 58, 89–90, 92, 122–23,
  150, 153, 168, 235, 259, 263
  Wild pigs, 4, 44, 147, 152, 205, 207,
    254
Pliable/soft (See mameu/matuw)
Pleiades (Gumeau), 285, 289–91, 340,
  346

Poison Fish Tree (Barrington aseatica),
  112
Pomentia pinnata (kaga), 183, 222,
  244n50
Posey, Darrell, 20, 22, 34n18, 42, 62
Potassium (K), 70–73
Potulski, C, 350n4
Power (See also tautoun), 25, 43, 138,
  197, 218, 247, 290, 326–28, 335, 345
Principle (kikun), 46, 90–91
Prow board (See also dabuiy/tabuiy,
  kunubwara & man), 101–102, 220, 230,
  267, 269, 306, 323, 330, 332–33, 350n5
Pwason (See also navel), 113, 267, 308

Region, 4, 85
Regional relationships, 4, 116, 130–32,
  151, 153, 165–174, 184, 208, 235, 279,
  296–98, 327, 333, 337–349, 352n33
  And exegesis—symbolism, 326
Reid, Shaw, 62–63, 69, 139, 145
Reversal, reverse (inversion), 121, 235,
  266–267, 271, 272, 283, 286, 290–91,
  300, 312, 349
Rhus taitensis (See also gwed), 38, 65,
  128, 254, 297
Rib (vilag), 312, 322–23
  boat ribs (See also geil, gulumom),
    193–94, 227, 300–01, 306–09, 314,
    333, 338, 350n6
Rice agriculture, 140
Robbins, Joel, 35n28, 75n10
Rooster (See kuk)
Ross, Malcolm, xiii, 119–20n36, 318
Rival, Laura, 20
Robbins, Joel, 35n28
Rudder (See also kavavis), 11, 49,
  110, 158, 178n36, 233, 235–36, 277,
  316–320
Ruddiman, William 34n16
Run Amok (kabalwein), 28, 81–83

Sacred grove, 133, 177n17
Sago, 50, 139–153, 154

Sago orchard, sago swamp, 121, 123, 126, 129, 166, 231, 259, 260, 327

Sago tree, 139

Sago trough, 105, 144, 177n23

Sago uses, 150–152, 270

Saido (*Terminalia catappa. See also* teili), 61, 81, 137, 155

Sails (*mweg*), 13–15, 106, 117, 123, 173, 190, 195–97, 200, 210, 214–18, 229–31, 248, 253, 256, 260–61, 263–64, 267–79, 282–83, 286–87, 292n3, 293n11, 294n19, 294n22, 308, 310, 312, 316–18, 320, 322, 327–28, 329, 337–39, 342, 347

    Sail burning (making a sail), 270–271, 342

    Sail and Kavavis (rudder), 318

Sail materials, 214, 250–256, 253, 262–63, 267–268, 270, 274, 293n16

    Aydinidin, 196–97, 214, 267

    "backside" (tapwan)/ "inside" (nuwan), 217

    Maylu lakatoi, 196, 202, 337

    Sail structure, 14–15, 102, 196–97, 267–279, 292n3

    Sail types,

    Sails, tying and wind, 83, 267–279, 318

    Seylau, 9, 196, 268, 279, 308

    Yabuloud, 196, 268, 274

Sailing village, 6, 151, 279, 293n14, 334, 336, 343

Sahlins, Marshall, 30, 235

Sasek (*See also* coral outcrop), 133, 222, 226, 236

Sather, Clifford, 23, 83, 113, 174, 181, 187, 241n10

Scoditti, G, 331–32

Seisuiy, 262, 276, 299–300, 304

Selau, 9, 308

Sexual imagery, 211, 237–238, 239, 257–58, 326–27, 335

SHANTI, xii

Shugart, H. Hank, 19, 33n15, 221

Silamuyuw (*See also* Dysoxylum), 56, 130, 178n36

Sinamat, 115, 169, 173

Sinasop (*See also* meadow), 109, 139–153, 208, 231, 291

Sinkwalay River, 127

Sip (*See also* knot), 119n35, 248, 263–282

Sipum, Amoen, xiv, 3, 4, 6, 22, 31–32, 37, 55, 84, 87, 97, 100, 149–50, 199, 206, 208, 212, 218, 238, 243–44n45, 247, 248, 250, 262, 291, 292n2, 324, 349

Skirt (*See also* dob), 5, 104, 150, 252, 256, 262, 288, 292, 341

Snakes (*mwatet*), 332–33, 351n27

Soil, 62–74, 106 119n26, 130, 145, 155, 226

    Dry, 106, 118n23 119n26 124–25, 176n12 190, 198–99 202 340, 344

    Wet, 95, 98, 106–08, 118n23 119n26 122, 124–25, 127 146, 148, 155, 166, 173 176n12, 177n20, 178n31, 190, 198–99, 202–03, 213, 229, 270, 312, 319 340, 344

Soul (See also Tying, method), 260–261

Spencer, Lee, 24, 33n8, 93

Sports cycles, 8

Springs, 6, 157, 212, 235, 266, 273, 277, 302–20

Stable isotope, x, 19, 30, 36, 40, 62–68, 76n22

Stanchions (*See also* watot), 264, 276, 299, 320

Star, constellation, 109, 138, 148, 218, 285, 289–91, 298, 321, 326, 328, 345–47, 352n39

Sterculiaceae, Sterculia sp. (*See also* ukw), 57, 106, 128, 144, 183, 249, 250, 260, 292n4

Stevens, Peter F., 20, 38, 119n25, n33, 176n9, 178n35, 180–82, 184, 185, 186, 188–91, 199–200, 206–07, 213, 240n2, 241n9, 241n11, 243n34, 243n35, 237, 294n19

Strake (*See also* budakay), 39, 45, 61, 127, 147, 152, 194, 220, 225, 233,

239–40, 257, 265, 300, 303, 305, 350–351n13
  Strake names, 306
Strangler figs, 55, 93–94
Streamers (*See also* bis, telltales), 12, 217, 262
String figures (*Kananik*), 282–291
  And boat tying, 283
  And gardens, 285
  Koyagaugau, 284, 294n30
  And Kula, 285, 287
  Nasikwabw, 284
  And stars, 285, 289–91, 348
  And time, 285, 289
  Wabunun, 288
Sulog, 12, 14, 15, 31, 37, 69–7, 108–9, 112, 121–23, 129–30, 175n4, 176n11, 176n13, 143, 154, 167, 170, 172–73, 208, 212–14, 219, 221, 274, 280, 294n26, 339, 345
Swamps (*See also demiavek*), 103, 129
Syzygium sp., 84, 96–99, 117n5, 134, 142, 166, 168, 221, 244n50, 299
Systematists, systematics, 20, 100, 118n22

Tabnayiyuw (Dysoxylum papuanum), 55, 66, 92, 126, 178n36
Tadob (*See* Masawa)
Tabuiy (*See also dabuiy* and prowboard)
Taiwan, xiii
Takon (as exchange), 167, 176n13, 253, 324
Takon (as chest), 274–75
Tan, potan (*See also* tide), 49, 224, 244n47
Taro, 36, 39, 53, 59, 66, 88, 117n9, 131, 321, 344
Tasim, 83, 132–139, 140, 147–8, 152, 155, 203, 221, 224, 226, 228, 229, 231, 237, 247, 291, 338, 345
Tatan (as base), 115–16, 165, 173, 325
Tautoun, 197, 247
Tawakw (Terminalia megalocarpa), 61
Tawan, tawala, 57–58, 60, 77n27, 125, 131, 339
Teili (*Terminalia catappa. See also* Saido), 81–83, 85, 100–101, 174, 305, 344

Telltales (*See also* bis, streamers), 12, 14, 262
Tibwelan (woven bands), 263
Tide/tidal, 18, 49, 96, 140192, 221, 224, 224n47, 256, 285, 288–89
Time, time system, 153, 163, 327
Totem, 90–91
Transformation, transforming, 4, 22, 24–25, 30, 42, 46, 53–54, 62, 74, 84, 116, 119n25, 140, 158, 165, 175, 196–97, 201–02, 219–20, 229–231, 237, 240, 244n55, 247, 249–250, 259, 268, 277, 282–84, 286–87, 289, 291–92, 293n11, 297, 299, 301, 327–28, 333, 348
Transition, 25, 27, 45, 59, 101, 113, 123, 203, 224, 229, 285, 316, 327
Trochus shell (*likilok*), 129
Tree form, 43, 88–91, 110, 114
Tree distribution, 87, 118n18, 127, 128, 202–05
Trobriand Is. (*Kilivil*), 17–18, 31, 45, 46, 53, 64–73, 76n21, 219, 244n52, 318
  T. Kavataria village, 76n17, 154
  T. Liluta village, 154
  T. Oburaku village, 154
  T. Okaibom village, 31, 36, 54, 134
Tulane Historical Ecology Conference, 20–22
Turner, Victor, 326
Turtle, 27, 89–90, 101, 125, 275, 328–29, 334, 346, 351n20
Tying, 195–6, 230, 251–55, 258–282, 300
  Kaynikw tying, 309–12
  Kinds of tying material, 262
  Method (*soul*), 260, 263–282

Ukw (*See also* Sterculia sp.), 57 105–6, 112, 144, 160, 250, 262, 293n10, 272, 273, 275
Ulakay (*See also* garden fallow), 42, 55, 57, 66–73, 97–99, 115, 129, 134–38, 155, 159
United Church, 28
Ungonam village, 144
Unmatan village, 51, 115, 118n, 124–25, 129, 142, 169–70, 173 175n3, 224, 227, 339

Urushiol, 38

Vakiya (*See also* kelkil), 51–52, 132
Vakuta, 175n4, 201
Vatul, 54, 89–90, 247–295
    Geliu's *vatul*, 320
    Names, 250–258
Vekway, 32, 341–42
Vayiel (*See also* mast), 209, 316–320
Venay, 44, 45, 127, 141, 152, 162, 175,
    293n7, 299, 346
Vesop (*See also* Alocasia sp.), 55, 81
Vibrations, vibrating boat parts, 12,
    209, 212, 216, 302–320
Village, 249
    Village form/order, 46–49, 114–5,
        288
    Villages and division of labor, 279,
        334, 336
Vine (*See also* vatul), 88–91, 146,
    247–295, 263

Wabunun, 15, 28, 56, 58–61, 71, 77n31,
    81, 91, 93, 100, 108, 118n15, 121,
    175n4, 125–26, 155–6, 158, 165, 167,
    169, 178n31, 195, 198, 227, 237,
    242n22, 270, 288, 302, 334, 344
    Origin point 28, 81
Wag, 152, 253, 289, 305, 314
Wagner, Roy, x, 22
Walker, Brian, 77n32
Wallace, Anthony F. C., xi, 282
Wamwan (*See also* Central Muyuw), 29,
    49, 176n7, 288
Wang, Mingming, 350n4
Water, freshwater lens, 17–18, 131, 137,
    140, 195, 224, 236, 242n20, 244n52
Watot (*See also* Stanchion), 157, 195,
    264, 276, 299
Waves (*kaysay*), 11, 13, 194, 228, 234,
    300, 303, 347
Wayavat village, 51, 115, 125, 169 288
Waviay, 15, 195
Website, xii
Weed, 89–90, 92
Western Muyuw (Nayem), 41, 45–46,
    51, 56, 59–60, 64, 68 74, 73, 77n3,

    85, 95, 101, 116, 118n12, 119n34,
    126–27, 129–32, 135–36, 138, 155, 158,
    161–162, 238–39, 248, 253, 257, 299,
    302, 322, 338, 341–42, 344, 346
    Tasim in Western Muyuw, 133,
        135
Wheatley, Paul, 33n10
Whistler, A., 188
Wills, Gary, 245
Wilson, E. O., 185–6
Winds, 9, 33n5, 141, 150, 157, 159, 192,
    195, 197, 201, 320
    Directions, 9, 43
    Directions when sailing, 11,
        214–217, 242n21, 277, 305, 318
    Wind and mast, 215, 230, 310
Wind reversals (*See also* kubum), 217,
    267
Witchcraft (*bwagau*), 5, 28
Winterhalder, Bruce, 33n15, 79, 83, 124,
    245
Wood, Emily, 293n18
Woodlark Island, 28, 41, 183
Wowun/dabwen (*See also* base/tip),
    113, 141, 237, 265, 299
Wot (*See* Fishing net)

Yalab, 193, 235, 333
Yams, 36, 46, 52–55, 64–73, 104, 131,
    142, 162, 247, 251–52, 256, 284–85,
    289, 344
    Yam pruning, 52, 86, 202, 337
Yam house, 41, 46, 158–165
Yam stakes, 61, 320
Yamwik, 115
Yard (*See* Kaley/kunay)
Yay tree (*See also* Casuarina littorale),
    83, 166
Yemga, 6, 77n31, 151, 201, 229, 334
    Tamdak as "rudder" 351n19
Yen, Douglas, 175n4, 177n20
Yoyit (Araceae, Raphidophora), 144,
    255, 259–60

Zhang, Bin, xiii, 350n4